Sourcebook of HVAC Specifications

Sourcebook of HVAC Specifications

Frank E. Beaty, Jr., P.E.

McGraw-Hill Book Company

New York St. Louis San Francisco Auckland Bogotá
Hamburg Johannesburg London Madrid Mexico
Montreal New Delhi Panama Paris São Paulo
Singapore Sydney Tokyo Toronto

Sourcebook of HVAC specifications.

Includes index.
1. Heating—Specifications. 2. Ventilation—
Specifications. 3. Air conditioning—Specifications.
I. Title.
TH7325.B43 1986 697'.00212 85-25634
ISBN 0-07-004192-X

1234567890 SEM/BKP 8932109876

ISBN 0-07-004192-X

The editor for this book was Betty Sun
and the production supervisor was Teresa F. Leaden

Printed by Semline, Inc. Bound by The Book Press.

Contents

Preface

This volume is intended as a guide for those in the business of specifying or using HVAC equipment. It is written with the assumption that the user of this guidebook is familiar with HVAC equipment and systems and is either preparing documents for the bidding of a project or is using this guide to see if such documents are reasonably complete.

This guidebook can be used as a brief reference to define HVAC equipment without the several pages usually required for an adequate description.

The *Sourcebook of HVAC Specifications* also relates to its companion volume: *The Sourcebook of HVAC Details.* The numbering system in both books is the same, and though not every section of this volume has details presented in the detail book, many do. The number of the detail is in the same series as the section number of the specifications. It is hoped that this cross reference will make the details easy to find for the specifier and the specification section easy to find for the detailer.

Each section of the specifications (except for Chapters 1 and 2, which relate to General and Special Conditions and how they are used) is divided into three parts.

Part One relates to the General and Special Conditions including the very necessary information about legal requirements, submittals, warranty, etc. This information is not used by everyone or every day in the life of a project and is included in each section by reference, not repeated for every section.

Part Two describes the equipment and the specific requirements of the equipment. The equipment supplier preparing a quotation for the project and the contractor receiving bids would be particularly interested in Part Two.

Part Three gives the specific requirements for the installation of the equipment, the nuts and bolts of how it is to be installed and supported and the related auxiliary materials and equipment involved in the installation.

Part Three details the requirements for the balancing and testing of the systems as they are installed. Before the contractor can consider the system ready for the owner's beneficial use, the requirements of this balancing and testing must be done. These requirements are included by reference and are not spelled out in each section.

These guide specifications are intended to include equipment by some manufacturers and are not intended to limit or restrict any manufacturer or supplier from participating in a project. This may be changed by a few words if the specification writer desires to limit the bidders.

identified in the "Description of Equipment" at the chapter heading with the name of the supplier given.

The body of the specifications allows the use of any equipment meeting the performance specifications. The specification heading "Acceptable Manufacturers" gives the user an opportunity to indicate on the drawings the equipment used as the basis of design which may or may not be one of the names listed.

Three names are listed in the specification under "Acceptable Manufacturers" to indicate at least the manufacturers that in the opinion of the author meet the specifications given. These three suppliers are named for the convenience of the guidebook user and have had no input into this author's work, other than having made their catalogs and other information available to the author over the more than 30 years he has been selecting and specifying HVAC equipment. The user of the guidebook may want to change the manufacturers named to those who have taken an active part in the market in the local area.

Options are noted in the sections by "(   )" where there are several different standards used in the industry. This is often indicated in the several types of controls used on equipment to meet differing requirements of the codes or the equipment.

The specifications are intended to be inclusive and may be considered "padded" by some specifiers or users. This is intentional on the part of the author, and it is up to the specifier to eliminate those added items that are not required for the specific needs of the project being addressed.

The author suggests that this book be used in conjunction with a copy machine and white-out materials. The user can make a copy of the desired section and amend it to meet his or her requirements by deleting those words or paragraphs that do not apply or by adding additional material where needed.

This guidebook is printed in typewriter print, rather than typeset, to make this type of addition and correction possible without being obvious. The type used is Prestige Elite 12, at 12 characters to the inch. This style of type is generally available on most office equipment. Pages are deliberately shorter than the space allows, so that there is additional room on most pages for more material to be added. A cut-and-paste approach with the modern copy machine can do a very nice job of editing the basic specifications printed here.

The author makes no representation that these specifications are accurate or complete or that they are fit for any specific application. They are presented as the type of material used by the author in his own practice for many years and seem to represent the information needed for the specification of HVAC equipment and systems.

The author regrets that every possible type of equipment is not covered, but it would take several volumes, each with more pages than permitted in this one, to make an attempt at that goal. The author has a sense of pride and surprise in the awesome number of products and materials that the HVAC contractor and engineer have at their disposal. Both must be familiar with and competent in the application and installation of a very large number of products in this industry. It is hoped that this guidebook will make the task a little easier.

FRANK E. BEATY, JR.

Acknowledgments

The author owes a debt of gratitude to all of those over the many years who have had a part in the instruction and experiences that have contributed to the education of the author as he has practiced in this field that is called *HVAC.* Some of this education came from books, some from the reams and reams of magazine articles, some suggested by friends, and much in the field where the contractor must make a profit to survive and the owner wants everything exactly right with all of the extras free.

The consulting engineer in private practice walks these paths daily, making decisions based on judgment and experience that will perhaps cost the contractor money not included in the bid, or deprive the owner of some feature that was not included but perhaps should have been. The pressure of clients, contractors, owners, suppliers, and his or her own morality all add to the pool of knowledge and experience.

To thank all of these individually or to single out any one would be a task beyond the author's ability and the scope of this book, for in the past, the present, and, hopefully, the future, there are those who will go out of their way to make life more pleasant for everyone connected with this industry that has been my life for more than a quarter of a century.

There are those who have encouraged me to proceed with this book and who have contributed in many ways to it and to its completion. Leading among those are friends and former employees Randall Jones and Greg White, who covered for me in my office while I spent more time than I should have on this project. Special among those that have contributed is my daughter Nancy who not only encouraged my efforts but also spent many hours at the keyboard of my limited word processor, correcting my spelling and adjusting my sentences to say what I intended to say.

Most of all, my appreciation to my wife Vinie who put up with a lot of nights alone and who missed out on many social and family outings because of my "work on the books." Her love and support have kept me going during those times when I would haved liked to have said "no more."

A special word of thanks to Betty Sun who edited this work and led a novice through this first experience at publication.

Sourcebook of HVAC Specifications

CHAPTER 1

General and Special Conditions of the Contract

CHAPTER ONE

****

USE OF PRINTED
GENERAL AND SPECIAL CONDITIONS OF THE CONSTRUCTION CONTRACT

1.1 GENERAL and SPECIAL CONDITIONS

a. The GENERAL CONDITIONS (of the contract) included in the construction contract cover the many items that are a necessary part of any contract that do not change substantially from one project to the next. As these do not change, they can be printed, saving time for both the design professional and the other parties of the contract.

b. The SPECIAL CONDITIONS (of the contract), on the other hand, tailor the General Conditions to fit the specific requirements of the owner by modification of the General Conditions through references.

1.2 STANDARD GENERAL CONDITIONS

a. The General Conditions should be standardized documents prepared by one of several agencies that produce contract documents.

b. One publisher of standard General Conditions is the American Institute of Architects (1735 New York Avenue, N.W., Washington, DC, 20006). The institute prepares a number of very helpful documents for the construction industry.

c. Another source of standard General Conditions is the Engineers Joint Contract Documents Committee (EJCDC). This committee represents the American Consulting Engineers Council (ACEC), the National Society of Professional Engineers, the American Society of Civil Engineers, and the Construction Specification Institute, Inc. EJCDC publishes a document named "Standard General Conditions of the Construction Contract."

d. Copies of "Standard General Conditions of the Construction Contract," as published by EJCDC, are available from any of the sponsoring societies. (The address of ACEC is 1015 15th Street, N.W., Washington, DC 20005.)

e. Some states, and even some municipalities, have their own General Conditions for use with work done for them. The design professional should read these carefully and be aware of any unusual wording or requirements they contain and how these differ from the requirements in the nationally recognized documents.

1.3 ADVANTAGE OF THE PRINTED DOCUMENTS

a. There are several advantages in the use of the printed documents:
b. First, since the document is printed (not reproduced from typed pages), any modification of the document would be noticeable. Therefore the user of the document knows that it is a standard form, and the user can refer to it in conversation with others by reference to the name, number, and date. This simplifies checking with attorney, insurance agent, bonding company, and others who provide the services to the contractor called for in the General Conditions.
c. The second advantage is the experience of those who write the standard documents. The authors of such documents are experts in every phase of the construction industry, so that the standard document carries many years of combined experience. These documents should not be changed unless absolutely necessary, and then only by the use of the Special Conditions. The following cautions are reprinted with permission from "Guide to the Preparation of Supplemental Conditions," prepared as a companion to the "Standard General Conditions" mentioned above.

THE GENERAL CONDITIONS ARE PRINTED AND USE CAREFULLY CHOSEN LANGUAGE. THEY SET FORTH THE BASIC RESPONSIBILITIES OF THE PARTIES IN RESPECT OF FUNDAMENTAL MATTERS AND LEGAL CONSEQUENCES. THEIR PROVISIONS SHOULD BE ALTERED ONLY WHERE MANDATED BY THE SPECIFIC REQUIREMENTS OF A GIVEN PROJECT AND THE CONSEQUENCES OF ANY MODIFICATION THOROUGHLY UNDERSTOOD.

A FURTHER WORD OF CAUTION ABOUT ANY CHANGE IN THE PRINTED FORMS. THEY HAVE BEEN CAREFULLY PREPARED, TERMS ARE USED UNIFORMLY THROUGHOUT, THEIR PROVISIONS CAREFULLY INTEGRATED AND THEY ARE QUITE DEPENDENT ON ONE ANOTHER. A CHANGE IN ONE DOCUMENT MAY NECESSITATE A CHANGE IN ANOTHER, AND A CHANGE IN ONE PARAGRAPH MAY NECESSITATE A CHANGE IN OTHER LANGUAGE OF THE SAME DOCUMENT. NO CHANGE SHOULD BE MADE UNTIL ITS FULL EFFECT ON THE REST OF THE GENERAL CONDITIONS AND OTHER CONTRACT DOCUMENTS HAS BEEN CONSIDERED.

LASTLY, ONE SHOULD BEAR IN MIND THAT AN ENGINEER IS NEITHER QUALIFIED NOR LICENSED TO GIVE ADVICE TO OTHERS ON THE LEGAL CONSEQUENCES OF THEIR CONTRACTS. ALL OF THE CONTRACT DOCUMENTS HAVE IMPORTANT LEGAL CONSEQUENCES. OWNERS SHOULD BE ENCOURAGED TO SEEK THE ADVICE OF AN ATTORNEY BEFORE ACCEPTING ANY MODIFICATION OF THE PRINTED FORMS, BEFORE THE DOCUMENTS ARE SENT OUT FOR BIDDING AND MOST ASSUREDLY BEFORE SIGNING ANY AGREEMENT.

d. The last and most important reason to use these standard documents is that the Specification writer does not have to compile all of the requirements for the project. This saves time and reduces the opportunities for mistakes. The omission of one vital requirement, which at the time seemed so trivial, can cost a great deal of time and money for all involved.

e. Printed standardized General Conditions are often included in the contract documents by reference rather than reproduction. This saves the Specification writer the cost of purchasing the printed copies and the additional cost in the assembly of the Specifications, but it is more difficult to answer questions as they arise. The design professional must be thoroughly familiar with all of the printed documents and have a copy at hand. There will be questions requiring a quick response.

1.4 SPECIAL CONDITIONS

a. Special (or Supplemental) Conditions are used to modify the requirements of the standard General Conditions by reference to that Section requiring modification, addition, or deletion.
b. Care should be taken in changing any part of the General Conditions. Ascertain that only the specific word or phrase is amended and that additional requirements are not deleted by reference in the part omitted. Modifications are often changes in the quantity of shop drawings, required amount of insurance, and similar items.
c. The engineering societies who publish the General Conditions recommended above also publish "Guide to the Preparation of Supplemental Conditions," a copy of which can be obtained from the sponsoring societies.
d. CAUTION! Changes in insurance and bonding requirements should be very carefully coordinated with the owner to ensure that the owner's legal and insurance counselors concur with these requirements as specified.
e. Duplication of insurance coverage serves no purpose and adds cost to the project, but omission of any insurance coverage should be made only with extreme care and a realization that insufficient insurance can be devastating to the owner, the contractor, and the design professional.
f. Some of the more common items covered by Special Conditions, in the author's experience, are:

1. Redefining the term "architect" to refer to engineer. (Many of the standard documents are written by architects who neglect the fact that engineers also design projects requiring Specifications.)
2. Clarifying the function of shop drawings and the engineer's responsibilities and actions in the review of such documents. (An excellent discussion of shop drawings and how they should be handled and stamped is included in Chapter 4 of UNTANGLING THE WEB OF PROFESSIONAL LIABILITY by Edward B. and Richard P. Howell, copyright 1980 by Design Professionals Insurance Company.)
3. It seems that General Conditions are usually written with new construction in mind. Often the author's work has been in existing buildings, sometimes occupied during the work, and the Special Conditions are used to spell out the responsibility of the contractor for protection of the public as well as owner's personnel and property.

4. The owner's right to recover all expenses and hidden costs may not be adequately covered in case of default by the contractor. The Special Conditions should spell out that in case of default on the contractor's part, the bond will cover cost of determination of exactly where the new contract must pick up. This should require that the determination be made by qualified experts, such as the professional responsible for the original design. The owner will usually have added legal cost and administrative cost that should be covered under the bond. The Special Conditions must ensure that the owner is protected in such a case.
5. In the case of existing facilities, it should be spelled out whether the contractor is to provide separate sanitary facilities for the workers or whether the toilet facilities in the building can be used.
6. If the building is in use by the owner, how often the contractor is to clean the area where he or she is working and how clean the contractor is required to leave the area at the end of a shift should be spelled out. Items such as this are very simple to clarify in the documents, but create hard feelings and involve much time if not spelled out.

g. The items above are examples from typical projects. Any project may have these or entirely different problems that need to be identified and clarified. Special Conditions should be used to clarify such items so there is no doubt as to what is required.

1.5 COMPLETENESS

a. It is in the completeness of items, such as the Special Conditions, that the expertise of the professional specification writer is revealed. How smoothly the project runs, or how many problems have to be resolved in the field or in the courtroom, largely depends on the ability and the care of the specification writer. Loopholes, ambiguities, and omissions lead to a problem job and give everyone connected a bad name and many sleepless nights. THE MORE COMPLETE AND TIGHT THE SPECIFICATIONS, THE HAPPIER THE JOB, THE GREATER THE PROFIT, AND THE MORE SATISFIED THE OWNER.

1.7 ADDITIONAL INFORMATION NEEDED FOR HVAC

a. Standard General Conditions as amended by the Special Conditions still leave much important information missing that is required by the Mechanical Sections. The mechanical work is very specialized and requires special instructions to the contractor who is bidding on and installing this type of work.

b. These additional instructions are usually incorporated in the beginning of the Technical Specification Section, where they are more closely identified with the work to be done under that Section.

1.8 ARCHITECTURALLY RELATED SYSTEMS

a. Where the HVAC Specifications are to be part of a document being prepared by another prime professional, they are usually allotted a section of those Specifications (usually Section 15000 or some part of that numbering series), and the additional information necessary for the HVAC Section is titled GENERAL and placed at the beginning of this Section.

b. Section 15100, contained in Chapter 2, is a guide for these additional requirements, as well as the introduction of the Technical Sections that follow.

c. Section 15101, in Chapter 2, is generally the same type of information for small jobs, and it is given in a short form.

d. It should be kept in mind by the Specification writer that the HVAC subcontractor is often not given a full set of specifications when asked to submit a bid. The subcontractor is not usually involved with much of the information contained in the General and Special Conditions directly, but the bid must reflect some of these requirements.

e. To be sure that the HVAC bidder is aware of the more important requirements, they are often repeated in an abbreviated form in the General Section of the HVAC Specifications.

f. The fact that the HVAC subcontractor is not given a full set of the contract documents is not to be condoned. It leads to many problems, but it is so common that it must be considered. The best way to live with this situation is to give the HVAC subcontractor enough information in the General Section to keep the subcontractor out of serious trouble.

1.9 DIRECT MECHANICAL WORK

a. Where the major part of the work under a contract is mechanical, rather than architectural, the General and Special Conditions more directly relate to the mechanical contractor, and the additional information required changes somewhat in nature and direction. Under these conditions the mechanical contractor is usually the prime contractor and may have other subcontractors performing part of the work.

b. Section 15102 of Chapter 2 is given as a guide to the additional information recommended in this situation.

END OF CHAPTER ONE

CHAPTER 2

Conditions for Mechanical Specifications

SECTION 15100
****

CONDITIONS FOR
HEATING, VENTILATION, AND AIR CONDITIONING SYSTEMS

****

THIS SECTION IS NORMALLY USED AS A PART OF A SET OF SPECIFICATIONS PREPARED BY AN ARCHITECT OR ANOTHER ENGINEER AND IS DESIGNED TO COVER A NORMAL HEATING, VENTILATING, AND AIR CONDITIONING SYSTEM TO BE INSTALLED IN A BUILDING.

THIS SECTION SHOULD FOLLOW THE GENERAL AND SPECIAL CONDITIONS OF THE OVERALL SET OF SPECIFICATIONS.

****

SECTION 15100
****
CONDITIONS FOR
HEATING, VENTILATION, AND AIR CONDITIONING SYSTEMS

PART ONE - GENERAL

1.1 These Conditions add to, and supplement, the GENERAL and SPECIAL CONDITIONS of the Contract Documents enumerated earlier. The General and Special Conditions are a part of this Section and shall apply as if written in full herein.

1.2 SCOPE

a. The work included in this Section shall consist of providing all materials, labor, tools, plant, and incidentals necessary to install and make ready for owner's use a complete HEATING, VENTILATION, AND AIR CONDITIONING SYSTEM (HVAC) for the proposed building as called for in the Contract Documents.

b. Review all drawings and visit the site; observe dimensions, construction, and details not shown on drawings.

c. The following sections of Technical Specifications are a part of the Contract Documents and are included as part of the HVAC system. The system shall include, but is not limited to, the following Sections:

SECTION	NAME	PAGES IN SECTION
15111	Balancing and Testing	4

SPECIFICATION WRITER SHOULD INSERT THE NUMBER, NAME, AND NUMBER OF PAGES OF EACH SECTION OF THE TECHNICAL SPECIFICATIONS. SECTION 15111, BALANCING AND TESTING, HAS BEEN ADDED AS AN EXAMPLE ONLY.

1.3 RESPONSIBILITY OF BIDDERS

a. Bidders shall examine all drawings and specifications issued and SHALL VISIT THE SITE of the work. Bidders must be familiar with the codes, rules, and regulations (and the local interpretations) in effect at the site of the work.

b. Where any of the above are at variance with the drawings and specifications, the code requirements shall take precedence, and any cost necessary to meet these shall be included in the bid price.

c. This contractor is assumed to be skilled in the trade and is solely responsible for compliance with health and safety regulations, performing the work in a safe and competent manner, and in installation procedures required for the work as outlined in these documents.

1.4 MECHANICAL PLANS

a. The mechanical plans are intended to be diagrammatic and are based on one manufacturer's equipment. They are not intended to show every item in its exact location, the exact dimensions, or all the details of the equipment. The contractor shall verify the actual dimensions of the equipment proposed to ensure that the equipment will fit in the available space.

b. Installation shall be within the limitations imposed by the architectural, structural, electrical, and plumbing requirements, with adequate space for maintenance.

1.5 QUESTIONS AND CLARIFICATIONS OF BID DOCUMENTS

a. Bidders shall not rely on any verbal clarification of the drawings or specifications. Any questions or clarifications shall be referred to the architect and/or engineer at least five working days prior to bidding to allow for issuance of an addendum. After the five-day deadline, bidder shall make a decision and qualify the bid, if the bidder feels it necessary.

1.6 TEN-DAY PRIOR APPROVAL

a. Any equipment or components proposed for this project, other than model numbers named in the bid documents, shall have pertinent submittal data and descriptive cover sheet submitted to the architect with a copy to the engineer 10 days prior to the bid date for inclusion in an addendum if, and when, reviewed and accepted for bidding.

b. This is for prebid review and is not to be regarded as submittals required for construction.

c. Bidder shall base the bid on items of equipment actually named in bid documents or addendums issued prior to bidding. Verbal acceptance will not be recognized unless verified in writing. It is the bidder's responsibility to ascertain that all equipment has been accepted by requiring copies of the written acceptance from suppliers.

1.7 GUARANTEES

a. All equipment, materials, and workmanship shall be guaranteed for a period of one year, beginning with the date of acceptance of the project in writing. Special warranties will be called for under some sections of EQUIPMENT. This warranty shall be in writing and shall include written copies of factory warranties with expiration dates on items of equipment where the warranty date might differ from the acceptance date, such as five-year warranty of sealed refrigerant systems. No warranty shall start before the acceptance date.

b. The contractor's warranty shall include at least two inspections of the system to repair and replace any items found to be defective during this period. The first shall be approximately six months after the acceptance of the system and the second at the end of the first year.

1.8 REQUIRED SUBMITTALS

a. Prior to starting any installation, submit no less than 5 nor more than 10 copies of items proposed for this work with necessary illustrations, drawings, and engineering data for review by the engineer. Submit in time to allow no less than 15 working days for checking and transmittal without delaying the construction schedule. Submit all items at one time no less than 30 days after award of the contract.

b. Submittals shall be clearly marked to show the intended item, with identification as to unit number or other marking to show location, service, and function. Submittals not marked to identify the equipment and application will be rejected.

c. Any equipment installed without prior acceptance shall be subject to rejection unless such items were identified by name and model number in the bid documents.

d. The supplier, by submitting, certifies that the materials or equipment proposed is satisfactory for the application intended, including adverse conditions that may prevail at the job site, and that the materials and equipment are in current production with no known plans to cease production.

e. Contractor agrees that submittals processed by the engineer are not change orders; that the purpose of submittals by the contractor is to demonstrate to the engineer that the contractor understands the design concept; and that this understanding is demonstrated by indicating which equipment and materials he or she intends to furnish and install and the fabrication and installation methods he or she intends to use.

f. Contractor further agrees that if deviations, discrepancies, or conflicts tetween submittals and contract documents are discovered either prior to or after submittals are processed by the engineer, the contract documents shall control and shall be followed.

g. Submittals shall include:

1. All equipment; cooling, heating, ventilation, etc.
2. Voltage, phase, and amps of each electrical item such as motors, heaters, etc.
3. All auxiliary equipment
4. Ductwork shop drawings, insulation, etc.
5. Supply and return registers, grilles, etc.
6. Pipe, valves, insulation, etc.

h. At the close of the job, prior to final review, five bound copies of the following shall be submitted by transmittal to the architect or engineer for review and acceptance.

1. Equipment warranties
2. Contractor's warranty
3. Parts list and manuals for all equipment
4. Balance and Test readings
5. Operating Instructions (in writing)
6. Written instructions on maintenance and care of the system

PART TWO - PRODUCTS

2.1 GENERAL

a. All products shall be first-line quality, of grade and type shown on the drawings and specified, or equivalents accepted by the architect or engineer in writing.

b. All products shall be in current production with no notice having been given that this product is to be drastically changed, modified, or discontinued from production.

c. The supplier, by submitting, certifies that equipment being proposed is proper for the application intended and that it has the capacity called for.

2.2 COMPLETE SYSTEM

a. All products, materials, and accessories shall be furnished and installed as required for a complete system ready for owner's beneficial use.

2.3 EQUIPMENT AND MATERIAL DEVIATIONS

a. When any material or equipment is identified on the plans or in the Specifications by reference to ONE manufacturer's name or model number, it is intended to establish a required standard of design and quality, and it is not intended to limit competition. It is understood that the phrase "or accepted equivalent" is hereby inserted following the one manufacturer's name, whether such phrase occurs or not.

b. When the drawings and/or specifications indicate ONE or TWO manufacturers' names for material or equipment, the bidder may submit a bid based on material or equipment of manufacturers not named but considered by the bidder to be equal to the standard of design and quality specified; however, such substitution must be accepted by the architect or engineer as equal. If the bidder elects to bid on a substitution without securing written consent of the architect or engineer prior to receipt of bids, then it will be understood that proof of compliance with specified requirements is the direct responsibility of the bidder and no such materials or equipment may be purchased or installed without written acceptance. Where THREE names are given for any equipment, THE CONTRACTOR SHALL FURNISH ONE OF NAMED MANUFACTURERS.

c. Bidders are advised to ascertain such acceptance from their suppliers by requesting copies of the acceptance in writing signed by the architect or engineer from their suppliers.

2.4 MOTORS AND STARTERS

a. All electric motors shall be high-efficiency type with maximum of 1750 rpm with open drip-proof enclosures. Motors located on air handling units shall be mounted in rubber supports, or the fan shall be independently supported on spring isolators. Motors located in the conditioned space shall be selected for quiet operation and shall not produce an objectionable "motor noise" in the space.

b. Electrical characteristics shall be determined from the Electrical Drawings and verified on the job from the electrical contractor.

c. Motor starters shall be sized by the National Electrical Code (NEC), and proper heater elements shall be provided and installed. Starters shall have overload trip element in each phase.

d. Larger motors and their starters shall meet the requirements of the utility company as to inrush allowable and the type of starting permitted.

2.5 LABELING

a. All electrical equipment and items consisting primarily of electrical components shall bear a label of an independent testing laboratory, such as Underwriters' Laboratories (UL).

b. Where such testing and labeling service is available for other products, such as fire dampers, boilers, etc., the equipment shall bear such a label.

PART THREE - EXECUTION

3.1 WORKMANSHIP

a. All work shall be performed by competent mechanics using proper tools and equipment to produce first-quality work. All work shall be neatly installed, accessible for maintenance, and complete with all accessories required.

3.2 ACCESSIBILITY

a. All equipment shall be installed in such a way that all components requiring access (such as drain pans, drains, fire dampers, control dampers, control operators, motors, drives, etc.) are so located and installed that they may be serviced, reset, replaced or recalibrated, etc., by service people with normal service tools and equipment. If any equipment or components are shown in such a position that this contractor cannot comply with the above, the contractor shall notify the general contractor and attempt to resolve the problem of access. If this consultation is not successful, the architect and engineer shall be notified in writing and a decision requested.

3.3 WORK BY OTHER TRADES

a. Cutting, patching, furring, painting, electrical, plumbing, etc., shall be done by the affected trade at this contractor's expense for changes required in work already installed or work required by other trades for changes made by this contractor in type or size of equipment purchased.

3.4 WORK NOT INCLUDED

a. Openings in floors, walls, and roof shall be furnished by the general contractor. This contractor shall inform the general contractor of the location and size required. This contractor shall furnish all sleeves, frames, including framing between joist unless shown on the Architectural or Structural drawings, access doors, prefabricated curbs, and other accessories necessary for a complete installation. Only those items specifically shown and/or specified in other Sections are excluded.

b Flashing of roof for curbs, pipes, stands, etc., shall be by the general contractor (roofer). (Curbs and counter flashing shall be by this contractor.)

c. Power wiring, including final connections, is by the electrical contractor. (This contractor shall install all motors and furnish the starting equipment to the electrical contractor for installation. Control wiring, including 115 volt from power source, conduit, switches, thermostats, interlocks, etc., shall be furnished by this contractor unless specifically shown on the Electrical Drawings. This contractor shall see that the electrical equipment does not block access to service areas of equipment, i.e., disconnect switches mounted on the compressor or control access doors of equipment.)

d. Power and fuel for testing. (General contractor shall provide power for buildings under construction. See General and Special Conditions.) (Owner shall provide power for existing buildings unless stated otherwise in General and Special Conditions.)
e. Door grille installation. (Grilles to be furnished by this contractor.)
f. Floor drains, hub drains, etc.

3.5 WORK FOR OTHER TRADES

a. This contractor shall furnish all vents required for water heaters, dryer vents, and other mechanical openings, as shown on the drawings.

3.6 FOUNDATIONS AND SPECIAL SUPPORTS

a. Furnish and install all special foundations and supports required for equipment installed under this Section, unless they are a part of the building structure and are shown in other Sections.

3.7 NOISE AND VIBRATION

a. Install vibration isolators, flexible connectors, expansion joints, and other safety measures to prevent noise and vibration from being transmitted to occupied areas. Equipment shall be selected to operate within the noise level recommended for the particular type installation in relation to its location.
b. Following installation, make proper adjustments to eliminate excessive noise and vibration.

3.8 PERMITS, CODES, AND LAWS

a. All work shall be in accordance with the following rules and regulations and any applicable laws.

National Fire Protection Association (NFPA)
Occupational Safety and Health Administration (OSHA)
State Building Code (SBC)
Local Building Codes (LBC)

b. Where any of the above are at variance with the drawings and specifications, the code requirements shall take precedence and any cost necessary to meet these shall be included in the contract.
c. This contractor is assumed to be skilled in the trade and is solely responsible for compliance with OSHA regulations, performing the work in a safe and competent manner, and in installation procedures required for this work. All supervision assigned to this project shall be experienced in this type of work. This contractor's superintendent shall be designated as Safety Inspector, unless the contractor designates another person and notifies the engineer of this change.

3.9 REVIEW BY ENGINEER

a. This contractor shall notify the engineer at the following stages of construction so that the engineer may visit the site for review and consultation:

1. When ductwork installation starts.
2. When equipment installation starts.
3. When refrigerant lines are to be run.
4. When the units are to be charged.
5. When ceiling installation will cover any work not reviewed.
6. When any lines or ducts are to be permanently concealed by construction.
7. When Balancing and Testing is started.

b. Should this contractor fail to notify the engineer at the times prescribed above, it shall then be his or her responsibility and cost to make ductwork accessible, expose any concealed lines or demonstrate the acceptability of any part of the system. Any extra cost, caused by the removal of work by other trades, shall be borne by this contractor.

3.10 EARLY START-UP

a. This contractor shall do all possible to see that the mechanical equipment is connected with electrical power as early as possible, so that final balancing and testing can be started. Should this contractor be ready for operation and power is not available, the general contractor and the architect or engineer shall be notified.

3.11 CLEANING AND PAINTING

a. Thoroughly clean all equipment and remove all trash, cartons, etc., from the area. Make any necessary corrections or repair/replace any damaged materials or equipment. Leave the entire system in a thoroughly clean and orderly manner.

b. Any finished surfaces that have been scratched or discolored shall be touched up or repainted with paint to match the original color. If any part has been bent, broken, or otherwise damaged, it shall be replaced prior to final review.

c. All metal items inside the building subject to rusting, and all ferrous metal exposed to weather, shall be given one coat of rust preventive primer as soon as installed.

END OF SECTION 15100

SECTION 15101
****
CONDITIONS FOR HEATING, VENTILATION, AND AIR CONDITIONING

****

THIS SECTION IS A "SHORT FORM", DESIGNED TO BE USED WITH SMALL JOBS WHERE THE AMOUNT OF WORK TO BE DONE DOES NOT WARRANT A FULL SET OF CONDITIONS, BUT IT IS STILL NECESSARY TO COVER THE ESSENTIALS.

THE SPECIFICATION WRITER SHOULD KEEP IN MIND THAT THE SIZE OF ANY JOB IS RELATIVE. IT MAY NOT BE LARGE IN DOLLAR VOLUME, BUT IT IS LARGE TO THE OWNER, SINCE IT IS THE OWNER'S MONEY AND HE OR SHE WANTS JUST AS MUCH FOR THE MONEY AS THE LARGE OWNER.

****

SECTION 15101
****
CONDITIONS FOR HEATING, VENTILATION, AND AIR CONDITIONING

PART ONE - GENERAL

1.1 These Specifications cover furnishing of all required equipment and the complete installation of the systems shown on the attached Drawings or defined in these Specifications.
1.2 These Specifications are abridged "short-form" specifications, due to the limited nature of the size of the system involved. All work shown or implied is to be performed in workmanlike manner, and all materials are to be first-line quality of the proper type, capacity, and size needed to make a complete and proper operating system ready for the owner's beneficial use, including the furnishing and installation of all utility connections required for the equipment.
1.3 All equipment is to be installed as recommended by the manufacturer, using all accessory equipment available from the manufacturer for supports, controls, etc., to make a complete system. All equipment or accessories needed and not shown or specified shall be furnished and installed by this contractor.
1.4 This contractor shall comply with all Codes and Regulations in effect at the job site, and obtain all approvals from regulatory agencies and deliver copies of such approvals to the architect or engineer. This contractor is solely responsible for the safety of his or her employees and compliance with all Health and Safety Regulations. No unsafe practices, equipment, or tools shall be allowed on the job site.

PART TWO - PRODUCTS

2.1 All products shall be the first-line equipment of recognized manufacturers. Submit seven copies of brochures, giving full details of the equipment and information on capacity, to the architect or engineer prior to purchasing of the equipment, and secure written consent for the use of any proposed equipment prior to the installation on the job.
2.2 All electrical equipment, or equipment containing electrical items, shall be UL-labeled. Check local codes for any other approvals required.

PART THREE - EXECUTION

3.1 All equipment and materials shall be installed in the best manner, as recommended by the manufacturer. Follow manufacturer's recommendations on the use of accessories and auxiliary equipment to provide for a complete first-class system ready for the owner's beneficial use.
3.2 Make all connections to equipment as recommended by the equipment manufacturer as far as traps, drains, flexible connections, etc., and as required by the equipment and the location.
3.3 Adjust the equipment for proper operation, check all controls, and verify that all safety devices are functioning properly.
3.4 Check all pipe and ductwork for leaks and excessive air loss. Correct any problems as soon as found.
3.5 Operate the system as a test and demonstrate to the owner and architect or engineer that the system is functioning properly.

PART FOUR - RELATED SECTIONS

4.1 The system shall include, but is not limited to, the labor, materials, and equipment required in the following sections of the Specifications and the Drawings:

SECTION	NAME	PAGES IN SECTION
15111	Balancing and Testing	4

(SPECIFICATION WRITER SHOULD INSERT THE NUMBER, NAME, AND NUMBER OF PAGES OF EACH SECTION OF THE TECHNICAL SPECIFICATIONS. SECTION 15111, BALANCING AND TESTING, HAS BEEN ADDED AS AN EXAMPLE ONLY.)

END OF SECTION 15101

SECTION 15102
****
CONDITIONS
FOR
MECHANICAL SYSTEMS

****

THIS SECTION IS DESIGNED TO BE USED FOR A MECHANICAL SYSTEM WHEN IT IS THE MAJOR PART OF THE CONTRACT.

THE MECHANICAL CONTRACTOR IS EXPECTED TO BE THE PRIME CONTRACTOR, AND OTHER WORK, SUCH AS BUILDING MODIFICATIONS, ELECTRICAL, PLUMBING, ETC., IS TO SUPPORT THE MECHANICAL.

****

SECTION 15102
****
CONDITIONS
FOR
MECHANICAL SYSTEMS

PART ONE - GENERAL

1.1 These Conditions add to, and supplement, the GENERAL and SPECIAL CONDITIONS of the Contract Documents enumerated earlier. The General and Special Conditions are a part of this Section and shall apply as if written in full herein.

1.2 SCOPE

a. The work included in this Section shall consist of providing all materials, labor, tools, plant, and incidentals necessary to install and make ready for owner's use a complete MECHANICAL SYSTEM as called for in the Contract Documents.

b. The following sections of Technical Specifications are a part of the Contract Documents and are included as part of the HVAC system. The system shall include, but is not limited to, the following Sections:

SECTION	NAME	PAGES IN SECTION
15111	Balancing and Testing	4

(SPECIFICATION WRITER SHOULD INSERT THE NUMBER, NAME, AND NUMBER OF PAGES OF EACH SECTION OF THE TECHNICAL SPECIFICATIONS. SECTION 15111, BALANCING AND TESTING, HAS BEEN ADDED AS AN EXAMPLE ONLY.)

1.3 RESPONSIBILITY OF BIDDERS

a. Bidders shall examine all drawings and specifications issued and SHALL VISIT THE SITE of the work. Bidders must be familiar with the codes, rules, and regulations in effect at the site of the work.

b. Where any of the above are at variance with the drawings and specifications, the code requirements shall take precedence and any cost necessary to meet these shall be included in the bid price.

c. This Contractor is assumed to be skilled in his trade and is solely responsible for compliance with health and safety regulations, performing the work in a safe and competent manner, and in installation procedures required for the work as outlined in these documents.

1.4 MECHANICAL PLANS

a. The mechanical plans are intended to be diagrammatic and are based on one manufacturer's equipment. They are not intended to show every item in its exact location, the exact dimensions, or all the details of the equipment. The Contractor shall verify the actual dimensions of the equipment proposed to assure himself that the equipment will fit in the available space.

b. Installation shall be within the limitations imposed by the architectural, structural, electrical, and plumbing requirements, with adequate space for maintenance.

1.5 QUESTIONS AND CLARIFICATIONS OF BID DOCUMENTS

a. Bidders shall not rely on any verbal clarification of the drawings or specifications. Any questions or clarifications shall be referred to the engineer at least five working days prior to bidding to allow for issuance of an addendum. After the five-day deadline, bidder shall make a decision and qualify the bid, if the bidder feels it necessary.

1.6 TEN-DAY PRIOR APPROVAL

a. Any equipment or components proposed for this project, other than manufacturer's equipment actually named in the bid documents, shall have pertinent submittal data and descriptive cover sheet submitted to the engineer 10 days prior to the bid date for inclusion in an addendum if, and when, reviewed and accepted for bidding.

b. This is for prebid review and is not to be regarded as submittals required for construction.

c. Bidder shall base the bid on items of equipment actually named in bid documents or addendums issued prior to bidding. Verbal acceptance will not be recognized unless verified in writing. It is the bidder's responsibility to ascertain that all equipment has been accepted by requiring copies of the written acceptance from suppliers.

1.7 GUARANTEES

a. All equipment, materials, and workmanship shall be guaranteed for a period of one year, beginning with the date of acceptance of the project in writing. Special warranties will be called for under some sections of EQUIPMENT. This warranty shall be in writing and shall include written copies of factory warranties with expiration dates on items of equipment where the warranty date might differ from the acceptance date, such as five-year warranty of sealed refrigerant systems. No warranty shall start before the date of acceptance in writing by the engineer.

b. The contractor's warranty shall include at least two inspections of the system to repair and replace any items found to be defective during this period. The first shall be approximately six months after the acceptance of the system and the second at the end of the first year. Engineer shall be provided with copies of every report of problems found and corrected. Any problems reported by the owner that were not corrected shall be called to the attention of the engineer and the reason for not correcting the problem shall be detailed.

1.8 REQUIRED SUBMITTALS

a. Prior to starting any installation, submit no less than 5 nor more than 10 copies of items proposed for this work with necessary illustrations, drawings, and engineering data for review by the engineer. Submit in time to allow no less than 15 working days for checking and transmittal without delaying the construction schedule. Submit all items at one time no less than 30 days after award of the contract.

b. Submittals shall be clearly designated as to the intended item with identification as to unit number or other marking to show location, service, and function. Submittals not marked to identify the equipment and application will be rejected.

c. Any equipment installed without prior acceptance shall be subject to rejection unless such items were identified by name and model number in the bid documents.

d. The supplier, by submitting, certifies that the materials or equipment proposed is satisfactory for the application intended, including adverse conditions that may prevail at the job site, and that the materials and equipment are in current production with no known plans to cease production.

e. Contractor agrees that submittals processed by the engineer are not change orders; that the purpose of submittals by the contractor is to demonstrate to the engineer that the contractor understands the design concept; and that the contractor demonstrates this understanding by indicating which equipment and materials he or she intends to furnish and install and the fabrication and installation methods he or she intends to use.

f. Contractor further agrees that if deviations, discrepancies, or conflicts between submittals and contract documents are discovered either prior to or after submittals are processed by the engineer, the contract documents shall control and shall be followed.

g. Submittals shall include:

1. All equipment; cooling, heating, ventilation, electrical motors, starters, controls, etc.
2. Voltage, phase, and amps of each electrical item, such as motors, heaters, etc.
3. All auxiliary equipment
4. Ductwork shop drawings, insulation, etc.
5. Supply and return registers, grilles, etc.
6. Pipe, valves, insulation, etc.

h. At the close of the job, prior to final review, five bound copies of the following shall be submitted by transmittal letter to the engineer for review and acceptance.

1. Equipment warranties
2. Contractor's warranty
3. Parts list and manuals for all equipment
4. Balance and Test readings
5. Operating Instructions (in writing)
6. Written instructions on maintenance and care of the system

PART TWO - PRODUCTS

2.1 GENERAL

a. All products shall be first-line quality, of grade and type shown on the drawings and specified, or equals accepted by the engineer in writing.

b. All products shall be in current production with no notice having been given that such product is to be drastically changed, modified, or discontinued from current production by the manufacturer proposed.

c. The supplier, by submitting, certifies that equipment being proposed is proper for the application intended and that it has the capacity called for.

2.2 COMPLETE SYSTEM

a. All products, materials, and accessories shall be furnished and installed as required for a complete system ready for owner's beneficial use.

2.3 EQUIPMENT AND MATERIAL DEVIATIONS

a. When any material or equipment is identified on the plans or in the specifications by reference to ONE manufacturer's name or model number, it is intended to establish a required standard of design and quality, and it is not intended to limit competition. It is understood that the phrase "or accepted equivalent" is hereby inserted following the one manufacturer's name, whether such phrase occurs or not.

b. When the drawings and/or specifications indicate ONE or TWO manufacturers' names for material or equipment, the bidder may submit a bid based on material or equipment of manufacturers not named but considered by the bidder to be equal to the standard of design and quality specified; however, such substitution must be accepted by the engineer as equal. If the bidder elects to bid on a substitution without securing written consent of the engineer prior to receipt of bids, then it will be understood that proof of compliance with specified requirements is the direct responsibility of the bidder and no such materials or equipment may be purchased or installed without written acceptance. Where THREE names are given for any equipment, the contractor shall furnish one of the named manufacturers.

c. Bidders shall ascertain such acceptance from their suppliers by requiring copies of the acceptance letter signed by the engineer.

2.4 MOTORS AND STARTERS

a. All electric motors shall be high-efficiency type with maximum of 1750 rpm with open drip-proof enclosures. Motors located on air handling units shall be mounted in rubber supports, or the fan shall be independently supported on spring isolators. Motors located in the conditioned space shall be selected for quiet operation and shall not produce an objectionable "motor noise" in the space.

b. Electrical characteristics shall be determined from the Electrical Drawings and verified on the job by the electrical contractor.

c. Motor starters shall be sized by the National Electrical Code (NEC), and proper heater elements shall be provided and installed as recommended by the manufacturer. Starters shall have overload trip element in each phase.

d. Larger motors and their starters shall meet the requirements of the utility company as to inrush allowable and the type of starting permitted.

2.5 LABELING

a. All electrical equipment and items consisting primarily of electrical components shall bear a label of a testing laboratory, such as Underwriters' Laboratories (UL).

b. Where such testing and labeling service is available for other products, such as fire dampers, boilers, etc., the equipment shall bear such a label.

PART THREE - EXECUTION

3.1 WORKMANSHIP

a. All work shall be performed by competent mechanics using proper tools and equipment to produce first-quality work. All work shall be neatly installed, accessible for maintenance, and complete with all accessories required.

3.2 ACCESSIBILITY

a. All equipment shall be installed in such a way that all components requiring access (such as drain pans, drains, fire dampers, control dampers, control operators, motors, drives, etc.) are so located and installed that they may be serviced, reset, replaced or recalibrated, etc., by service people with normal service tools and equipment. If any equipment or components are shown in such a position that this contractor cannot comply with the above, the contractor shall notify the engineer in writing and a decision requested.

3.3 WORK FOR OTHER TRADES

a. This contractor shall furnish all vents required for water heaters, dryer vents, and other mechanical openings, as shown on the drawings.

3.4 FOUNDATIONS AND SPECIAL SUPPORTS

a. Furnish and install all special foundations and supports required for equipment installed.

3.5 NOISE AND VIBRATION

a. Install vibration isolators, flexible connectors, expansion joints, and other safety measures to prevent noise and vibration from being transmitted to occupied areas. Equipment shall be selected to operate within the noise level recommended for the particular type installation in relation to its location.

b. After installation, make proper adjustments to eliminate excessive noise and vibration.

3.6 PERMITS, CODES, AND LAWS

a. All work shall be in accordance with the following rules and regulations and any applicable laws:

National Fire Protection Association (NFPA)
Occupational Safety and Health Administration (OSHA)
State Building Code (SBC)
Local Building Codes (LBC)

b. Where any of the above are at variance with the drawings and specifications, the code requirements shall take precedence, and any cost necessary to meet these shall be included in the contract.

c. This contractor is assumed to be skilled in the trade and is solely responsible for compliance with OSHA regulations, performing the work in a safe and competent manner, and in installation procedures required for this work. All supervision assigned to this project shall be experienced in this type of work. This contractor's superintendent shall be designated as the "safety inspector" unless the contractor appoints another person and notifies the engineer of this change.

d. Should asbestos be encountered in small amounts in any existing system (such as built-up insulation on fittings of otherwise acceptable systems), this contractor shall be responsible for the proper handling of any that might be disturbed. This is not an asbestos removal contract. Should substantial amounts of asbestos be encountered, this contractor shall notify the engineer and the owner prior to disturbing the asbestos.

3.7 REVIEW BY ENGINEER

a. This contractor shall notify the engineer at the following stages of construction so that the engineer may visit the site for review and consultation:

1. When ductwork installation starts
2. When equipment installation starts
3. When refrigerant lines are to be run
4. When the units are to be charged
5. When ceiling installation is started
6. When any lines or ducts are to be permanently concealed by construction.
7. When Balancing and Testing is started
8. When major equipment is set and connected

b. Should this contractor fail to notify the engineer at the times prescribed above, it shall then be the contractor's responsibility and cost to make ductwork accessible, expose any concealed lines, or demonstrate the acceptability of any part of the system. Any extra cost caused by the removal of such work shall be borne by this contractor or the subcontractor responsible.

3.8 EARLY START-UP

a. This contractor shall do all possible to see that the mechanical equipment is connected with electrical power as early as possible so that final balancing and testing can be started. Any reason for delay beyond the control of the contractor shall be reported to the engineer.

3.9 CLEANING AND PAINTING

a. Thoroughly clean all equipment and remove all trash, cartons, etc., from the area. Make any necessary corrections or repair/replace any damaged materials or equipment. Leave the entire work area in a thoroughly clean and orderly manner.

b. Any finished surfaces that have been scratched or discolored shall be touched up or repainted with paint to match the original color. If any part has been bent, broken, or otherwise damaged, it shall be replaced prior to final review.

c. All metal items inside the building subject to rusting, and all ferrous metal exposed to weather, shall be given one coat of rust preventive primer as soon as installed.

END OF SECTION 15102

SECTION 15111
****
BALANCING AND TESTING OF AIR AND WATER SYSTEMS

****

THIS SECTION ON BALANCING AND TESTING SHOULD BE USED WITH ANY SPECIFICATIONS THAT INCLUDE AIR CONDITIONING EQUIPMENT OR HEATING EQUIPMENT WHERE WATER OR AIR IS USED AS A TRANSFER MEDIUM. IT WOULD NOT BE APPROPRIATE FOR A STEAM RADIATOR HEATING SYSTEM, BUT APPLIES TO MOST OTHER SYSTEMS.

****

SECTION 15111

****

BALANCING AND TESTING OF AIR AND WATER SYSTEMS

PART ONE - GENERAL

1.1 The GENERAL and SPECIAL CONDITIONS, Section 15100, are included as a part of this Section as though written in full in this document.

1.2 Scope of the Work shall include the furnishing and complete installation of the equipment covered by this Section, with all auxiliaries, ready for owner's use.

PART TWO - PRODUCTS

2.1 EVALUATION OF SYSTEM

a. This contractor shall furnish all materials and equipment necessary to properly measure the air capacity of the system, the electrical voltage and current, fan speeds, static pressures, air velocity, water pressure drops, refrigeration pressures, and all other readings normally necessary to evaluate the performance of a system, adjust the quantities to those called for, and test the system.

2.2 SYSTEM PERFORMANCE

a. This contractor is responsible for the performance of the equipment and the system he or she installs. Contractor cannot assume that supplier will ship equipment adjusted to meet the job requirements.

2.3 EQUIPMENT OPERATION

a. All equipment shall be checked for proper operation as soon as electrical power is available to do so. Any malfunction shall be reported to the manufacturer, and corrective action taken as soon as possible to prevent delay of the acceptance of the work.

2.4 EQUIPMENT PROBLEMS AND ADJUSTMENTS

a. Required adjustments and minor problems with mechanical equipment are to be expected to some extent, and it is this contractor's responsibility to determine if there are any in the work and to correct them without causing any undue alarm on the part of the owner and without delay of the job.

PART THREE - EXECUTION

3.1 INITIAL BALANCING - AIR SYSTEMS

a. As soon as electrical power is available, the contractor shall check all equipment for electrical problems, check rotation of motors, read voltage and current in each leg of each motor, heater, etc., and check the readings against the nameplate.

b. Complete ductwork as soon as possible and operate the evaporator fan(s) (with filters in place); adjust the units for maximum air supply by reading motor power. Supply outlets shall be adjusted to the required air quantity. If air quantity is in great excess, the fan speed shall be reduced until the noise level is acceptable. If the air quantity at this point is not up to the design, the contractor shall notify the manufacturer and the engineer!

c. The return air system shall then be adjusted to design capacity with the proper outside air.

d. Check and balance each exhaust system to the design air quantity. Excess exhaust air will not be permitted since this wastes energy.

e. After supply and return air are in balance and the quantity correct, adjust the outside air dampers to the air quantity shown on the drawings (if not shown, use 10% of the supply air quantity). If economizer control is specified, check for proper setting of the controls and for proper operation of the dampers (outside air and relief).

3.2 RESPONSIBILITY FOR PROPER BALANCING AND TESTING

a. The GENERAL CONTRACTOR is responsible for the performance of the entire building, including the work in this Section. After this contractor has completed the installation, the superintendent for the general contractor shall monitor the Balancing and Testing of the system and shall certify that the readings required under this Section have actually been made and that all systems are in actual operation. The Test and Balance data shall be signed by the general superintendent. At time of final review, if it is apparent that these readings have not been made, or that equipment is not in operation, the expense for the return of the engineer and/or the architect shall be billed to the general contractor.

3.3 READINGS REQUIRED TO BE REPORTED

a. The following readings shall be made and reported to the engineer after the building is balanced and all equipment is operating properly.

b. All readings shall be recorded on a print of the Mechanical system giving the actual raw data read for each supply and return opening, including exhaust hoods and openings. All readings made shall be recorded, and if any readings are invalid, they shall be identified as such. Any readings out of line shall be explained by a note on the print. (The original print shall be submitted to the engineer for review.) If additional copies are required, they may be transcribed from this print on to other copies.

c. Air quantity readings shall include:

1. Actual measured air quantity of each supply and return outlet shall be read and recorded. Measurement shall be made with a cone with a calibrated outlet and velometer equal to Alnor.
2. Same for each return or exhaust inlet.
3. Same for each hood, giving supply (if any) and exhaust.

d. Temperature readings required as above are:
   1. Outside air at equipment.
   2. Return air at unit.
   3. Supply air leaving unit.
   4. Mixture of outside and return air BEFORE entering the cooling or heating coil or heater. Readings 1, 2, and 4 allow the determination of the outside air/return air ratio.

e. Electrical readings required are:
   1. Measured voltage and amps on EACH phase of each major motor (compressor, evaporator fan, condenser fan(s), roof exhaust fans, etc.) while the equipment is under maximum normal load.
   2. The nameplate voltage and current for each of the above motors.

f. Refrigeration readings required are:
   1. Suction and discharge pressure of each compressor or, in the case of packaged condensing units, the suction and liquid line pressure.

g. Water readings required are:
   1. Water pressure on the inlet and outlet of each pump as close to the pump impeller as is possible.
   2. Water pressure drop across the suction strainer of each pump.
   3. Water pressure drop across the outlet valve and check valve on the pump discharge.
   4. Water pressure drop across each chiller, or boiler producing energy for the system. These readings shall be at the flanges of the equipment.
   5. Water pressure drop across every coil, heating element, etc., using energy from the system.
   6. The pressure drop calculations shall be applied against the manufacturers' rating sheets to determine the actual flow through the system and the equipment. These readings give the owner the basic information to determine at a later time if the equipment is fouled and if the water flow is being maintained at the design conditions.
   7. Any flow-measuring devices in the system shall be read and reported.
   8. Should any readings made above indicate that the flow in the system is below design, the contractor shall determine the reason for the difference and correct the problem so that the flow will be up to design conditions when the system is turned over to the owner.

3.4 SYSTEM DIFFICULTIES

a. The above readings shall be made on each unit or piece of equipment and these readings sent to the engineer for review as early as possible so that any apparent difficulties can be resolved before the anticipated close of the job and before such problems are called to the attention of the owner. Minor problems, such as the necessityi to adjust a fan sheave, often raise questions and doubts in the owner's mind about the system. Such problems are normal, and if corrected without delay lead to a much happier owner.

3.5 REVIEW BY ENGINEER

a. After the above information is received by the engineer, it will be reviewed and compared against the design. The engineer will generally review the job for the owner and recommend final acceptance or the holding of funds pending additional work. Such review will not be scheduled until the above information can be reviewed and accepted. The work required under this contract is not complete until this information is accepted as accurate and complete.

END OF SECTION 15111

CHAPTER 3

Central Station Refrigeration Equipment

SECTION 15150
****
RECIPROCATING COMPRESSOR UNIT
HERMETIC

PART ONE - GENERAL

1.1 The GENERAL and SPECIAL CONDITIONS, Section 15100, are included as a part of this Section as though written in full in this document.

1.2 Scope of the Work shall include the furnishing and complete installation of the equipment covered by this Section, with all auxiliaries, ready for owner's use.

PART TWO - PRODUCTS

2.1 COMPRESSOR

a. Compressor unit shall consist of reciprocating refrigeration compressor with factory-wired control panel. Compressor capacity and EER shall be as shown on the drawings at the conditions given. All electrical components shall be UL-labeled.
b. Compressor shall be reciprocating serviceable semihermetic unit with suction and discharge valves, automatically reversible oil pump, and with the operating oil charge.
c. Compressors in excess of 30 tons shall be provided with unloaders to allow capacity modulation in reasonable steps, to prevent short-cycling under reduced load. Unloading may be electrically operated or electrically controlled and operated by suction gas or oil pressure.
d. Compressor head assemblies shall be protected from liquid slugging.
e. Compressor suction and discharge valves shall be designed to prevent the flexing of the valves during operation.
f. Compressor shall be provided with a crankcase emersion-type heater to control dilution of the oil during shutdown.
g. Compressor motor shall not exceed 1750 rpm and shall be cooled by suction gas. Motor shall be protected by internal sensors from high temperature. Motor power rating shall not exceed that shown on the drawings.
h. Provide suction gas strainer to trap oil mist, liquid, and any foreign matter in the return gas.

2.2 CONTROL PANEL

a. Control panel shall be factory-wired and -installed. The electrical controls shall be separated from the refrigeration controls.
b. The electrical section shall contain terminal blocks for all wiring connections, including power and interlock and control wiring. This section shall contain the motor starter for part winding or across-the-line starter (as required by the local utility company), control transformer, control fuse, pumpdown control relay, compressor starter relay, reset relay, and nonrecycling compressor overload relay.

c. The refrigeration section shall contain the high-pressure control, low-pressure control, and solid-state motor protector with oil pressure cutout.

2.3 AUXILIARY EQUIPMENT

a. Provide gauges with refrigerant valve for suction, discharge, and oil pressure. Mount on the control panel and pipe to the compressor.

b. Provide hot gas muffler tuned to the compressor and mounted on the unit.

c. Provide rubber-in-shear isolators for mounting under the unit. Isolators shall be matched to the unit weight.

d. Provide acoustically lined compartment of heavy-gauge metal to reduce the amount of noise reaching adjacent spaces.

e. Provide suction line and hot gas line flexible connectors to match the unit capacity.

2.4 CONTROL SEQUENCE

a. Evaporator fan or pump shall be operating before compressor unit can be started. Condenser fan shall be cycled by head pressure control.

2.5 ACCEPTABLE MANUFACTURERS

a. Compressor units shall be the make and model number shown on the drawings or approved equivalents by Carrier (06 Series), Trane (Model CUAB), York.

PART THREE - EXECUTION

3.1 INSTALLATION

a. Install the unit in place on the vibration isolators specified, with clearances required for proper maintenance.

b. Install and connect the compressor to the condenser and the evaporator with lines of the size shown on the drawings and as recommended by the manufacturer. Install suction and hot gas flexible connectors. Provide traps and loops as required for proper oil return and to prevent slugging.

c. Make proper electrical connections for power, control, and interlock wiring as required by the codes.

d. Pressure-test the entire system with dry nitrogen and check for leaks.

e. Evacuate the system to a vacuum and hold for four hours at the low pressure.

f. Charge the system with the proper refrigerant and check for leaks using a freon detector.

g. Operate the system and check suction and discharge pressures for proper readings. Report these readings and the power draw of the motor as required in Balancing and Testing, Section 15111.

END OF SECTION 15150

SECTION 15151

****

WATER-COOLED CONDENSING UNIT
HERMETIC RECIPROCATING COMPRESSOR

PART ONE - GENERAL

1.1 The GENERAL and SPECIAL CONDITIONS, Section 15100, are included as a part of this Section as though written in full in this document.

1.2 Scope of the Work shall include the furnishing and complete installation of the equipment covered by this Section, with all auxiliaries, ready for owner's use.

PART TWO - PRODUCTS

2.1 CONDENSING UNIT

a. Condensing unit shall consist of reciprocating refrigeration compressor, water-cooled shell and tube condenser and factory-wired control panel. The unit capacity and EER shall be as shown on the drawings at the conditions given. All electrical components shall be UL-labeled.

2.2 COMPRESSOR

a. Compressor shall be reciprocating serviceable semihermetic unit with suction and discharge valves, automatically reversible oil pump, and with the operating oil charge.
b. Compressors in excess of 30 tons shall be provided with unloaders to allow capacity modulation in reasonable steps to prevent short-cycling under reduced load. Unloading may be electrically operated or electrically controlled and operated by suction gas or oil pressure.
c. Compressor head assemblies shall be protected from liquid slugging.
d. Compressor suction and discharge valves shall be designed to prevent the flexing of the valves during operation.
e. Compressor shall be provided with a crankcase emersion-type heater to control dilution of the oil during shutdown.
f. Compressor motor shall not exceed 1750 rpm and shall be cooled by suction gas. Motor shall be protected from high temperature by internal sensors. Motor power rating shall not exceed that shown on the drawings.
g. Provide suction gas strainer to trap oil mist, liquid, and any foreign matter in the return gas.

2.3 WATER-COOLED CONDENSER

a. Condenser shall have a steel shell with straight copper tubes with removable heads on both ends for cleaning and replacement of tubes. Tubes shall be integral finned type with refrigerant-side working pressure of 300 psi (2070 kPa/m) and water-side working pressure of 150 psi (1035 kPa/m). Condenser shall be ASME-stamped.
b. Provide safety relief valve set at the maximum working pressure.
c. Water pressure drop shall be maximum shown on the drawings. Low-pressure loss is required for maximum energy savings.

d. Provide a back-seated refrigerant valve on the liquid connection of the condenser, purge connection, and other standard connections required by the system.

2.4 CONTROL PANEL

a. Control panel shall be factory-wired and -installed. The electrical controls shall be separated from the refrigeration controls.

b. The electrical section shall contain terminal blocks for all wiring connections, including power and interlock and control wiring. This section shall contain the motor starter for part winding or across-the-line starter (as required by the local utility company), control transformer, control fuse, pumpdown control relay, compressor starter relay, reset relay, and nonrecycling compressor overload relay.

c. The refrigeration section shall contain the high-pressure control, low-pressure control, and solid-state motor protector with oil pressure cutout.

2.5 AUXILIARY EQUIPMENT

a. Provide gauges and refrigerant valves for suction, discharge, and oil pressure. Mount on the control panel and pipe to the compressor.

b. Provide hot gas muffler between the compressor and the condenser.

c. Provide rubber-in-shear isolators for mounting under the unit. Isolators shall be matched to the unit weight.

d. Provide acoustically lined compartment of heavy-gauge metal to reduce the noise of the unit reaching adjacent spaces.

e. Provide suction line flexible connector to match the unit capacity.

f. Provide flex connector in condenser piping connections.

2.6 CONTROL SEQUENCE

a. Evaporator fan or pump shall be operating before compressor unit can be started. Condenser water pump shall be started by system temperature controls, and the condensing unit controls will be energized after the water flow is proved by a flow switch or pressure differential switch.

b. See manufacturer's standard controls for safety controls.

2.7 ACCEPTABLE MANUFACTURERS

a. Compressor units shall be the make and model number shown on the drawings or approved equivalents by Carrier (07 Series), Trane (Model RWUB), York (LCH).

PART THREE - EXECUTION

3.1 INSTALLATION

a. Install the unit in place on the vibration isolators specified, with the clearances required for proper maintenance.

b. Install and connect the compressor to the condenser and the evaporator with lines of the size shown on the drawings and as recommended by the manufacturer. Install braided-type flex connector in suction line. Provide traps and loops as required for proper oil return and to prevent slugging.

c. Make proper electrical connections for power, control, and interlock wiring as required by the codes.

d. Pressure-test the entire system with dry nitrogen and check for leaks.

e. Evacuate the system to a vacuum and hold for four hours at the low pressure.

f. Charge the system with the proper refrigerant and check for leaks using a freon detector.

g. Operate the system and check suction and discharge pressures of the compressor for proper readings. Record the pressure readings as required in Balance and Testing, Section 15111.

h. Check the water quantity for proper flow. Report the water flow and the water pressure drop as measured across the heads, as required in Balancing and Testing, Section 15111.

i. Measure the power draw in each phase and the actual voltage of the motor, and report these readings as required in Balancing and Testing, Section 15111.

END OF SECTION 15151

SECTION 15152

****

WATER-COOLED LIQUID CHILLER
HERMETIC RECIPROCATING COMPRESSOR

PART ONE - GENERAL

1.1 The GENERAL and SPECIAL CONDITIONS, Section 15100, are included as a part of this Section as though written in full in this document.

1.2 Scope of the Work shall include the furnishing and complete installation of the equipment covered by this Section, with all auxiliaries, ready for owner's use.

PART TWO - PRODUCTS

2.1 LIQUID CHILLER

a. Chiller unit shall consist of reciprocating refrigeration compressor, water-cooled shell and tube condenser, evaporator, and factory-wired control panel. The unit capacity and EER shall be as shown on the drawings at the conditions given. All electrical components shall be UL-labeled.

2.2 COMPRESSOR

a. Compressor shall be reciprocating serviceable semihermetic unit with suction and discharge valves, automatically reversible oil pump, and with the operating oil charge.

b. Compressors in excess of 30 tons shall be provided with unloaders to allow capacity modulation in reasonable steps to prevent short-cycling under reduced load. Unloading may be electrically operated or electrically controlled and operated by suction gas or oil pressure.

c. Compressor head assemblies shall be provided with protection from liquid slugging.

d. Compressor suction and discharge valves shall be designed to prevent the flexing of the valves during operation.

e. Compressor shall be provided with a crankcase emersion-type heater to control dilution of the oil during shutdown.

f. Compressor motor shall not exceed 1750 rpm and shall be cooled by suction gas. Motor shall be protected from high temperature by internal winding sensors. Motor power rating shall not exceed that shown on the drawings.

g. Provide suction gas strainer to trap oil mist, liquid, and any foreign matter in the return gas.

2.3 WATER-COOLED CONDENSER

a. Condenser shall have a steel shell with straight copper tubes with removable heads on both ends for cleaning and replacement of tubes. Tubes shall be integral finned type with refrigerant-side working pressure of 300 psi (2070 kPa) and water-side working pressure of 150 psi (1035 kPa). Condenser shall be ASME-stamped.

b. Provide safety relief valve set at the maximum working pressure.
c. Water pressure drop shall be maximum shown on the drawings. Low pressure drop is required for maximum energy savings.
d. Provide a back-seated refrigerant valve on the liquid connection of the condenser, purge connection, and other standard connections required by the system.

2.4 EVAPORATOR

a. Evaporator shall be shell and tube type with straight copper tubes with integral external fins. Evaporator shall be designed for 225 psi (1552 kPa) refrigerant working pressure and 150 psi (1035 kPa) water-side working pressure. Evaporator shall be ASME-stamped.
b. Water pressure drop shall not exceed that shown on the drawings. Low pressure drop is required for maximum energy savings.
c. Provide water pressure drop connections on each head, drain provisions, and necessary control wells for safety and operating controls.
d. Evaporator shall be factory-insulated with closed-cell insulation having a "K" factor of not less than 0.26. Insulation on heads shall be removable and replaceable for access to the heads for service.

2.5 CONTROL PANEL

a. Control panel shall be factory-wired and -installed. The electrical controls shall be separated from the refrigeration controls.
b. The electrical section shall contain terminal blocks for all wiring connections, including power and interlock and control wiring. This section shall contain the motor starter for part winding or across-the-line starter (as required by the local utility company), control transformer, control fuse, pumpdown control relay, compressor starter relay, reset relay, and nonrecycling compressor overload relay.
c. The refrigeration section shall contain the high-pressure control, low-pressure control, and solid-state motor protector with oil pressure cutout.

2.6 AUXILIARY EQUIPMENT

a. Provide gauge with refrigerant valve for suction, discharge, and oil pressure. Mount on the control panel and pipe to the compressor.
b. Provide hot gas muffler between the compressor and the condenser.
c. Provide rubber-in-shear isolators for mounting under the unit. Isolators shall be matched to the unit weight.
d. Provide acoustically lined compartment of heavy-gauge metal to reduce the noise of the unit reaching adjacent spaces.

2.7 CONTROL SEQUENCE

a. Chilled liquid pump shall be operating and the flow proved by a flow switch or pressure differential switch before the condenser water pump can be energized. Condenser water pump shall be started by chilled liquid return temperature controls, and the condensing unit controls will be energized after the water flow is proved by a flow switch or pressure differential switch.

b. See manufacturer's standard controls for safety and interlock controls.

2.8 ACCEPTABLE MANUFACTURERS

a. Liquid chiller shall be the make and model number shown on the drawings or approved equivalents by Carrier (30HK Series), Trane (Model CGW), York (LCH).

PART THREE - EXECUTION

3.1 INSTALLATION

a. Install the unit in place on the vibration isolators specified, with clearances required for proper maintenance.

b. Install and connect the chilled liquid lines to the evaporator and the condenser water lines to the condenser with pipe of the size shown on the drawings and as recommended by the manufacturer. (See piping details shown on the drawings.) All connections shall allow for easy removal of the heads for servicing of the condenser and the evaporator. Allow room for tube removal if necessary.

c. Make proper electrical connections for power, control, and interlock wiring as required by the codes.

d. Operate the system and check suction and discharge pressures of the compressor for proper readings. Read and record refrigerant suction and discharge pressure readings, as required in Balance and Testing, Section 15111.

e. Check the water quantity for proper flow through the evaporator and the condenser. Report the water flow and the water pressure drop as measured across the heads, as required in Balancing and Testing, Section 15111.

f. Measure the power draw in each phase and the actual voltage of the motor and report these readings, as required in Balancing and Testing, Section 15111.

END OF SECTION 15152

SECTION 15153

****

WATER-COOLED LIQUID CHILLER
DUAL HERMETIC RECIPROCATING COMPRESSORS

PART ONE - GENERAL

1.1 The GENERAL and SPECIAL CONDITIONS, Section 15100, are included as a part of this Section as though written in full in this document.

1.2 Scope of the Work shall include the furnishing and complete installation of the equipment covered by this Section, with all auxiliaries, ready for owner's use.

PART TWO - PRODUCTS

2.1 LIQUID CHILLER

a. Chiller unit shall consist of dual reciprocating refrigeration compressors, water-cooled shell and tube condenser, shell and tube evaporator, and factory-wired control panel.

b. Chiller shall require only one power feed supplying both compressors through the control panel.

c. The unit capacity and EER shall be as shown on the drawings at the conditions given. All electrical components shall be UL-labeled.

2.2 COMPRESSORS

a. Compressors shall be reciprocating serviceable semihermetic type with suction and discharge back-seated refrigerant valves, automatically reversible oil pump, and with the operating oil and freon charge.

b. Compressors in excess of 30 tons shall be provided with unloaders to allow capacity modulation in reasonable steps in order to prevent short-cycling under reduced load. Unloading may be electrically operated or electrically controlled and operated by suction gas or oil pressure.

c. Compressor head assemblies shall be provided with protection from liquid slugging.

d. Compressor suction and discharge valves shall be designed to prevent the flexing of the valves during operation.

e. Compressor shall be provided with a crankcase emersion-type heater to control dilution of the oil during shutdown.

f. Compressor motor shall not exceed 1750 rpm and shall be cooled by suction gas. Motor shall be protected from high temperature by internal winding sensors. Motor power rating shall not exceed that shown on the drawings.

g. Provide suction gas strainer to trap oil mist, liquid, and any foreign matter in the return gas.

2.3 WATER-COOLED CONDENSER

a. Dual condensers shall have a steel shell, with straight copper tubes with removable heads on both ends for cleaning and replacement of tubes. Tubes shall be integral finned type, with refrigerant-side working pressure of 300 psi (2070 kPa) and water-side working pressure of 150 psi (1035 kPa). Condenser shall be ASME-stamped.
b. Provide safety relief valves set at the maximum working pressure.
c. Water pressure drop shall be maximum shown on the drawings. Low pressure drop is required for maximum energy savings.
d. Provide a back-seated refrigerant valve in the liquid and suction line connections of each circuit to the compressor, purge connection, and other standard connections required by the system.

2.4 EVAPORATOR

a. Evaporator shall be dual-circuited shell and tube type, with straight copper tubes with integral external fins. Evaporator shall be designed for 225 psi (1552 kPa) refrigerant working pressure and 150 psi (1035 kPa) water-side working pressure. Evaporator shall be ASME-stamped.
b. Water pressure drop shall not exceed that shown on the drawings. Low-pressure drop is required for maximum energy savings.
c. Provide connection on each head to monitor the water pressure drop; provide drain and necessary control wells for safety and operating controls.
d. Evaporator shall be factory-insulated with closed-cell insulation having a "K" factor of not less than 0.26. Insulation on heads shall be removable and replaceable for access to the heads for service.

2.5 CONTROL PANEL

a. Control panel shall be factory-wired and -installed. The electrical controls shall be separated from the refrigeration controls.
b. The electrical section shall contain terminal blocks for all wiring connections, including power and interlock and control wiring. This section shall contain the motor starters for part winding or across-the-line starter (as required by the local utility company), control transformer, control fuse, pumpdown control relays, compressor starter relay, reset relay, and nonrecycling compressor overload relay.
c. The refrigeration section shall contain the high-pressure control, low-pressure control, and solid-state motor protector with oil pressure cutout for each compressor.

2.6 AUXILIARY EQUIPMENT

a. Provide suction, discharge, and oil pressure gauges with valve for each. Mount on the control panel and pipe to the compressor.
b. Provide hot gas muffler between the compressor and the condenser.
c. Provide rubber-in-shear isolators for mounting under the unit. Isolators shall be matched to the unit weight.
d. Provide acoustically lined compartment of heavy-gauge metal to reduce the noise of the unit reaching adjacent spaces.
e. Provide unit-mounted disconnect switch mounted in the control compartment at the factory.

f. Provide a load limit thermostat to unload the compressor if the chilled fluid temperature is too high and the compressor tends to overload.

g. Provide a factory-mounted alarm package with loud alarm horn and pilot lights to indicate loss of evaporator flow, low-temperature thermostat operation, or compressor malfunction. Lights shall indicate that power is on and indicate which compressor is in operation.

h. Provide factory-installed hot gas bypass valves and controls to allow operation of unit below the minimum step of unloading.

i. Provide cycle counter and hour meter to indicate operating time of unit.

j. Provide factory-installed power supply monitor to protect against phase loss, imbalance, incorrect sequence, and low voltage.

k. Provide factory-mounted timer to allow the compressor to periodically pump out the chiller, if required.

2.7 CONTROL SEQUENCE

a. Chilled liquid pump shall be operating, and the flow proved by a flow switch or pressure differential switch, before the condenser water pump can be energized. Condenser water pump shall be started by chilled liquid return temperature controls, and the condensing unit controls will be energized after the water flow is proved by a flow switch or pressure differential switch in the condenser water circuit.

b. See manufacturer's standard controls for safety and interlock controls.

2.8 ACCEPTABLE MANUFACTURERS

a. Liquid chiller shall be the make and model number shown on the drawings or approved equivalents by Carrier (30HK Series), Trane (Model CGW), York (LCH).

PART THREE - EXECUTION

3.1 INSTALLATION

a. Install the unit in place on the vibration isolators specified, with clearances required for proper maintenance.

b. Install and connect the chilled liquid lines to the evaporator and the condenser water lines to the condenser, with pipe of the size shown on the drawings and as recommended by the manufacturer. (See piping details shown on the drawings.) All connections shall allow for easy removal of the heads for servicing of the condenser and the evaporator. Allow room for tube removal if it should be necessary.

c. Make proper electrical connections for power, control, and interlock wiring as required by the codes.

d. Operate the system and check suction and discharge pressures of each of the compressors for proper readings. Read and record refrigerant suction and discharge pressure readings as required in Balance and Testing, Section 15111.

e. Check the water quantity for proper flow through the evaporator and the condenser. Report the water flow and the water pressure drop as measured across the heads as required in Balancing and Testing, Section 15111.

f. Measure the power draw in each phase and the actual voltage of the motor, and report these readings as required in Balancing and Testing, Section 15111.

END OF SECTION 15153

SECTION 15160
****
CENTRIFUGAL LIQUID CHILLER
HERMETIC

PART ONE - GENERAL

1.1 The GENERAL and SPECIAL CONDITIONS, Section 15100, are included as a part of this Section as though written in full in this document.

1.2 Scope of the Work shall include the furnishing and complete installation of the equipment covered by this Section, with all auxiliaries, ready for owner's use.

PART TWO - PRODUCTS

2.1 DESIGN CHANGES

a. The design and layout shown on the drawings are based on the manufacturer shown in the equipment schedule. If equipment other than that of the manufacturer shown is submitted to the engineer for consideration as an equal, it shall be the responsibility of the bidder wishing to make the substitution to submit with the request a revised drawing of the mechanical room layout acceptable to the engineer. This revised drawing shall show the proposed location of the substitute and the unit and the area required to pull the tubes, compressor, and motor. This drawing shall also show clearances of adjacent equipment and service area required by that equipment.

b. Changes in the architectural, structural, electrical, mechanical, and plumbing requirements for the substitution shall be the responsibility of the bidder wishing to make the substitution. This shall include the cost of redesign by the affected designer(s). Any additional cost incurred by affected subcontractors shall be the responsibility of this bidder and not the owner.

2.2 CHILLER

a. Chiller shall consist of motor, compressor, evaporator (cooler), condenser, purge unit, isolation assembly, machine control center, and starter, with controls for automatic operation mounted on the chiller.

b. Chiller shall be a fully packaged unit complete in all details to chill fluid shown on drawings in indicated quantities from and to the temperatures shown, when supplied with condenser water quantity shown and with a rise from 85 deg. F (30 deg. C) to 95 deg. F (35 deg. C).

c. Unit power required shall not exceed that shown on the drawings in kW/ton (kW/watt) at full load (based on load required, not on machine total capacity). Power required at 25, 50, 75 and 100% of rated equipment capacity shall be submitted for evaluation.

d. Pressure drop through the evaporator and condenser shall not exceed that shown on the drawings. Units submitted having less pressure drop than shown are encouraged but should be called to the attention of the engineer.

e. Chiller shall be shipped factory assembled with all refrigerant piping and control wiring factory-installed.

f. Chiller using refrigerant at a pressure higher than R-11 shall be provided with a pump-out system for removal of the refrigerant. Provide compressor, drive, piping, wiring, etc., as required for removal and external storage of the refrigerant. Storage vessel shall have capacity equal to that of the largest unit in the installation, plus 20% excess storage. Only one pump-out system is required for multiple units, but system shall have capacity of the largest system.
g. Provide positive metering of the refrigerant flow in the machine. Chiller shall be capable of operating with entering condenser water of 55 deg. F (13 deg. C) or a cooling tower bypass valve, controls, and associated piping, all of which shall be provided by the contractor whether shown on the drawings or not.

2.3 COMPRESSOR AND DRIVE

a. Motor, transmission (if used), and compressor shall be hermetically sealed into a common assembly.
b. Compressor shall be single stage or multistage with an interstage economizer between stages. Impellers shall be overspeed-tested to a minimum of 15% above operating conditions at the factory.
c. Compressor motor shall be cooled by refrigerant in contact with all internal motor components. Motor stator shall be arranged to permit service with only minor compressor disassembly and without requiring the breaking of main refrigerant piping connections.
d. Transmission gears (if used) shall be arranged to allow visual inspection without disassembly or removal of the compressor casing or impeller.

2.4 LUBRICATION SYSTEM

a. Lubrication system shall be factory-installed to deliver oil under pressure to all bearings and gears. System shall consist of hermetic motor-driven oil pump with starter and controls, oil cooler, pressure regulator, oil filter, automatic water control valve, thermostatically controlled oil heater, and oil reservoir with temperature gauge.

2.5 COMPRESSOR MOTOR CONTROLS

a. Motor starter shall be reduced-voltage type, unless across-the-line starting is allowed by the utility company.
b. Provide devices to limit current draw to full-load amps. Provide demand-limiting device so that maximum current may be set between 40 and 100% of full load.

2.6 COOLER AND CONDENSER

a. Cooler and condenser may be unishell or separate shell and tube vessels with ASME stamp where applicable.
b. Tubes shall be integrally finned copper tubing rolled into tube sheets and support sheets and individually replaceable. Pipe connections and water boxes shall be designed for 150 psig (1035 kPa) unless noted otherwise and shall be provided with vents and drains. Provide tappings in water boxes for control sensors, gauges, and thermometers.

2.7 PURGE SYSTEM

a. Purge system shall be furnished for chillers operating in a vacuum.
b. Purge system shall evacuate air and water vapor from the system and shall separate the refrigerant and return it to the system.

2.8 CONTROLS

a. Controls shall be solid-state, fully automatic, and fail-safe. The The chiller shall shut down for motor overcurrent, high bearing temperature, high condenser pressure, high motor temperature, and low oil pressure. These controls shall each have an individual manual reset feature.
b. Provide a leaving chilled water low-temperature shutdown with automatic reset.
c. Provide a low-limit, manual reset, freeze protection thermostat in the leaving chilled water.
d. Capacity control shall be by means of variable inlet guide vanes in the compressor suction to modulate the capacity from 100 to 10% of full load (ARI conditions) without the use of hot gas bypass.
e. Controls shall be factory-prepiped and -prewired to terminal strips where interlocks to other equipment can be easily field-connected.
f. Controls shall include a program timer to ensure prelube and postlube needs prior to start and during coast-down. The control shall prevent restart for a field-adjustable time period.

2.9 SUPPORTS

a. Chiller manufacturer shall furnish soleplates and isolation pad assembly for mounting and leveling chiller on concrete base.
b. Spring-type isolators shall be provided for chillers not installed on a grade-supported concrete base. Isolators shall be selected by the chiller manufacturer for the conditions involved in order to prevent noise and vibration transmission to the structural frame.

2.10 INSULATION

a. The compressor motor purge chamber, and all miscellaneous piping, shall be factory-insulated by the manufacturer.
b. The evaporator (cooler), the suction elbow between the compressor and the evaporator, and the cooler water box covers shall be field-insulated.
c. Insulation shall be closed-cell, foamed, fireproof plastic, 3/4 in. (20 mm) thick, with thermal conductivity as recommended by the manufacturer to prevent condensation on the surface.

2.11 COMPRESSOR MOTOR STARTER

a. Compressor manufacturer shall furnish a wye-delta closed-transition starter in NEMA I enclosure for each chiller. Overload relays shall be provided in each leg of the starter.
b. Provide ammeter for each power leg and mount the meters in the door of the starter enclosure.
c. Chiller manufacturer shall furnish control wiring diagrams for the installation of chiller and associated equipment.

PART THREE - EXECUTION

3.1 INSTALLATION

a. The contractor shall set the chiller in place as shown on the submittal drawings and in such a position that all tubes, compressor, and motor can be pulled and removed from the equipment room.

b. The contractor shall install and leave in place necessary rigging to lift the motor, compressor, and transmission. The rigging shall be supported by structural members of adequate size, or additional support shall be provided.

c. The chiller shall be set level on soleplates or isolators as required above.

d. Piping connections for chilled fluid, condenser water, and miscellaneous small piping shall be installed as shown on the drawings and required by the manufacuturer. Note that piping to the unit shall provide for removal and replacement of the equipment by the use of flanges, couplings, etc.

e. The chiller starter and related pump motor starters shall be mounted, wired, and interlocked with controls.

f. Flow or pressure differential switches shall be provided in the chilled liquid and in the condenser water circuits as recommended by the manufacturer.

g. Power wiring shall be provided for the oil pump starter, oil heater, purge unit (if required), and other equipment requiring power. This power shall be independent of the main compressor starter.

h. Provide small water piping for the oil cooler and any other locations required. Provide drain and vent connections for all water connections.

i. Provide refrigerant vent connection to the outside from the rupture disk and from other locations, where refrigerant might be expelled or released from the chiller.

j. Install thermometers and gauges as recommended by the chiller manufacturer and as shown on the drawings.

3.2 START-UP AND TESTING

a. The chiller manufacturer shall provide the services of a trained engineer to check the installation and report to the contractor and the engineer any changes required in the installation for proper operation and servicing of the chiller. Changes recommended by the start-up engineer shall be made as soon as recommended, so as to avoid any delay in the use of the equipment. This review and necessary changes shall be at the contractor's expense. Changes required shall be discussed with the contractor and, if necessary, the engineer (not the owner).

b. After the equipment installation has been checked, the start-up engineer shall check out the equipment. If all factory wiring, etc., is in order, the chiller shall be charged and placed in service.

c. The operation of the chiller shall be observed and all safeties set and checked for proper operation.

d. The refrigerant pressure, water pressure, water flow, temperature, and power shall be read, recorded, and submitted to the owner as required under Balancing and Testing, Section 15111.

e. The start-up engineer shall submit a full report to the contractor, engineer and the owner, certifying that the system was thoroughly checked and placed in service. This report shall include the items required above. Comments should not be directed to the owner unless the engineer and the contractor are participating in the discussion.

3.3 TRAINING

a. The contractor and the manufacturer's engineer shall provide three full sets of parts and operating instructions to the owner's operating personnel.

b. These people shall be fully briefed in the normal start-up of the system, operation under light load and full load, and normal and emergency shutdown of the chiller and associated equipment.

c. Routine maintenance, yearly maintenance, winterization, and spring start-up shall be fully discussed and documented.

d. Names of those instructed and dates, as well as a list of information turned over to the owner, shall be included in the start-up report.

END OF SECTION 15160

SECTION 15162

****

CENTRIFUGAL LIQUID CHILLER

HIGH-SPEED GEAR TYPE

PART ONE - GENERAL

1.1 The GENERAL and SPECIAL CONDITIONS, Section 15100, are included as a part of this Section as though written in full in this document.

1.2 Scope of the Work shall include the furnishing and complete installation of the equipment covered by this Section, with all auxiliaries, ready for owner's use.

PART TWO - PRODUCTS

2.1 DESIGN CHANGES

a. The design and layout shown on the drawings are based on the manufacturer shown in the equipment schedule. If equipment other than that of the manufacturer indicated is submitted to the engineer for consideration as an equal, it shall be the responsibility of the bidder wishing to make the substitution to submit with the request a revised drawing of the mechanical room layout acceptable to the engineer. This revised drawing shall show the proposed location of the substitute unit, and the area required to pull the tubes, compressor, and motor. This drawing shall also show clearances of adjacent equipment and service area required by that equipment.

b. Changes in architectural, structural, electrical, mechanical, and plumbing requirements for the substitution shall be the responsibility of the bidder wishing to make the substitution. This shall include the cost of redesign by the affected designer(s). Any additional cost incurred by affected subcontractors shall be the responsibility of this bidder and not the owner.

2.2 CHILLER

a. Chiller shall consist of motor, compressor, evaporator (cooler), condenser, purge unit, isolation assembly, machine control center, and starter, with controls for automatic operation mounted on the chiller.

b. Chiller shall be a fully packaged unit complete in all details to chill fluid shown on drawings in quantities indicated from and to the temperatures shown, when supplied with condenser water quantity indicated and with a rise from 85 deg F (30 deg. C) to 95 deg. F (35 deg. C).

c. Unit power required shall not exceed that shown on the drawings in kW/ton or kW/watt at full load (based on load required), not on machine total capacity. Power required at 25, 50, 75, and 100% of rated equipment capacity shall be submitted for evaluation.

d. Pressure drop through evaporator and condenser shall not exceed that shown on the drawings. Units submitted having less pressure drop than shown are encouraged but should be called to the attention of the engineer.

e. Chiller shall be shipped factory-assembled with all refrigerant piping and control wiring factory-installed.

f. Chiller using refrigerant at a pressure higher than R-11 shall be provided with a pump-out system for removal of the refrigerant. Provide compressor, drive, piping, wiring, etc., as required for removal and external storage of the refrigerant. Storage vessel shall have capacity equal to that of the largest unit in the installation, plus 20% excess storage. Only one pump-out system is required for multiple units, but system shall have capacity of the largest system.

g. Provide positive metering of the refrigerant flow in the machine. Chiller shall be capable of operating with entering condenser water of 55 deg. F (13 deg. C) or a cooling tower bypass valve, controls, and associated piping. These shall be provided by the contractor whether shown on the drawings or not.

2.3 COMPRESSOR AND DRIVE

a. Motor, transmission, and compressor shall be hermetically sealed into a common assembly.

b. Compressor shall be single-stage or two-stage with an interstage economizer. Impellers shall be overspeed-tested to 20% above operating conditions at the factory.

c. Compressor motor shall be cooled by subcooled liquid refrigerant in intimate contact with all internal motor components. Motor stator shall be arranged to permit service with only minor compressor disassembly and without requiring the breaking of main refrigerant piping connections.

d. Transmission gears shall be double helical type, arranged to allow visual inspection without disassembly or removal of the compressor casing or impeller.

2.4 LUBRICATION SYSTEM

a. Lubrication system shall be factory-installed to deliver oil under pressure to all bearings and gears. System shall consist of hermetic motor-driven oil pump with starter and controls, oil cooler, pressure regulator, oil filter, automatic water control valve, thermostatically controlled oil heater, and oil reservoir with temperature gauge.

2.5 COMPRESSOR MOTOR CONTROLS

a. Motor starter shall be wye-delta-type reduced-voltage starter, unless across-the-line starting is allowed by the utility company.

b. Provide devices to limit current draw to full-load amps. Provide demand-limiting device so that maximum current may be set between 40 and 100% of full load.

2.6 COOLER AND CONDENSER

a. Cooler and condenser may be unishell or separate shell and tube vessels, with ASME stamp where applicable.

b. Tubes shall be integrally finned copper tubing rolled into tube sheets and support sheets and individually replaceable. Pipe connections and water boxes shall be designed for 150 psig (1035 kPa) unless noted otherwise and shall be provided with vents and drains. Provide tappings in water boxes for control sensors, gauges, and thermometers.

2.7 PURGE SYSTEM

a. Purge system shall be furnished for chillers operating in a vacuum.

b. Purge system shall be self-contained thermal type with necessary devices for evacuating air and water vapor from the system and with a means of separating the refrigerant and returning it to the system.

c. If other types of purge systems are proposed, all piping and wiring shall be included and noted in the submittal as required in Section 2.1 of the Specifications.

2.8 CONTROLS

a. Controls shall be a micro-computer center in a locked enclosure, factory mounted wired and tested. The control center shall include a 40 character alpha-numeric display that shall spell out the system parameters in English.

b. Keypad shall allow programming of setpoints as follows: leaving chilled water temperature; percent (%) current limit; pulldown demand limiting; seven day time clock with holiday schedule for chiller, pumps, and tower; and remote reset temperature range.

c. All safety controls shall be annunciated through the display to show day, time of shutdown, cause of shutdown, and type of restart required.

d. Controls shall provide adjustable rate at which the chiller is allowed to load, ranging from one minute to four hours. Pulse rate of pre-rotation vanes shall be indicated on LED to show loading/unloading condition.

e. Controls shall not allow compressor to cycle until a safe period of time has passed. If cycling continues, controls shall shut the chiller down automatically and the display shall show the cause of the shutdown.

f. Keyswitch shall provide for operation in the following modes: local; program; remote; and service.

g. Any improper input to the controls shall be rejected and the operator informed through the display.

h. Backup battery shall be provided to preserve all programming in memory for not less than 30 days.

i. Heat efficiency indicating need for tube cleaning or water treatment shall be displayed on the contol center.

j. Numbered terminals shall be provided to accept a remote pulse width signal for remote reset of the leaving chilled water temperature and for power demand limiting.

k. Interlocks shall be provided to indicate if a shutdown is because of safeties or normal cycling and to indicate if all safety and cycle controls will allow chiller to start.

l. System evaporator and condenser refrigerant pressures shall be indicated on the display along with oil pressure and purge pressure.

m. System temperatures for return and leaving chilled water, return and leaving condenser water, evaporator and condenser refrigerant saturation, compressor discharge, and oil shall be displayed on the control center display.

2.9 SUPPORTS

a. Chiller manufacturer shall furnish soleplates and isolation pad assembly for mounting and leveling chiller on concrete base.

b. Spring-type isolators shall be provided for chillers not installed on a grade-supported concrete base. Isolators shall be selected by the chiller manufacturer for the conditions involved in order to prevent noise and vibration transmission to the structural frame.

2.10 INSULATION

a. The compressor motor purge chamber and all miscellaneous piping shall be factory-insulated by the manufacturer.

b. The evaporator (cooler), the suction elbow between the compressor and the evaporator, and the cooler water box covers shall be field-insulated.

c. Insulation shall be closed-cell, foamed, fireproof plastic, 3/4 in. (20 mm) thick, with thermal conductivity as recommended by the manufacturer to prevent condensation on the surface.

2.11 COMPRESSOR MOTOR STARTER

a. Compressor manufacturer shall furnish a wye-delta closed-transition starter in NEMA I enclosure for each chiller. Overload relays shall be provided in each leg of the starter.

b. Provide ammeter for each power leg of the starter and mount meters in the door of the starter enclosure.

c. Chiller manufacturer shall furnish control wiring diagrams for the installation of chiller and associated equipment.

PART THREE - EXECUTION

3.1 INSTALLATION

a. The contractor shall set the chiller in place as shown on the submittal drawings and in such a position that all tubes, compressor, and motor can be pulled and removed from the equipment room.

b. The contractor shall install, and leave in place, necessary rigging to lift the motor, compressor, and transmission. The rigging shall be supported by structural members of adequate size, or additional support shall be provided.

c. The chiller shall be set level on soleplates or isolators as required above.

d. Piping connections for chilled fluid, condenser water, and all miscellaneous small piping required shall be installed as shown on the drawings and required by the manufacturer. Note that piping to the unit shall provide for removal and replacement for the servicing of the equipment by the use of flanges, couplings, etc.

e. The chiller starter and related pump motor starters shall be mounted and wired for power and interlocked for controls.

f. Flow or pressure differential switches shall be provided in the chilled liquid and in the condenser water circuits as recommended by the manufacturer.

g. Power wiring shall be provided for the oil pump starter, oil heater, purge unit (if required), and other equipment requiring power. This power shall be independent of the main compressor starter.

h. Provide small water piping for the oil cooler and any other locations required. Provide drain and vent connections for all water connections.
i. Provide refrigerant vent connection to the outside from the rupture disk and any other location where refrigerant might be expelled or released from the chiller.
j. Install thermometers and gauges as recommended by the chiller manufacturer and as shown on the drawings.

3.2 START-UP AND TESTING

a. The chiller manufacturer shall provide the services of a factory-trained engineer to check the installation and report to the contractor and the engineer any changes required in the installation to ensure the proper operation and servicing of the chiller. Changes recommended by the start-up engineer shall be made as soon as recommended so as to avoid delay in the use of the equipment. This review and necessary changes shall be at the contractor's expense. Any necessary changes shall be made by the contractor and reported to the engineer, not to the owner.
b. After the equipment installation has been checked, the start-up engineer shall check out the equipment. If all factory wiring, etc., is in order, the chiller shall be charged, if necessary, and placed in service.
c. The operation of the chiller shall be observed and all safeties set and checked for proper operation.
d. The refrigerant pressure, water pressure, water flow, temperature, and power shall be read, recorded, and submitted to the owner as required under Balancing and Testing, Section 15111.
e. The start-up engineer shall submit a full report to the contractor, engineer and owner confirming that the system was thoroughly checked and placed in service. This report shall include the items required above. Verbal comments and reports should not be made to the owner or the owner's representative unless the contractor and engineer are present.

3.3 TRAINING

a. The contractor and the manufacturer's engineer shall provide three full sets of parts and operating instructions to the owner's selected operating personnel.
b. Owner's people shall be fully briefed in the normal start-up of the system, operation under light load and full load, and normal and emergency shutdown of the chiller and associated equipment.
c. Routine maintenance, yearly maintenance, winterization, and spring start-up shall be fully discussed and documented.
d. Names of those instructed and dates, as well as a list of information handed over to the owner, shall be included in the start-up report.

END OF SECTION 15162

SECTION 15163

****

CENTRIFUGAL HEAT RECOVERY LIQUID CHILLER
HIGH SPEED, GEAR DRIVEN

PART ONE - GENERAL

1.1 The GENERAL and SPECIAL CONDITIONS, Section 15100, are included as a part of this Section as though written in full in this document.

1.2 Scope of the Work shall include the furnishing and complete installation of the equipment covered by this Section, with all auxiliaries, ready for owner's use.

PART TWO - PRODUCTS

2.1 DESIGN CHANGES

a. The design and layout shown on the drawings are based on the manufacturer indicated in the equipment schedule. If equipment other than that of the manufacturer shown is submitted to the engineer for consideration as an equal, it shall be the responsibility of the bidder wishing to make the substitution to submit with the request a revised drawing of the mechanical room layout acceptable to the engineer. This revised drawing shall show the proposed location of the substitute unit and the area required to pull the tubes, compressor, and motor. This drawing shall also show clearances of adjacent equipment and service area required by that equipment.

b. Changes in architectural, structural, electrical, mechanical, and plumbing requirements for the substitution shall be the responsibility of the bidder wishing to make the substitution. This shall include the cost of redesign by the affected designer(s). Any additional cost incurred by affected subcontractors shall be the responsibility of this bidder and not the owner.

2.2 CHILLER

a. Chiller shall consist of motor, compressor, evaporator (cooler), condenser, auxiliary condenser, purge unit, isolation assembly, machine control center, and starter, with controls for automatic operation mounted on the chiller.

b. Chiller shall be a fully packaged unit complete in all details to chill the fluid shown on the drawings in quantities shown from and to the temperatures shown, when supplied with quantity of condenser water shown, and with a rise from 85 deg. F (30 deg. C) to 95 deg. F (35 deg. C).

c. Unit power required shall not exceed that shown on the drawings in kW/ton or kW/watt at full load, based on load required, not on machine total capacity. Power required at 25, 50, 75, and 100% of rated equipment capacity shall be submitted for evaluation.

d. Pressure drop through evaporator and condensers shall not exceed that shown on the drawings. Units submitted having less pressure drop than shown are encouraged but should be called to the attention of the engineer.

e. Chiller shall be shipped factory-assembled with all refrigerant piping and control wiring factory-installed.
f. Chiller using refrigerant at a pressure higher than R-11 shall be provided with a pump-out system for removal of the refrigerant. Provide compressor, drive, piping, wiring, etc., as required for removal and external storage of the refrigerant. Storage vessel shall have capacity equal to that of the largest unit in the installation, plus 20% excess storage. Only one pump-out system is required for multiple units, but system shall have capacity of the largest system.
g. Provide positive metering of the refrigerant flow in the machine. Chiller shall be capable of operating with entering condenser water of 55 deg. F (13 deg. C), or a cooling tower bypass valve, controls, and associated piping shall be provided by the contractor whether shown on the drawings or not.

2.3 COMPRESSOR AND DRIVE

a. Motor, transmission, and compressor shall be hermetically sealed into a common assembly.
b. Compressor shall be multiple-stage with an interstage economizer between each stage. Impellers shall be overspeed-tested to 20% above operating conditions at the factory.
c. Compressor motor shall be cooled by subcooled liquid refrigerant in intimate contact with all internal motor components. Motor stator shall be arranged to permit service with only minor compressor disassembly and without requiring the breaking of main refrigerant piping connections.
d. Transmission gears shall be double helical type arranged to allow visual inspection without disassembly or removal of the compressor casing or transmission.

2.4 LUBRICATION SYSTEM

a. Lubrication system shall be factory-installed to deliver oil under pressure to all bearings. System shall consist of hermetic motor-driven oil pump with starter and controls, oil cooler, pressure regulator, oil filter, automatic water control valve, thermostatically controlled oil heater, and oil reservoir with temperature gauge.

2.5 COMPRESSOR MOTOR CONTROLS

a. Motor starter shall be wye-delta-type reduced-voltage starter, unless across-the-line starting is allowed by the utility company.
b. Provide electronic devices to limit current draw to full-load amps. Provide demand-limiting device so that maximum current may be set between 40 and 100% of full load.

2.6 COOLER AND CONDENSER

a. Cooler, auxiliary condenser, and condenser may be unishell or separate shell and tube vessels with ASME stamp where applicable.
b. Tubes shall be integrally finned copper tubing rolled into tube sheets and support sheets and individually replaceable. Pipe connections and water boxes shall be designed for 150 psig (1035 kPa) unless noted otherwise and shall be provided with vents and drains. Provide tappings in water boxes or adjacent piping for control sensors, gauges, and thermometers.

2.7 PURGE SYSTEM

a. Purge system shall be furnished for chillers operating in a vacuum.

b. Purge system shall be self-contained thermal type with necessary devices for evacuating air and water vapor from the system and means of separating the refrigerant and returning it to the system.

c. If other types of purge systems are proposed, all changes in piping and wiring shall be noted in the submittal as required in 2.1 of this Section of the Specifications.

2.8 CONTROLS

a. Electronic temperature control system, including temperature sensor, vane actuator, and integrated circuit microprocessor, shall be provided.

b. Controls shall control the rate at which the chiller is allowed to load (ramp function) from 2 to 45 minutes (field adjustable).

c. Individual lights shall indicate when machine is loading and unloading or if automatic current limiting is occurring.

d. Inlet guide vanes shall close in case of low refrigerant temperature and at start-up.

e. Provide a 30-minute timer to prevent recycling.

f. Provide double-break, first-out indication and manual reset for low evaporator temperature, high condenser pressure, high motor temperature, low oil pressure, and starter fault.

g. Controls shall be solid-state, fully automatic, and fail-safe. These controls shall each have an individual manual reset.

h. Provide a leaving chilled water low-temperature shutdown with automatic reset.

i. Provide a low-limit, manual reset, freeze protection thermostat in the leaving chilled water.

j. Capacity control shall be by means of variable inlet guide vanes in the compressor suction to modulate the capacity from 100 to 10% of full load (ARI conditions) without the use of hot gas bypass.

k. Provide precise control of chilled water reset based on building return water temperature or a discrete external signal such as building humidity, ambient temperature, etc.

l. Controls shall be factory-prepiped and -prewired to terminal strips so that interlocks to other equipment can be easily field-connected. Terminals and wires shall be individually identified.

m. Controls shall include a program timer to ensure prelube and postlube needs prior to start and during coast-down (under power or for power failure conditions).

2.9 SUPPORTS

a. Chiller manufacturer shall furnish soleplates and isolation pad assembly for mounting and leveling chiller on concrete base.

b. Spring-type isolators shall be provided for chillers not installed on a grade-supported concrete base. Isolators shall be selected by the chiller manufacturer for the conditions involved in order to prevent noise and vibration transmission to the structural frame.

2.10 INSULATION

a. The compressor motor purge chamber and miscellaneous piping shall be factory-insulated.
b. The evaporator (cooler), the suction elbow between the compressor and the evaporator, the cooler water box covers, and all items that are cold enough to sweat shall be factory-insulated.
c. All small water piping and incidental items subject to condensation shall be field-insulated.
d. Insulation shall be closed-cell, foamed, fireproof plastic, 3/4 in. (20 mm) thick, with thermal conductivity as recommended by the manufacturer to prevent condensation on the surface.

2.11 COMPRESSOR MOTOR STARTER

a. Compressor manufacturer shall furnish a wye-delta closed-transition starter in NEMA I enclosure mounted on each chiller. Overload protection shall be provided in each leg of the starter.
b. Provide distribution fault protection that will shut the chiller down if a fault of 1-1/2 electrical cycle duration occurs. A fault indicator shall indicate that a "distribution fault has occurred."
c. Provide ammeter and voltmeter for each power leg of the starter. Mount meters in the door of the starter enclosure or another equally visible location. Single meter and switch are not acceptable.
d. Chiller manufacturer shall furnish control wiring diagrams for the installation of chiller and associated equipment.

PART THREE - EXECUTION

3.1 INSTALLATION

a. The contractor shall set the chiller in place as shown on the submittal drawings and in such a position that all tubes, compressor, and motor can be pulled and removed from the equipment room.
b. The contractor shall install, and leave in place, necessary rigging to lift the motor and compressor. The rigging shall be supported by structural members of adequate size, or additional support shall be provided.
c. The chiller shall be set level on soleplates or isolators as required above.
d. Piping connections for chilled fluid, condenser water, and miscellaneous small piping required shall be installed as shown on the drawings and required by the manufacturer. Note that piping to the unit shall provide for removal and replacement for the servicing of the equipment. Flanges, couplings, etc., shall be provided. All pipe connections to the chiller shall be valved.
e. The chiller starter and related pump motor starters shall be mounted and wired for power and interlocked for controls.
f. Flow or pressure differential switches shall be provided in the chilled liquid and in each condenser water circuit as recommended by the manufacturer.
g. Power wiring shall be provided for the oil pump starter, oil heater, purge unit, and other equipment requiring power. This power shall be independent of the main compressor starter.

h. Provide small water piping for the oil cooler and any other locations required. Provide drain and vent connections for all water connections.

i. Provide refrigerant vent connection to the outside from the rupture disk and any other location where refrigerant might be expelled or released from the chiller.

j. Install thermometers and gauges as recommended by the chiller manufacturer and as shown on the drawings.

3.2 START-UP AND TESTING

a. The chiller manufacturer shall provide the services of a factory-trained engineer to check the installation and report to the contractor and the engineer any changes required in the installation to ensure proper operation and servicing of the chiller. Changes recommended by the start-up engineer shall be made as soon as recommended, so as to avoid delay in the use of the equipment. This review and necessary changes shall be at the contractor's expense.

b. After the equipment installation has been checked, the start-up engineer shall check out the equipment. If all factory wiring, etc., is in order, the chiller shall be charged, if necessary, and placed in service.

c. The operation of the chiller shall be observed and all safeties set and checked for proper operation.

d. The refrigerant pressure, water pressure, water flow, temperature, and power shall be read, recorded, and submitted to the owner, as required under Balancing and Testing, Section 15111.

e. The start-up engineer shall submit a full report to the contractor, engineer and the owner to the effect that the system was checked and placed in service. This report shall include the items required above. Verbal comments and reports should not be directed to the owner, or the owner's representative, except in the presence of the contractor and engineer.

3.3 TRAINING

a. The contractor and the manufacturer's engineer shall provide three full sets of parts and operating instructions to the owner's selected operating personnel.

b. These persons shall be fully briefed in the normal start-up of the system, operation under light load and full load, and normal and emergency shutdown of the chiller and associated equipment.

c. Routine maintenance, yearly maintenance, winterization, and spring start-up shall be fully discussed and documented by written instructions.

d. Names of those instructed and dates, as well as a list of information handed over to the owner, shall be included in the start-up report.

END OF SECTION 15163

SECTION 15164
****
CENTRIFUGAL LIQUID CHILLER
DIRECT DRIVE

PART ONE - GENERAL

1.1 The GENERAL and SPECIAL CONDITIONS, Section 15100, are included as a part of this Section as though written in full in this document.

1.2 Scope of the Work shall include the furnishing and complete installation of the equipment covered by this Section, with all auxiliaries, ready for owner's use.

PART TWO - PRODUCTS

2.1 DESIGN CHANGES

a. The design and layout shown on the drawings are based on the manufacturer shown in the equipment schedule. If equipment other than that of the manufacturer shown is submitted to the engineer for consideration as an equal, it shall be the responsibility of the bidder wishing to make the substitution to submit with the request a revised drawing of the mechanical room layout acceptable to the engineer. This revised drawing shall show the proposed location of the substitute unit and the area required to pull the tubes, compressor, and motor. This drawing shall also show clearances of adjacent equipment and service area required by that equipment.

b. Changes in architectural, structural, electrical, mechanical, and plumbing requirements for the substitution shall be the responsibility of the bidder wishing to make the substitution. This shall include the cost of redesign by the affected designer(s). Any additional cost incurred by affected subcontractors shall be the responsibility of this bidder and not the owner.

2.2 CHILLER

a. Chiller shall consist of motor, compressor, evaporator (cooler), condenser, purge unit, isolation assembly, machine control center, and starter, with the controls for automatic operation mounted on the chiller.

b. Chiller shall be a fully packaged unit complete in all details to chill fluid shown on drawings. Capacity shall be no less than that noted in the schedules on the drawings.

c. Unit power required shall not exceed that shown on the drawings in kW/ton (kW/watt) at full load, based on load required, not on machine total capacity. Power required at 25, 50, 75, and 100% of rated equipment capacity shall be submitted for evaluation.

d. Pressure drop through evaporator and condenser shall not exceed that shown on the drawings. Units submitted having less pressure drop than shown are encouraged but should be called to the attention of the engineer.

e. Chiller shall be shipped factory-assembled with all refrigerant piping and control wiring factory-installed.

f. Chiller using refrigerant at a pressure higher than R-11 shall be provided with a pump-out system for removal of the refrigerant. Provide compressor, drive, piping, wiring, etc., as required for removal and external storage of the refrigerant. Storage vessel shall have capacity equal to that of the largest unit in the installation, plus 20% excess storage. Only one pump-out system is required for multiple units, but system shall have capacity of the largest system.

g. Provide positive metering of the refrigerant flow in the machine. Chiller shall be capable of operating with entering condenser water of 55 deg. F (13 deg. C), or a cooling tower bypass valve, controls, and associated piping shall be provided by the contractor, whether shown on the drawings or not.

2.3 COMPRESSOR AND DRIVE

a. Motor and compressor shall be hermetically sealed into a common assembly.

b. Compressor shall be single-stage or multiple stage with an interstage economizer. Impellers shall be overspeed-tested to 20% above operating conditions at the factory.

c. Compressor motor shall be cooled by subcooled liquid refrigerant in intimate contact with all internal motor components. Motor stator shall be arranged for service with only minor compressor disassembly and without requiring the breaking of main refrigerant piping connections.

d. Compressor wheel(s) shall be mounted on the motor shaft and supported by the motor bearings. Motor speed shall be 3600 rpm.

2.4 LUBRICATION SYSTEM

a. Lubrication system shall be factory-installed to deliver oil under pressure to all bearings. System shall consist of hermetic motor-driven oil pump, with starter and controls, oil cooler, pressure regulator, oil filter, automatic water control valve, thermostatically controlled oil heater, and oil reservoir with temperature gauge.

2.5 COOLER AND CONDENSER

a. Cooler and condenser shall be separate shell and tube vessels with ASME stamp where applicable.

b. Tubes shall be integrally finned copper tubing rolled into tube sheets and support sheets and individually replaceable. Pipe connections and water boxes shall be designed for 150 psig (1035 kPa) unless noted otherwise and shall be provided with vents and drains. Provide tappings in water boxes or adjacent piping for control sensors, gauges, and thermometers.

2.6 PURGE SYSTEM

a. Purge system shall be furnished for chillers operating in a vacuum.

b. Purge system shall be self-contained compressor type with necessary devices for evacuating air and water vapor and with means of separating the refrigerant and returning it to the system.

c. If other types of purge systems are proposed, all changes in piping and wiring shall be noted in the submittal, as required in 2.1 of this Section of the Specifications.

2.7 CONTROLS

a. Controls shall be solid-state, fully automatic, and fail-safe. The chiller shall shut down in the event of motor overcurrent, high bearing temperature, high condenser pressure, high motor temperature, and low oil pressure. Each of the above controls shall have an individual manual reset.

b. Provide a leaving chilled water low-temperature shutdown with automatic reset.

c. Provide a low-limit, manual reset, freeze protection thermostat in the leaving chilled water.

d. Provide devices to limit current draw to full-rated amps. Provide demand-limiting device so that maximum current may be set between 40 and 100% of full load.

e. Capacity control shall be by means of variable inlet guide vanes in the compressor suction in order to modulate the capacity from 100 to 10% of full load (ARI conditions) without the use of hot gas bypass.

f. Controls shall be factory-prepiped and -prewired to terminal strips where interlocks to other equipment can be easily field-connected. Terminals and wires shall be individually identified.

g. Controls shall include a program timer to ensure prelube and postlube needs prior to start and during coast-down (under power for power failure conditions). The control shall prevent restart for a field-adjustable time period.

2.8 SUPPORTS

a. Chiller manufacturer shall furnish soleplates and isolation pad assembly for mounting and leveling chiller on concrete base.

b. Spring-type isolators shall be provided for chillers not installed on a grade-supported concrete base. Isolators shall be selected by the chiller manufacturer for the conditions involved to prevent noise and vibration transmission to the structural frame.

2.9 INSULATION

a. The evaporator (cooler), the suction elbow between the compressor and the evaporator, and the cooler water box covers shall be factory-insulated.

b. All small water piping and incidental items shall be field-insulated.

c. Insulation shall be closed-cell, foamed, fireproof plastic, 3/4 in. (20 mm) thick, with thermal conductivity as recommended by the manufacturer to prevent condensation on the surface.

2.10 COMPRESSOR MOTOR STARTER

a. Compressor manufacturer shall furnish a wye-delta closed-transition starter in NEMA I enclosure for each chiller. Overload relays shall be provided in each leg of the starter.

b. Provide ammeter for each power leg of the starter. Mount ammeters in the door of the starter enclosure or another equally visible location.

c. Chiller manufacturer shall furnish control wiring diagrams for the installation of chiller and associated equipment.

PART THREE - EXECUTION

3.1 INSTALLATION

a. The contractor shall set the chiller in place as shown on the submittal drawings and in such a position that all tubes, compressor, and motor can be pulled and removed from the equipment room.

b. The contractor shall install, and leave in place, necessary rigging to lift the motor and compressor. The rigging shall be supported by structural members of adequate size, or sufficient members shall be provided.

c. The chiller shall be set level on soleplates or isolators as required above.

d. Piping connections for chilled fluid, condenser water, and miscellaneous small piping required shall be installed as shown on the drawings and required by the manufacturer. Note that piping to the unit shall provide flanges, couplings, etc., for removal and replacement of the equipment.

e. The chiller starter and related pump motor starters shall be mounted, wired for power, and interlocked with the controls.

f. Flow or pressure differential switches shall be provided in the chilled liquid and in the condenser water circuits as recommended by the manufacturer.

g. Power wiring shall be provided for the oil pump starter, oil heater, purge unit, and other equipment requiring power. This power shall be independent of the main compressor starter.

h. Provide small water piping for the oil cooler and any other equipment required. Provide drain and vent connections for all water connections.

i. Provide refrigerant vent connection to the outside from the rupture disk and any other location where refrigerant might be expelled or released from the chiller.

j. Install thermometers and gauges as recommended by the chiller manufacturer and as shown on the drawings.

3.2 START-UP AND TESTING

a. The chiller manufacturer shall provide the services of a factory-trained engineer to check the installation and report to the contractor and the engineer any changes required in the installation to ensure proper operation and servicing of the chiller. Changes recommended by the start-up engineer shall be made as soon as recommended, so as to avoid delay in the use of the equipment. This review and necessary changes shall be at the contractor's expense.

b. After the equipment installation has been checked, the start-up engineer shall check out the equipment. If all factory wiring, etc., is in order, the chiller shall be charged as necessary and placed in service.

c. The operation of the chiller shall be observed and all safeties set and checked for proper operation.

d. The refrigerant pressure, water pressure, water flow, temperature, and power shall be read and recorded and submitted to the engineer, as required under Balancing and Testing, Section 15111.

e. The start-up engineer shall submit a full report to the contractor, engineer, and owner to the effect that the system was thoroughly checked and placed in service. This report shall include the items required above. Verbal comments and reports should not be directed to the owner, or the owner's representative, unless contractor and engineer are present in the conference.

3.3 TRAINING

a. The contractor and the manufacturer's engineer shall provide three full sets of parts and operating instructions to the owner's operating personnel.

b. These persons shall be fully briefed in the normal start-up of the system, operation under light load and full load, and normal and emergency shutdown of the chiller and associated equipment.

c. Routine maintenance, yearly maintenance, winterization, and spring start-up shall be fully discussed and documented by written instructions.

d. Names of those instructed and dates, as well as a list of information handed over to the owner, shall be included in the start-up report.

END OF SECTION 15164

SECTION 15165

****

CENTRIFUGAL HEAT RECOVERY LIQUID CHILLER
DIRECT DRIVE

PART ONE - GENERAL

1.1 The GENERAL and SPECIAL CONDITIONS, Section 15100, are included as a part of this Section as though written in full in this document.

1.2 Scope of the Work shall include the furnishing and complete installation of the equipment covered by this Section, with all auxiliaries, ready for owner's use.

PART TWO - PRODUCTS

2.1 DESIGN CHANGES

a. The design and layout shown on the drawings are based on the manufacturer indicated in the equipment schedule. If equipment other than that of the manufacturer shown is submitted to the engineer for consideration as an equal, it shall be the responsibility of the bidder wishing to make the substitution to submit with the request a revised drawing of the mechanical room layout acceptable to the engineer. This revised drawing shall show the proposed location of the substitute unit and the area required to pull the tubes, compressor, and motor. This drawing shall also show clearances of adjacent equipment and service area required by that equipment.

b. Changes in architectural, structural, electrical, mechanical, and plumbing requirements for the substitution shall be the responsibility of the bidder wishing to make the substitution. This shall include the cost of redesign by the affected designer(s). Any additional cost incurred by affected subcontractors shall be the responsibility of this bidder and not the owner.

2.2 CHILLER

a. Chiller shall consist of motor, compressor, evaporator (cooler), condenser, auxiliary condenser, purge unit, isolation assembly, machine control center, and starter, with controls for automatic operation mounted on the chiller.

b. Chiller shall be a fully packaged unit complete in all details to chill the fluid shown on the drawings in quantities shown from and to the temperatures shown, when supplied with quantity of condenser water shown and with a rise from 85 deg. F (30 deg. C) to 95 deg. F (35 deg. C).

c. Unit power required shall not exceed that shown on the drawings in kW/ton (kW/watt) at full load, based on load required, not on machine total capacity. Power required at 25, 50, 75, and 100% of rated equipment capacity shall be submitted for evaluation.

d. Pressure drop through evaporator and condensers shall not exceed that shown on the drawings. Units submitted having less pressure drop than shown are encouraged but should be called to the attention of the engineer.
e. Chiller shall be shipped factory-assembled with all refrigerant piping and control wiring factory-installed.
f. Chiller using refrigerant at a pressure higher than R-11 shall be provided with a pump-out system for removal of the refrigerant. Provide compressor, drive, piping, wiring, etc., as required for removal and external storage of the refrigerant. Storage vessel shall have capacity equal to that of the largest unit in the installation, plus 20% excess storage. Only one pump-out system is required for multiple units, but system shall have the capacity of the largest system.
g. Provide positive metering of the refrigerant flow in the machine. Chiller shall be capable of operating with entering condenser water of 55 deg. F (13 deg. C) or a cooling tower bypass valve, controls, and associated piping. These shall be provided by the contractor whether shown on the drawings or not.

2.3 COMPRESSOR AND DRIVE
a. Motor and compressor shall be hermetically sealed into a common assembly.
b. Compressor shall be multiple-stage with an interstage economizer between each stage. Impellers shall be overspeed-tested to 15% above operating conditions at the factory.
c. Compressor motor shall be cooled by subcooled liquid refrigerant in intimate contact with all internal motor components. Motor stator shall be arranged for service with only minor compressor disassembly and without requiring the breaking of main refrigerant piping connections.
d. Compressor wheel(s) shall be mounted on the motor shaft and supported by the motor bearings. Motor speed shall be 3600 rpm.

2.4 LUBRICATION SYSTEM
a. Lubrication system shall be factory-installed to deliver oil under pressure to all bearings. System shall consist of hermetic motor-driven oil pump with starter and controls, oil cooler, pressure regulator, oil filter, automatic water control valve, thermostatically controlled oil heater, and oil reservoir with temperature gauge.

2.5 COMPRESSOR MOTOR CONTROLS
a. Motor starter shall be wye-delta-type reduced-voltage starter, unless across-the-line starting is allowed by the utility company.
b. Provide electronic devices to limit current draw to full-load amps. Provide demand-limiting device so that maximum current may be set between 40 and 100% of full load.

2.6 COOLER AND CONDENSER
a. Cooler, auxiliary condenser, and condenser shall be separate shell and tube vessels, with ASME stamp where applicable.
b. Tubes shall be integrally finned copper tubing rolled into tube sheets and support sheets and individually replaceable. Pipe connections and water boxes shall be designed for 150 psig (1035 kPa) unless noted otherwise and shall be provided with vents and drains. Provide tappings in water boxes or adjacent piping for control sensors, gauges, and thermometers.

2.7 PURGE SYSTEM

a. Purge system shall be furnished for chillers operating in a vacuum.

b. Purge system shall be self-contained compressor type with necessary devices for evacuating air and water vapor from the system and with means of separating the refrigerant and returning it to the system.

c. If other types of purge systems are proposed, all changes in piping and wiring shall be noted in the submittal as required in 2.1 of this Section of the Specifications.

2.8 CONTROLS

a. Electronic temperature control system, including temperature sensor, vane actuator, and integrated circuit microprocessor, shall be provided.

b. Controls shall control the rate at which the chiller is allowed to load (ramp function) from 2 to 45 minutes. (field adjustable).

c. Individual lights shall indicate when machine is loading and unloading or if automatic current limiting is occurring.

d. Inlet guide vanes shall close in case of low refrigerant temperature and at start-up.

e. Provide a 30-minute timer to prevent recycling.

f. Provide double-break, first-out indication and manual reset for low evaporator temperature, high condenser pressure, high motor temperature, low oil pressure, and starter fault.

g. Controls shall be solid-state, fully automatic, and fail-safe. Each of these controls shall have an individual manual reset.

h. Provide a low temperature shutdown with automatic reset in the leaving chilled water.

i. Provide a low-limit, manual reset, freeze protection thermostat in the leaving chilled water.

j. Capacity control shall be by means of variable inlet guide vanes in the compressor suction to modulate the capacity from 100 to 10% of full load (ARI conditions) without the use of hot gas bypass.

k. Provide precise control of chilled water reset based on building return water temperature or a discrete external signal such as building humidity, ambient temperature, etc.

l. Controls shall be factory-prepiped and -prewired to terminal strips so that interlocks to other equipment can be easily field-connected. Terminals and wires shall be individually identified.

m. Controls shall include a program timer to ensure prelube and postlube needs prior to start and during coast-down (under power or power failure conditions).

2.9 SUPPORTS

a. Chiller manufacturer shall furnish soleplates and isolation pad assembly for mounting and leveling chiller on concrete base.

b. Spring-type isolators shall be provided for chillers not installed on a grade-supported concrete base. Isolators shall be selected by the chiller manufacturer for the conditions involved in order to prevent noise and vibration transmission to the structural frame.

2.10 INSULATION

a. The evaporator (cooler), the suction elbow between the compressor and the evaporator, the cooler water box covers, and all items that are cold enough to sweat shall be factory-insulated.

b. All small water piping and incidental items shall be field-insulated.

c. Insulation shall be closed-cell, foamed, fireproof plastic, 3/4 in. (20 mm) thick, with thermal conductivity as recommended by the manufacturer to prevent condensation on the surface.

2.11 COMPRESSOR MOTOR STARTER

a. Compressor manufacturer shall furnish a wye-delta closed-transition starter in NEMA I enclosure mounted on each chiller. Overload protection shall be provided in each leg of the starter.

b. Provide distribution fault protection that will shut the chiller down if a fault of 1-1/2 electrical cycle duration occurs. A fault indicator shall indicate "distribution fault has occurred."

c. Provide ammeter and voltmeter for each power leg of the starter. Mount meters in the door of the starter enclosure or another equally visible location. Single meters and switch are not acceptable.

d. Chiller manufacturer shall furnish control wiring diagrams for the installation of chiller and associated equipment.

PART THREE - EXECUTION

3.1 INSTALLATION

a. The contractor shall set the chiller in place as shown on the submittal drawings and in such a position that all tubes, compressor, and motor can be pulled and removed from the equipment room.

b. The contractor shall install, and leave in place, necessary rigging to lift the motor and compressor. The rigging shall be supported by structural members of adequate size, or sufficient members shall be provided.

c. The chiller shall be set level on soleplates or isolators as required above.

d. Piping connections for chilled fluid, condenser water, and miscellaneous small piping required shall be installed as shown on the drawings and required by the manufacturer. Note that piping to the unit shall provide for removal and replacement for the servicing of the equipment by the use of flanges, couplings, etc. All pipe connections to the chiller shall be valved.

e. The chiller starter and related pump motor starters shall be mounted and wired for power and interlocked for controls.

f. Flow or pressure differential switches shall be provided in the chilled liquid and in each condenser water circuit as recommended by the manufacturer.

g. Power wiring shall be provided for the oil pump starter, oil heater, purge unit, and other equipment requiring power. This power shall be independent of the main compressor starter.

h. Provide small water piping for the oil cooler and any other locations required. Provide drain and vent connections for all water connections.

i. Provide refrigerant vent connection to the outside from the rupture disk and any other location where refrigerant might be expelled or released from the chiller.

j. Install thermometers and gauges as recommended by the chiller manufacturer and as shown on the drawings.

3.2 START-UP AND TESTING

a. The chiller manufacturer shall provide the services of a factory-trained engineer to check the installation and report to the contractor and the engineer any changes required in the installation to ensure proper operation and servicing of the chiller. Changes recommended by the start-up engineer shall be made as soon as recommended, so as not to delay the use of the equipment. This review and necessary changes shall be at the contractor's expense.

b. After the equipment installation has been checked, the start-up engineer shall checkout the equipment. If all factory wiring, etc., is in order, the chiller shall be charged as necessary and placed in service.

c. The operation of the chiller shall be observed and all safeties set and checked for proper operation.

d. The refrigerant pressure, water pressure, water flow, temperature, and power shall be read, recorded, and submitted to the Owner, as required under Balancing and Testing, Section 15111.

e. The start-up engineer shall submit a full report to the contractor, engineer, and owner to the effect that the system was thoroughly checked and placed in service. This report shall include the items required above. Verbal comments and reports should not be directed to the owner, or the owner's representative, except in the presence of the contractor and engineer.

3.3 TRAINING

a. The contractor and the manufacturer's engineer shall provide three full sets of parts and operating instructions to the owner's operating personnel.

b. These persons shall be fully briefed in the normal start-up of the system, operation under light load and full load, and normal and emergency shutdown of the chiller and associated equipment.

c. Routine maintenance, yearly maintenance, winterization, and spring start-up shall be fully discussed and documented by written instructions.

d. Names of those instructed and dates, as well as alist of information handed over to the owner, shall be included in the start-up report.

END OF SECTION 15165

SECTION 15169
****

AIR-COOLED CENTRIFUGAL HEAT RECOVERY LIQUID CHILLER
HIGH SPEED, GEAR DRIVEN

PART ONE - GENERAL

1.1 The GENERAL and SPECIAL CONDITIONS, Section 15100, are included as a part of this Section as though written in full in this document.

1.2 Scope of the Work shall include the furnishing and complete installation of the equipment covered by this Section, with all auxiliaries, ready for owner's use.

PART TWO - PRODUCTS

2.1 DESIGN CHANGES

a. The design and layout shown on the drawings are based on the manufacturer shown in the equipment schedule. If equipment other than that of the manufacturer shown is submitted to the engineer for consideration as an equal, it shall be the responsibility of the bidder wishing to make the substitution to submit with the request a revised drawing of the mechanical room layout acceptable to the engineer. This revised drawing shall show the proposed location of the substitute unit, and the area required to pull the tubes, compressor, and motor. This drawing shall also show clearances of adjacent equipment and service area required by that equipment.

b. Changes in architectural, structural, electrical, mechanical, and plumbing requirements for the substitution shall be the responsibility of the bidder wishing to make the substitution. This shall include the cost of redesign by the affected designer(s). Any additional cost incurred by affected subcontractors shall be the responsibility of this bidder and not the owner.

2.2 CHILLER

a. Air-cooled centrifugal chiller shall consist of motor, compressor, evaporator (cooler), air-cooled condenser, water-cooled heat recovery auxiliary condenser, purge unit, isolation assembly, machine control center, and starter, with controls for automatic operation mounted on the chiller.

b. Chiller shall be a fully packaged unit and shall be operated and tested at the factory. The chiller shall be complete in all details to chill the fluid shown on the drawings in quantities specified from and to the temperatures shown.

c. Unit power required shall not exceed that shown on the drawings in kW/ton (kW/watt) at full load, based on load required, not on machine total capacity. Power required at 25, 50, 75, and 100% of rated equipment capacity shall be submitted for evaluation.

d. Pressure drop through evaporator and heat recovery condenser shall not exceed that shown on the drawings. Units submitted having less pressure drop than shown are encouraged but should be called to the attention of the engineer.

2.3 COMPRESSOR AND DRIVE

a. Motor, transmission, and compressor shall be hermetically sealed into a common assembly.

b. Compressor shall be multiple-stage with an interstage economizer between each stage. Impellers shall be overspeed-tested to 20% above operating conditions at the factory.

c. Compressor motor shall be cooled by subcooled liquid refrigerant in intimate contact with all internal motor components. Motor stator shall be arranged to permit service with only minor compressor disassembly and without requiring the breaking of main refrigerant piping connections.

d. Transmission gears shall be arranged to allow visual inspection without disassembly or removal of the compressor casing or transmission.

2.4 LUBRICATION SYSTEM

a. Lubrication system shall deliver oil under pressure to all bearings and gears. System shall consist of hermetic motor-driven oil pump with starter and controls, oil cooler, pressure regulator, oil filter, automatic water control valve, thermostatically controlled oil heater, and oil reservoir with thermometer.

2.5 COOLER AND CONDENSER

a. Cooler, condenser, and auxiliary condenser shall be shell and tube vessels with ASME stamp.

b. Tubes shall be integrally finned copper tubing rolled into tube and support sheets and individually replaceable. Pipe connections and water boxes shall be designed for 150 psig (1035 kPa) unless noted otherwise and shall be provided with vents and drains. Provide tappings in water boxes or adjacent piping for control sensors, gauges, and thermometers.

2.6 PURGE SYSTEM

a. Purge system shall be furnished for chillers operating in a vacuum.

b. Purge system shall be self-contained thermal type with necessary devices for evacuating air and water vapor from the system and with means of separating the refrigerant and returning it to the system.

c. If other types of purge systems are proposed, all changes in piping and wiring shall be noted in the submittal as required in 2.1 of this section of the Specifications.

2.7 CONTROLS

a. Electronic temperature control system, including temperature sensor, vane actuator, and integrated circuit microprocessor shall be provided.

b. Controls shall control the rate at which the chiller is allowed to load (ramp function) from 2 to 45 minutes (field adjustable).

c. Individual lights shall indicate when machine is loading and unloading or if automatic current limiting is occurring.

d. Provide electronic devices to limit current draw to full-rated amps. Provide demand-limiting device so that maximum current may be set between 40 and 100% of full load.

e. Inlet guide vanes shall close in case of low refrigerant temperature and at start-up.

f. Provide a 30-minute timer to prevent recycling.

g. Provide double-break, first-out indication and manual reset for: low evaporator temperature, high condenser pressure, high motor temperature, low oil pressure, and starter fault.

h. Controls shall be solid-state, fully automatic, and fail-safe. These controls shall each have an individual manual reset.
i. Provide a leaving chilled water low-temperature shutdown with automatic reset.
j. Provide a low-limit, freeze protection thermostat with manual reset in the leaving chilled water.
k. Capacity control shall be by means of variable inlet guide vanes in the compressor suction to modulate the capacity from 100 to 10% of full load (ARI conditions) without the use of hot gas bypass.
l. Provide precise control of chilled water reset based on building return water temperature or a discrete external signal such as building humidity, ambient temperature, etc.
m. Controls shall be factory-prepiped and -prewired to terminal strips, so that interlocks to other equipment can be easily field-connected. Terminals and wires shall be individually identified.
n. Controls shall include a program timer to ensure prelube and postlube needs prior to start and during coast-down (under power or for power failure conditions).

2.8 SUPPORTS

a. Chiller shall be mounted on structural steel frame ready for mounting and leveling on concrete base.
b. Spring-type isolators shall be provided for chillers not installed on a grade-supported concrete base. Isolators shall be selected by the chiller manufacturer for the conditions involved in order to prevent noise and vibration transmission to the structural frame.

2.9 INSULATION

a. The compressor motor purge chamber and miscellaneous piping shall be factory-insulated.
b. The evaporator (cooler), the suction elbow between the compressor and the evaporator, the cooler water box covers, and all items that are cold enough to sweat shall be factory-insulated.
c. All small water piping and incidental items subject to condensation shall be field-insulated.
d. Insulation shall be closed-cell, foamed, fireproof plastic, 3/4 in. (20 mm) thick, with thermal conductivity as recommended by the manufacturer to prevent condensation on the surface.

2.10 COMPRESSOR MOTOR STARTER

a. Compressor manufacturer shall furnish a wye-delta closed-transition starter in NEMA I enclosure mounted on each chiller. Overload protection shall be provided in each leg of the starter.
b. Provide distribution fault protection that will shut the chiller down if a fault of 1-1/2 electrical cycle duration occurs. A fault indicator shall indicate "distribution fault has occurred."
c. Provide ammeter and voltmeter for each power leg of the starter. Mount meters in the enclosure. Single meter and switch are not acceptable.
d. Chiller manufacturer shall furnish control wiring diagrams for the installation of chiller and associated equipment.

PART THREE - EXECUTION

3.1 INSTALLATION

a. The contractor shall set the chiller in place as shown on the submittal drawings and in such a position that all tubes, compressor, and motor can be removed from the equipment compartment.

b. The chiller shall be set level on supports as required above.

c. Piping connections for chilled fluid, condenser heat recovery water, and miscellaneous small piping required shall be installed as shown on the drawings and required by the manufacturer. Note that piping to the unit shall provide for removal and replacement for the servicing of the equipment by the use of flanges, couplings, etc. All pipe connections to the chiller shall be valved.

d. The systems shall be charged with glycol of the amount required on the drawings to prevent freezing during operation and during winter shutdown. Both chiller and heat recovery system shall be protected by glycol.

e. The chiller starter and related pump motor starters shall be mounted, wired, and factory-tested, ready for interlocking with controls.

f. Flow or pressure differential switches shall be provided in the chilled liquid as recommended by the manufacturer.

g. Power wiring shall be included for the oil pump starter, oil heater, purge unit, and other equipment requiring power. This power shall be independent of the main compressor starter.

h. Provide small water piping for the oil cooler and any other locations required. Provide drain and vent connections for all water connections.

i. Install thermometers and gauges as recommended by the chiller manufacturer and as shown on the drawings.

3.2 START-UP AND TESTING

a. The chiller manufacturer shall provide the services of a factory-trained engineer to check the installation and report to the contractor and the engineer any changes required in the installation for proper operation and servicing of the chiller. Changes recommended, by the start-up engineer shall be made as soon as recommended so as to avoid delay in the use of the equipment. This review and necessary changes shall be at the contractor's expense.

b. After the equipment installation has been checked, the start-up engineer shall check the equipment and charge if necessary. If all factory wiring, etc., is in order, the chiller shall be placed in service.

c. The operation of the chiller shall be observed and all safeties set and checked for proper operation.

d. The refrigerant pressure, water pressure, and flow for both chiller and the heat recovery condenser shall be read and recorded. Temperature and power shall be read, recorded, and submitted to the engineer also, as required under Balancing and Testing, Section 15111.

e. The start-up engineer shall submit a full report to the contractor, engineer, and owner to the effect that the system was thoroughly checked and placed in service. This report shall include the items required above. Verbal comments and reports should not be directed to the owner, or the owner's representative, except in the presence of the contractor and engineer.

3.3 TRAINING

a. The contractor and the manufacturer's engineer shall provide three full sets of parts and operating instructions to the owner's operating personnel.

b. These persons shall be fully briefed in the normal start-up of the system, operation under light load and full load, and normal and emergency shutdown of the chiller and associated equipment.

c. Routine maintenance, yearly maintenance, winterization, and spring start-up shall be fully discussed and documented by written instructions.

d. Names of those instructed and dates, as well as a list of information handed over to the owner, shall be included in the start-up report.

END OF SECTION 15169

SECTION 15175
****
AIR-COOLED CONDENSER
PROPELLER TYPE

PART ONE - GENERAL

1.1 The GENERAL and SPECIAL CONDITIONS, Section 15100, are included as a part of this Section as though written in full in this document.

1.2 Scope of the Work shall include the furnishing and complete installation of the equipment covered by this Section, with all auxiliaries, ready for owner's use.

PART TWO - PRODUCTS

2.1 AIR-COOLED CONDENSER

a. Air-cooled condenser shall be packaged type for outdoor mounting, with condenser coils in one or more circuits to match the refrigeration equipment and with propeller-type vertical discharge fan(s).

b. Casing shall be heavy-gauge steel with metal surfaces given a protective treatment and all surfaces given a heavy, protective, baked, enamel-finished coat.

c. Coils shall be plate type, designed for this service, and tested at no less than 425 psig (2930 kPa). Coils shall be divided into entirely separate circuits to match the refrigeration compressor circuits. Each circuit shall have an integral subcooler.

d. Fans shall be quiet-type propeller fans with heavy steel hub mounted on the fan motor shaft. Fan blades and hub shall be protected from rust by protective treatment. Provide heavy-gauge guard over each fan.

e. Fan motors shall be slow-speed and designed for this service. Provide weathertight slinger over the shaft. Motor shall be UL-labeled for outdoor mounting. Fan shall be mounted on rubber isolators.

f. Provide head pressure controls for operation of the system down to 40 deg. F (4 deg. C) by cycling the fans.

2.2 AUXILIARY EQUIPMENT

a. Condenser shall provide for low ambient operation down to 0 deg. F (-18 deg. C) by dampers or other means.

b. Provide heavy-gauge guards over all surfaces where the coil might be subjected to damage.

c. Provide vibration pad under each leg, except where condenser is mounted on concrete pad outside.

2.3 ACCEPTABLE MANUFACTURERS

a. Air-cooled condenser shall be model number shown on the drawings or approved equivalent by Trane (Model CAU), Carrier (Model 09D), York.

PART THREE - EXECUTION

3.1 INSTALLATION

a. Mount the air-cooled condenser on concrete slab on grade as shown on the drawings. Concrete pad on grade shall extend 4 in. (10 cm) above the highest adjacent grade and extend out 18 in. (45 cm) on all sides beyond the edge of the condenser.

b. Condensers mounted on roof shall be provided with supports as detailed on the drawings or with stub columns and plastic flashing caps to make the entire installation watertight. Pitch cups are not acceptable.

c. Make proper refrigerant connections and power connections required. Roof-mounted units shall be provided with curbs and plastic flashing caps.

d. Install back-seated refrigerant valves at each connection to the condenser, unless in sight of the compressor(s). Provide refrigerant line traps and loops as recommended by the manufacturer for mounting of the condenser above or below the compressor. Size the lines for oil return to the compressor, but with a maximum pressure drop recommended by the manufacturer.

3.2 OPERATION AND TEST

a. Operate the system and check for proper fan operation, head pressure control, and refrigerant pressure drop.

b. Record the motor current and refrigerant pressures and outside air temperature as required by Balancing and Testing, Section 15111 of these Specifications.

END OF SECTION 15175

SECTION 15176

****

AIR-COOLED CONDENSER
CENTRIFUGAL BLOWER

PART ONE - GENERAL

1.1 The GENERAL and SPECIAL CONDITIONS, Section 15100, are included as a part of this Section as though written in full in this document.

1.2 Scope of the Work shall include the furnishing and complete installation of the equipment covered by this Section, with all auxiliaries, ready for owner's use.

PART TWO - PRODUCTS

2.1 AIR COOLED CONDENSER

a. Capacity of the air-cooled condenser shall match the capacity of the refrigeration compressor system as shown on the drawings, with air entering at 95 deg. F (35 deg. C).

b. Air-cooled condenser shall be packaged type for outdoor or indoor mounting, with condenser coils in one or more circuits to match the refrigeration equipment. The fan shall be centrifugal cabinet type with air quantity as shown on the drawings.

c. Casing shall be heavy-gauge steel with metal surfaces given a protective treatment and all surfaces given a heavy, protective coat of baked enamel.

d. Coils shall be plate type, designed for this service, and tested at no less than 425 psig (2930 kPa). Coils shall be divided into entirely separate circuits to match the refrigeration compressor circuits. Each circuit shall have an integral subcooler.

e. Fan(s) shall be belt-driven centrifugal type with motor external to the cabinet.

f. Cabinet shall be heavy-gauge steel with protective treatment to prevent rust spread and given a final heavy coat of baked-enamel finish.

g. Fan drive shall be provided with adjustable drive to supply the air quantity shown on the drawings against the external static pressure shown.

h. Provide head pressure controls of the system shown, with face and bypass dampers around the coil and with the external duct- mounted dampers as shown on the drawings.

2.2 AUXILIARY EQUIPMENT

a. See the drawings and the control section of the Specifications for related equipment included with this system.

b. Provide heavy-gauge guard over the belts and drive pulley.

c. Provide rubber-in-shear vibration eliminator for each support of the condenser.

2.3 ACCEPTABLE MANUFACTURERS

a. Air-cooled condenser shall be model number shown on the drawings or approved equivalent by Trane (Model TC), Carrier, or York.

PART THREE - EXECUTION

3.1 INSTALLATION

a. Mount the air-cooled condenser as shown on the drawings.

b. Condensers mounted on roof shall be provided with supports as detailed on the drawings or with stub columns and plastic flashing caps to make the entire installation watertight. Pitch cups are not acceptable.

c. Make proper refrigerant and power connections required. Roof-mounted units shall be provided with curbs and plastic flashing caps.

d. Install back-seated refrigerant valves at each connection to the condenser, unless in sight of the compressor(s). Provide refrigerant line traps and loops as recommended by the manufacturer for mounting of the condenser above or below the compressor. Size the lines for oil return to the compressor, but with a maximum pressure drop recommended by the manufacturer.

3.2 OPERATION AND TEST

a. Operate the system and check for proper fan operation, airflow, motor current, control operators operation, head pressure control, and refrigerant pressure drop.

b. Record the airflow, air pressure drop, fan motor current, and refrigerant pressures, as well as outside air temperature, all as required by Balancing and Testing, Section 15111 of these Specifications.

END OF SECTION 15176

SECTION 15180
****
COOLING TOWER
PROPELLER-TYPE CROSS FLOW

PART ONE - GENERAL

1.1 The GENERAL and SPECIAL CONDITIONS, Section 15100, are included as a part of this Section as though written in full in this document.

1.2 Scope of the Work shall include the furnishing and complete installation of the equipment covered by this Section, with all auxiliaries, ready for owner's use.

PART TWO - PRODUCTS

2.1 COOLING TOWER

a. Cooling tower shall be a packaged type of heavy steel construction with noncrossive fill. Air inlet shall be on one end and discharge out the opposite end with propeller-type fan. Tower capacity shall be as shown on the drawings to cool water from and to the temperatures shown with the wet bulb temperature shown.

b. Cooling tower casing shall be constructed of no less than 14-gauge, heavy galvanized steel.

c. Splash box, hot water basin, top sheet, and fan sheet shall be no less than 14-gauge, heavy galvanized steel.

d. Tower fill shall be film type of fire-retardant PVC material.

e. Fan cylinder and fan guard shall be very heavy galvanized steel to adequately support and protect the fan. Fan guard shall protect against personnel coming in contact with the fan blades.

f. Inlet louvers of heavy galvanized steel shall be provided to prevent back-splash and to protect the fill.

g. Fan shall be constructed on heavy-gauge steel mounted on heavy steel hub, all galvanized. Fan shaft shall be mounted on long-life bearings with external lubricating lines and grease fittings. Fan drive shall be heavy-duty V-belt(s) with variable pitch drive pulley.

h. Fan motor shall be of power shown, high-efficiency, single- or two-speed (as shown on the drawings), for mounting outdoors and exposed to weather. All wiring and junction boxes shall be weathertight. All electrical equipment shall be UL-labeled.

i. Tower base shall be no less than 12-gauge galvanized steel, constructed to support the tower from four support points on steel framework supplied by the installing contractor.

j. Provide bronze float-type water make-up valve and suction screen, as are standard with the tower.

2.2 AUXILIARY EQUIPMENT

a. Provide heavy-gauge galvanized steel guard on intake to tower.

b. Provide two-speed motor as called for on drawings. Provide controls necessary for two-speed operation.

c. Provide heavy-gauge galvanized steel covers for the hot water basin.

d. Provide float-operated or probe-type controls for electric solenoid-operated water make-up valve.

e. Provide a manual bleed valve in the supply to the tower and an adequate point to waste the water.

2.3 ACCEPTABLE MANUFACTURERS

a. Cooling tower shall be the make and model number shown on the drawings or approved equivalents by Marley (Model 47000) or Baltimore Aircoil (Model FXT).

PART THREE - EXECUTION

3.1 INSTALLATION

a. Erect the proper support grillage to elevate the tower above the pump suction and as shown on the drawings.

b. Mount the tower as directed by the manufacturer and secure to the grillage.

c. Make the proper pipe connections for supply and return. Provide drainage and water make-up for the tower. Provide connections for washing the tower down with the supply to the tower valved off. Carry the tower drain to an adequate point of disposal.

d. Install the motor and the drive pulley and align as recommended by the manufacturer. Install the proper type of belts and adjust. Readjust the belts after two weeks of operation.

e. Fill the tower with water, add commercial cleaner, and clean the tower and connected piping with cleaner to remove mill grease and scale. Drain and flush the tower and refill to proper operating level.

f. Check the fan for proper operation and record motor current draw as required under Balancing and Testing, Section 15111 of these Specifications.

END OF SECTION 15180

SECTION 15182
****
COOLING TOWER
VERTICAL DISCHARGE CROSS FLOW

PART ONE - GENERAL

1.1 The GENERAL and SPECIAL CONDITIONS, Section 15100, are included as a part of this Section as though written in full in this document.

1.2 Scope of the Work shall include the furnishing and complete installation of the equipment covered by this Section, with all auxiliaries, ready for owner's use.

PART TWO - PRODUCTS

2.1 COOLING TOWER

a. Cooling tower shall be a packaged type of heavy steel construction with noncrossive fill. Air inlet shall be on both ends and vertical discharge out top center with propeller-type fan(s). Tower capacity shall be as shown on the drawings to cool water from and to the temperatures shown with the wet bulb temperature shown.

b. Cooling tower casing shall be constructed of heavy galvanized steel, with sheets installed vertically with watertight connections.

c. Distribution basin shall be open gravity type with plastic metering orifices, constructed of heavy galvanized steel.

d. Tower fill shall be film type of fire-retardant PVC material with integral eliminators. Drift loss shall not exceed 0.2% of water circulated.

e. Fan cylinder and fan guard shall be very heavy galvanized steel to adequately support and protect the fan. Fan guard shall protect against personnel coming in contact with the fan blades.

f. Fan shall be propeller-type with cast aluminum individually adjustable blades, mounted on an aluminum hub and driven by a motor mounted out of the air-stream through a right-angle gear reducer. Motor power shall not exceed that shown on the drawings.

g. Fan motor shall be of power shown, high-efficiency, single- or two-speed as shown on the drawings, for mounting outdoors and exposed to weather. All wiring and junction boxes shall be weathertight. All electrical equipment shall be UL-labeled.

h. Tower base shall be of heavy, bolted galvanized steel construction, with depressed sump for side suction connection. Provide screen and anti-cavitation device in tower basin. Provide bronze float-type water make-up valve.

2.2 AUXILIARY EQUIPMENT

a. Provide heavy-gauge galvanized steel guards on inlets to tower.

b. Provide flow control valves from each supply connection to the tower.

c. Provide two-speed motor as called for on drawings. Provide controls necessary for two-speed operation.

d. Provide heavy-gauge galvanized steel covers for the hot water basin.

e. Provide float-operated or probe-type controls for electric solenoid-operated water make-up valve.
f. Provide a manual bleed valve in the supply to the tower and an adequate point to waste the water.
g. Provide adequate point of discharge for the tower drain and connect the drain and overflow to this point.

2.3 ACCEPTABLE MANUFACTURERS
a. Cooling tower shall be the make and model number shown on the drawings or approved equivalent by Marley (Model 8800).

PART THREE - EXECUTION

3.1 INSTALLATION
a. Erect the proper support grillage to elevate the tower above the pump suction and as shown on the drawings.
b. Mount the tower as directed by the manufacturer and secure to the grillage.
c. Make the proper pipe connections for supply and return. Provide drainage and water make-up for the tower. Provide connections for washing the tower down with the supply to the tower valved off. Carry the tower drain to an adequate point of disposal.
d. Install the motor and the drive and align as recommended by the manufacturer. Install the proper type and quantity of lubricants in the gear reducer.
e. Fill the tower with water, add commercial cleaner, and clean the tower and connected piping to remove mill grease and scale. Drain and flush the tower and refill to proper operating level.
f. Check the fan for proper operation and record motor current draw as required under Balancing and Testing, Section 15111 of these Specifications.

END OF SECTION 15182

SECTION 15185
****
COOLING TOWER
CENTRIFUGAL FAN TYPE

PART ONE - GENERAL

1.1 The GENERAL and SPECIAL CONDITIONS, Section 15100, are included as a part of this Section as though written in full in this document.

1.2 Scope of the Work shall include the furnishing and complete installation of the equipment covered by this Section, with all auxiliaries, ready for owner's use.

PART TWO - PRODUCTS

2.1 COOLING TOWER

a. Cooling tower shall be a packaged type of heavy steel construction with noncrossive fill. Air inlet shall be on one or both sides as shown on the drawings with vertical discharge. Fan(s) shall be centrifugal type. Tower capacity shall be as shown on the drawings to cool water from and to the temperatures shown with the wet bulb temperature shown.
b. Cooling tower casing shall be constructed of heavy-gauge, hot-dipped galvanized steel with watertight connections.
c. Distribution system shall consist of schedule 40 PVC pipe, with large-diameter PVC nozzles held into the headers by rubber grommets.
d. Tower fill shall be film type of fire-retardant PVC material with hot-dipped galvanized eliminators. Drift loss shall not exceed 0.2% of water circulated.
e. Fan section shall consist of hot-dipped galvanized steel section with forward-curved centrifugal fans mounted on a large-diameter fan shaft, belt driven by a high-efficiency single- or two-speed motor as shown on the drawings. Fan power shall not exceed that shown on the schedule on the drawings. Fan bearings shall be grease-lubricated by externally mounted grease fittings and connecting tubes. Fan drive shall be rated at no less than 150% of the motor rating. All electrical equipment shall be UL-labeled.
f. Provide float-operated, all brass water make-up valve.
g. Provide access doors for areas of tower needing cleaning.

2.2 AUXILIARY EQUIPMENT

a. Provide heavy-gauge galvanized steel guards on inlet(s) and outlet.
b. Provide sound-attenuating plenum on the discharge of the tower.
c. Provide two-speed motor as called for on drawings. Provide controls necessary for two-speed operation.
d. Provide float-operated or probe-type controls for electric solenoid-operated water make-up valve.
e. Provide a manual bleed valve in the supply to the tower.

2.3 ACCEPTABLE MANUFACTURERS

a. Cooling tower shall be the make and model number shown on the drawings or approved equivalent by Baltimore Aircoil (Model VX).

PART THREE - EXECUTION

3.1 INSTALLATION

a. Erect the proper support grillage to elevate the tower above the pump suction and as shown on the drawings.

b. Mount the tower as directed by the manufacturer and secure to the gillage.

c. Make the proper pipe connections for supply and return. Provide drainage and water make-up for the tower. Provide connections for washing the tower down with the supply to the tower valved off. Carry the tower drain to an adequate point of disposal.

d. Install the motor and the drive and align as recommended by the manufacturer.

e. Fill the tower with water, add commercial cleaner, and clean the tower and connected piping to remove mill grease and scale. Drain and flush the tower and refill to proper operating level.

f. Place the tower in operation and adjust the bleed rate or install automatic water solids control and chemical treatment.

g. Check the fan for proper operation and record motor current draw as required under Balancing and Testing, Section 15111 of these Specifications.

END OF SECTION 15185

SECTION 15186
****
COOLING TOWER
MULTISTAGE AXIAL FANS

PART ONE - GENERAL

1.1 The GENERAL and SPECIAL CONDITIONS, Section 15100, are included as a part of this Section as though written in full in this document.

1.2 Scope of the Work shall include the furnishing and complete installation of the equipment covered by this Section, with all auxiliaries, ready for owner's use.

PART TWO - PRODUCTS

2.1 COOLING TOWER

a. Cooling tower shall be a packaged type of heavy steel construction with noncrossive fill. Air inlet shall be on one or both sides as shown on the drawings with vertical discharge. Fan(s) shall be multistage axial type. Tower capacity shall be as shown on the drawings to cool water from and to the temperatures shown with the wet bulb temperature shown.
b. Cooling tower casing shall be constructed of heavy-gauge hot-dipped galvanized steel with watertight connections.
c. Distribution system shall consist of schedule 40 PVC pipe, with large-diameter PVC nozzles held into the headers by rubber grommets.
d. Tower fill shall be film type of fire-retardant PVC material with hot-dipped galvanized eliminators. Drift loss shall not exceed 0.2% of water circulated.
e. Fan section shall consist of hot-dipped galvanized steel section with multistage axial-type fans, belt driven by a high-efficiency single- or two-speed motor as shown on the drawings. Fan power shall not exceed that shown on the schedule on the drawings. Fan bearings shall be grease-lubricated by externally mounted grease fittings and connecting tubes. Fan drive shall be rated at not less than 150% of the motor rating. All electrical equipment shall be UL-labeled.
f. Provide float-operated, all brass water make-up valve.
g. Provide access doors for areas of tower needing cleaning.

2.2 AUXILIARY EQUIPMENT

a. Provide heavy-gauge galvanized steel guards on inlet(s) and outlet.
b. Provide sound-attenuating plenum on the discharge of the tower.
c. Provide two-speed motor as called for on drawings. Provide controls necessary for two-speed operation.
d. Provide float-operated or probe-type controls for electric solenoid-operated water make-up valve.
e. Provide a manual bleed valve in the supply to the tower.

2.3 ACCEPTABLE MANUFACTURERS

a. Cooling tower shall be the make and model number shown on the drawings or approved equivalent by Baltimore Aircoil (Model VXMT).

PART THREE - EXECUTION

3.1 INSTALLATION

a. Erect the proper support grillage to elevate the tower above the pump suction and as shown on the drawings.

b. Mount the tower as directed by the manufacturer and secure to the grillage.

c. Make the proper pipe connections for supply and return. Provide drainage and water make-up for the tower. Provide connections for washing the tower down with the supply to the tower valved off. Carry the tower drain to an adequate point of disposal.

d. Install the motor and the drive and align as recommended by the manufacturer.

e. Fill the tower with water, add commercial cleaner, and clean the tower and connected piping to remove mill grease and scale. Drain and flush the tower and refill to proper operating level.

f. Place the tower in operation and adjust the bleed rate or install automatic water solids control and chemical treatment.

g. Check the fan(s) for proper operation and record motor current draw as required under Balancing and Testing, Section 15111 of these Specifications.

END OF SECTION 15186

SECTION 15187
****
EVAPORATIVE CONDENSER
MULTISTAGE AXIAL FANS

PART ONE - GENERAL

1.1 The GENERAL and SPECIAL CONDITIONS, Section 15100, are included as a part of this Section as though written in full in this document.

1.2 Scope of the Work shall include the furnishing and complete installation of the equipment covered by this Section, with all auxiliaries, ready for owner's use.

PART TWO - PRODUCTS

2.1 EVAPORATIVE CONDENSER

a. Evaporative condenser shall be a packaged type of heavy steel construction with condensing coil of steel tubing. Air inlet shall be on side shown on the drawings with vertical discharge. Fan(s) shall be multistage axial type. Evaporative condenser capacity shall be as shown on the drawings with the wet bulb temperature shown.

b. Evaporative condenser casing shall be constructed of heavy-gauge, hot-dipped galvanized steel with watertight connections. Condenser coil shall be hot-dipped galvanized with the coil section as a unit.

c. Distribution system shall consist of schedule 40 PVC pipe, with large-diameter PVC nozzles held into the headers by rubber grommets.

d. Recirculation pump and spray system shall circulate necessary water quantity over the condenser coil to produce the capacity called for. Drift loss shall not exceed 0.2% of water circulated.

e. Fan section shall consist of hot-dipped galvanized steel section with multistage axial-type fans, belt driven by a high-efficiency single- or two-speed motor as shown on the drawings. Fan power shall not exceed that shown on the schedule on the drawings. Fan bearings shall be grease-lubricated by externally mounted grease fittings and connecting tubes. Fan drive shall be rated at not less than 150% of the motor rating. All electrical equipment shall be UL-labeled.

f. Provide float-operated, all brass water make-up valve.

g. Provide access doors for areas of evaporative condenser needing cleaning.

2.2 AUXILIARY EQUIPMENT

a. Provide heavy-gauge galvanized steel guards on inlet(s) and outlet.

b. Provide two-speed motor as called for on drawings. Provide controls necessary for two-speed operation.

c. Provide float-operated or probe-type controls for electric solenoid-operated water make-up valve.

d. Provide a manual bleed valve in the supply to the evaporative condenser.

2.3 ACCEPTABLE MANUFACTURERS

a. Evaporative condenser shall be the make and model number shown on the drawings or approved equivalent by Baltimore Aircoil (Model VX).

PART THREE - EXECUTION

3.1 INSTALLATION

a. Erect the proper support grillage as shown on the drawings.
b. Mount the evaporative condenser as directed by the manufacturer and secure to the grillage.
c. Make the proper refrigerant pipe connections including the traps and loops recommended by the manufacturer of the compressor.
d. Provide drainage and water make-up for the evaporative condenser. Provide connections for washing the evaporative condenser down with the supply to the evaporative condenser valved off. Carry the evaporative condenser drain to an adequate point of disposal.
e. Install the motor and the drive and align as recommended by the manufacturer.
f. Fill the evaporative condenser with water, add commercial cleaner, and clean the evaporative condenser of mill grease and scale. Drain and flush the evaporative condenser and refill to proper operating level.
g. Place the evaporative condenser in operation and adjust the bleed rate or install automatic water solids control and chemical treatment.
h. Check the evaporative condenser for refrigerant leaks. After the unit is in service, read and record the refrigerant pressure readings required by Balancing and Testing, Section 15111.
i. Check the fan(s) for proper operation and record motor current draw as required under Balancing and Testing, Section 15111 of these Specifications.

END OF SECTION 15187

SECTION 15188
****
FLUID COOLER
MULTISTAGE AXIAL FANS

PART ONE - GENERAL

1.1 The GENERAL and SPECIAL CONDITIONS, Section 15100, are included as a part of this Section as though written in full in this document.

1.2 Scope of the Work shall include the furnishing and complete installation of the equipment covered by this Section, with all auxiliaries, ready for owner's use.

PART TWO - PRODUCTS

2.1 FLUID COOLER

a. Fluid cooler shall be a packaged type of heavy steel construction with fluid coil of steel or copper tubing, as called for on drawings. Air inlet shall be on side shown on the drawings with vertical discharge. Fan(s) shall be multistage axial type. Fluid cooler capacity shall be as shown on the drawings, with the wet bulb temperature shown.

b. Fluid cooler casing shall be constructed of heavy-gauge hot-dipped galvanized steel. Fluid coil shall be hot-dipped galvanized, with the coil section as a unit if steel is required, or coil section shall be copper if noted on drawings.

c. Distribution system shall consist of schedule 40 PVC pipe, with large-diameter PVC nozzles held into the headers by rubber grommets.

d. Recirculation pump and spray system shall circulate necessary water over the coil to produce the capacity called for.

e. Fan section shall consist of hot-dipped galvanized steel section with multistage axial-type fan(s), belt driven by a high-efficiency single- or two-speed motor(s), as shown on the drawings. Fan power shall not exceed that in the schedule on the drawings. Fan bearings shall be grease-lubricated by externally mounted grease fittings and connecting tubes. Fan drive shall be rated at not less than 150% of the motor rating. All electrical equipment shall be UL-labeled.

f. Provide float-operated, all brass water make-up valve.

g. Provide access doors for areas of fluid cooler needing cleaning.

2.2 AUXILIARY EQUIPMENT

a. Provide heavy-gauge galvanized steel guards on inlet(s) and outlet.

b. Provide two-speed motor as called for on drawings. Provide controls necessary for two-speed operation.

c. Provide float-operated or probe-type controls for electric solenoid-operated water make-up valve.

d. Provide a manual bleed valve in the recirculation to the fluid cooler.

2.3 ACCEPTABLE MANUFACTURERS

a. Fluid cooler shall be the make and model number shown on the drawings or approved equivalent by Baltimore Aircoil (Model VX).

PART THREE - EXECUTION

3.1 INSTALLATION

a. Erect the proper support grillage as shown on the drawings.

b. Mount the fluid cooler as directed by the manufacturer and secure to the grillage.

c. Make the proper pipe connections for the fluid to be cooled. Provide drain connections and valves as shown on the drawings.

d. Provide drainage and water make-up for the fluid cooler. Provide connections for washing the fluid cooler down, with the supply to the make-up valve turned off.

e. Install the motor and the drive and align as recommended by the manufacturer.

f. Fill the fluid cooler sump with water, add commercial cleaner, and clean the unit of mill grease and scale. Drain and flush the unit and refill to proper operating level.

g. Place the fluid cooler in operation and adjust the bleed rate or install automatic water solids control and chemical treatment.

h. Check the fluid cooler for leaks. After the unit is in service, read and record the pressure drop readings required by Balancing and Testing, Section 15111.

i. Check the fan(s) for proper operation and record motor current draw as required under Balancing and Testing, Section 15111 of these Specifications.

END OF SECTION 15188

CHAPTER 4

Packaged Equipment

SECTION 15200
****
ROOFTOP HEATING AND COOLING UNITS
ELECTRIC COOLING - GAS HEATING

PART ONE - GENERAL

1.1 The GENERAL and SPECIAL CONDITIONS, Section 15100, are included as a part of this Section as though written in full in this document.

1.2 Scope of the Work shall include the furnishing and complete installation of the equipment covered by this Section, with all auxiliaries, ready for owner's use.

PART TWO - PRODUCTS

2.1 ROOFTOP UNIT

a. Rooftop unit (RTU) shall be packaged and include electric cooling and gas-fired heat, with capacity and steps of cooling and heating as shown on the drawings.

b. Unit shall be factory-charged and -tested, shall be UL-labeled and ARI-certified by Standard 210 and 270, and shall be AGA-certified.

c. Unit casing shall be heavy-gauge galvanized steel or heavy-gauge aluminum with protective coat of baked enamel. Weatherproof access panels shall be provided for access to all parts requiring service.

d. Compressor(s) shall be sealed or serviceable hermetic type and shall be resiliently mounted to avoid vibration and noise. Compressor shall be provided with antislugging protection, crankcase heater, and time delay on recycling of the compressor. Two internal compressor motor thermal cutouts and a hot gas cutout shall protect the compressor in addition to high- and low-pressure safeties. Standard controls shall permit operation down to 35 deg. F (2 deg. C), and compressor shall be locked out below this temperature.

e. Condenser fan(s) shall be direct-driven for the shaft of the slow-speed motor, which shall be designed for operation exposed to the weather.

f. Condenser coils shall have a subcooling section.

g. Refrigerant circuit shall include filter dryer, moisture indicator, sight glass, and gauge ports.

h. Filter rack shall be provided for filters 1 in.(25 mm) thick and shall filter both outdoor air and return air. See Section 15299 of these Specifications for type of filters and the number of filter changes to be furnished with the equipment.

i. Evaporator fan shall be quiet-type centrifugal blower, directly connected to an adjustable-speed motor or belt driven with an adjustable-pitch pulley on the motor.

j. Heat exchanger shall be aluminized steel, designed for long life and quiet operation. Burner shall provide dependable and quiet ignition in the stages as called for.

k. Gas burner controls shall provide automatic safety pilot, dual automatic gas valves, manual gas cock, and pressure regulator. Ignition shall be electric for the intermittent pilot with 100% shutoff when the unit is off.

l. Induced draft blower shall provide prepurge and shall be provided with a proving switch to prevent burner operation if venter is not in operation.

m. Provide fan switch and limit control to delay the fan until heat is available and to continue fan operation until heat is dispersed. Limit switch shall shut the burner down in case of failure of operating controls.

2.2 ACCESSORY EQUIPMENT

a. Condenser coil guards shall be provided for all units mounted within reach of the public.

b. Roof mounting frame shall be provided for all units mounted on the roof. Frame shall be approved by the National Roofing Contractors Association. Provide all necessary flashing and counterflashing.

c. Provide "power saver" dampers and controls to provide "free cooling" from 0 to 100% outdoor air (OA) when the outside air humidity and temperature are acceptable. Provide OA, return air, and relief air dampers in a factory-provided enclosure. All air shall be filtered and bird screen shall be installed.

d. Provide a warm-up thermostat to prevent the OA dampers from opening if the return air temperature is below the set point (65 deg. F) (18 deg. C).

e. Provide necessary controls for operation of the compressor below the normal temperature of the compressor cutout. Operation shall be permitted down to temperature specified on drawings.

f. Provide factory-trained service person to check out the system, calibrate the controls, and see that the RTU is operating properly. The service person making the settings shall make a written report to the engineer and the owner with all set points listed for future reference.

g. Rooftop units mounted on slabs or other fixed locations shall be provided with adapters for end discharge and return to the unit.

h. Provide thermostat and other controls required to produce the control functions called for.

2.3 ACCEPTABLE MANUFACTURERS

a. RTU shall be the make and model number shown on the schedule on the drawings, or acceptable equivalents by Carrier, Lennox, Trane, or York.

PART THREE - EXECUTION

3.1 INSTALLATION

a. Install the curb as required by the job conditions and as recommended by the manufacturer, and install proper flashing and counterflashing. See details on the drawings.

b. Set the unit in place, taking care to protect the adjacent roofing, and connect the supply and return ductwork.

c. Make electrical and gas line connections, taking care that these do not block access to any part of the equipment requiring service.

d. Have the factory service person check out the unit and make a written report. Place the unit in service.

3.2 BALANCING AND TEST

a. Operate the RTU and check for proper supply air quantity, noise, and proper operation.

b. Report the airflow, static pressure, voltage and current draw of each item, refrigerant pressure readings, etc., as required by Section 15111 of these Specifications. This system is not complete until these readings have been made, submitted to the engineer, and accepted.

END OF SECTION 15200

SECTION 15205

****

ROOFTOP HEATING AND COOLING UNITS
ELECTRIC COOLING - ELECTRIC HEAT

PART ONE - GENERAL

1.1 The GENERAL and SPECIAL CONDITIONS, Section 15100, are included as a part of this Section as though written in full in this document.

1.2 Scope of the Work shall include the furnishing and complete installation of the equipment covered by this Section, with all auxiliaries, ready for owner's use.

PART TWO - PRODUCTS

2.1 ROOFTOP UNIT

a. Rooftop unit (RTU) shall be packaged and include electric cooling and electric heat with capacity and steps of cooling and heating as shown on the drawings.

b. Unit shall be factory-charged and -tested, shall be UL-labeled and ARI-certified by Standard 210 and 270, and shall be AGA-certified.

c. Unit casing shall be heavy-gauge galvanized steel or heavy-gauge aluminum with protective coat of baked enamel. Weatherproof access panels shall be provided for access to all parts requiring service.

d. Compressor(s) shall be sealed or serviceable hermetic type and shall be resiliently mounted to avoid vibration and noise. Compressor shall be provided with antislugging protection, crankcase heater, and time delay on recycling of the compressor. Two internal compressor motor thermal cutouts and a hot gas cutout shall protect the compressor in addition to high- and low-pressure safeties. Standard controls shall permit operation down to 35 deg. F (95 deg. C) and compressor shall be locked out below this temperature.

e. Condenser fan(s) shall be direct-driven on the shaft of the slow-speed motor, which shall be designed to operate exposed to the weather.

f. Condenser coils shall have a subcooling section.

g. Refrigerant circuit shall include filter dryer, moisture indicator, sight glass, and gauge ports.

h. Filter rack shall be provided for filters 1 in.(25 mm) thick and shall filter both outdoor air and return air. See Section 15299 of these Specifications for type of filters and the number of filter changes to be furnished with the equipment.

i. Evaporator fan shall be quiet-type centrifugal blower, directly connected to an adjustable-speed motor or belt driven with an adjustable-pitch pulley on the motor.

j. Electric heat section shall be installed in the unit and served by the same power source as the rest of the unit. Only one power feed shall be required for the unit.

2.2 ACCESSORY EQUIPMENT

a. Condenser coil guards shall be provided for all units mounted within reach of the public.

b. Roof mounting frame shall be provided for all units mounted on the roof. Frame shall be approved by the National Roofing Contractors Association. Provide all necessary flashing and counterflashing.

c. Provide "power saver" dampers and controls to provide "free cooling" from 0 to 100% outdoor air (OA) when the outside air humidity and temperature are acceptable. Provide OA, return air, and relief air dampers in a factory-provided enclosure. All air shall be filtered and bird screen shall be installed.

d. Provide a warm-up thermostat to prevent the OA dampers from opening if the return air temperature is below the set point (65 deg. F) (18 deg. C).

e. Provide necessary controls for operation of the compressor below the normal temperature of the compressor cutout. Operation shall be permitted down to temperature specified on drawings.

f. Provide factory-trained service person to check out the system, calibrate the controls, and see that the RTU is operating properly. The service person making the settings shall make a written report to the engineer and the owner with all set points listed for future reference.

g. Rooftop units mounted on slabs or other fixed locations shall be provided with adapters for end discharge and return to the unit.

h. Provide thermostat and other controls required to produce the control functions called for.

2.3 ACCEPTABLE MANUFACTURERS

a. RTU shall be the make and model number shown on the schedule on the drawings, or acceptable equivalents by Carrier, Lennox, Trane, or York.

PART THREE - EXECUTION

3.1 INSTALLATION

a. Install the curb as required by the job conditions and as recommended by the manufacturer, and install proper flashing and counterflashing. See details on the drawings.

b. Set the unit in place, taking care to protect the adjacent roofing, and connect the supply and return ductwork.

c. Make electrical connections, taking care that these do not block access to any part of the equipment requiring service.

d. Have the factory service person check out the unit and make a written report. Place the unit in service.

3.2 BALANCING AND TEST

a. Operate the RTU and check for proper supply air quantity, noise, and correct operation.

b. Report the airflow, static pressure, voltage and current draw of each item, refrigerant pressure readings, etc., as required by Section 15111 of these Specifications. This system is not complete until these readings have been made, submitted to the engineer, and accepted.

END OF SECTION 15205

SECTION 15210

****

ROOFTOP COOLING UNITS
ELECTRIC COOLING

PART ONE - GENERAL

1.1 The GENERAL and SPECIAL CONDITIONS, Section 15100, are included as a part of this Section as though written in full in this document.

1.2 Scope of the Work shall include the furnishing and complete installation of the equipment covered by this Section, with all auxiliaries, ready for owner's use.

PART TWO - PRODUCTS

2.1 ROOFTOP UNIT

a. Rooftop unit (RTU) shall be packaged and include electric cooling only with capacity and steps of cooling as shown on the drawings.

b. Unit shall be factory-charged and -tested, shall be UL-labeled and ARI-certified by Standard 210 and 270, and shall be AGA-certified.

c. Unit casing shall be heavy-gauge galvanized steel or heavy-gauge aluminum with protective coat of baked enamel. Weatherproof access panels shall be provided for access to all parts requiring service.

d. Compressor(s) shall be sealed or serviceable hermetic type and shall be resiliently mounted to avoid vibration and noise. Compressor shall be provided with antislugging protection, crankcase heater, and time delay on recycling of the compressor. Two internal compressor motor thermal cutouts and a hot gas cutout shall protect the compressor in addition to high- and low-pressure safeties. Standard controls shall permit operation down to 35 deg. F (95 deg. C) and compressor shall be locked out below this temperature.

e. Condenser fan(s) shall be direct-driven on the shaft of the slow-speed motor, which shall be designed to operate exposed to the weather.

f. Condenser coils shall have a subcooling section.

g. Refrigerant circuit shall include filter dryer, moisture indicator, sight glass, and gauge ports.

h. Filter rack shall be provided for filters 1 in.(25 mm) thick and shall filter both outdoor air and return air. See Section 15299 of these Specifications for type of filters and the number of filter changes to be furnished with the equipment.

i. Evaporator fan shall be quiet-type centrifugal blower directly connected to an adjustable-speed motor or belt driven with an adjustable-pitch pulley on the motor.

2.2 ACCESSORY EQUIPMENT

a. Condenser coil guards shall be provided for all units mounted within reach of the public.

b. Roof mounting frame shall be provided for all units mounted on the roof. Frame shall be approved by the National Roofing Contractors Association. Provide all necessary flashing and counterflashing.

c. Provide "power saver" dampers and controls to provide "free cooling" from 0 to 100% outdoor air (OA) when the outside air humidity and temperature are acceptable. Provide OA, return air, and relief air dampers in a factory-provided enclosure. All air shall be filtered and bird screen shall be installed.
d. Provide a warm-up thermostat to prevent the OA dampers from opening if the return air temperature is below the set point (65 deg. F) (18 deg. C).
e. Provide necessary controls for operation of the compressor below the normal temperature of the compressor cutout. Operation shall be permitted down to temperature specified on drawings.
f. Provide factory-trained service person to check out the system, calibrate the controls, and see that the RTU is operating properly. The service person making the settings shall make a written report to the engineer and the owner with all set points listed for future reference.
g. Rooftop units mounted on slabs or other fixed locations shall be provided with adapters for end discharge and return to the unit.
h. Provide thermostat and other controls required to produce the control functions called for.

2.3 ACCEPTABLE MANUFACTURERS

a. RTU shall be the make and model number shown on the schedule on the drawings, or acceptable equivalents by Carrier, Lennox, Trane, or York.

PART THREE - EXECUTION

3.1 INSTALLATION

a. Install the curb as required by the job conditions and as recommended by the manufacturer, and install proper flashing and counterflashing. See details on the drawings.
b. Set the unit in place, taking care to protect the adjacent roofing, and connect the supply and return ductwork.
c. Make electrical connections, taking care that these do not block access to any part of the equipment requiring service.
d. Have the factory service person check out the unit and make a written report. Place the unit in service.

3.2 BALANCING AND TEST

a. Operate the RTU and check for proper supply air quantity, noise, and correct operation.
b. Report the airflow, static pressure, voltage and current draw of each item, refrigerant pressure readings, etc., as required by Section 15111 of these Specifications. This system is not complete until these readings have been made, submitted to the engineer, and accepted.

END OF SECTION 15210

SECTION 15215

****

ROOFTOP SINGLE-PACKAGED HEAT PUMP

PART ONE - GENERAL

1.1 The GENERAL and SPECIAL CONDITIONS, Section 15100, are included as a part of this Section as though written in full in this document.

1.2 Scope of the Work shall include the furnishing and complete installation of the equipment covered by this Section, with all auxiliaries, ready for owner's use.

PART TWO - PRODUCTS

2.1 ROOFTOP UNIT

a. Rooftop unit (RTU) shall be single packaged heat pump unit, complete with reversing valve, defrost controls, auxiliary heat, and steps of cooling and heating as shown on the drawings.

b. Unit shall be factory-charged and -tested, shall be UL-labeled and ARI-certified by Standard 240 and 270.

c. Unit casing shall be heavy-gauge galvanized steel or heavy-gauge aluminum with protective coat of baked enamel. Weatherproof access panels shall be provided for access to all parts requiring service.

d. Compressor(s) shall be heavy-duty type, rated for heat pump service, sealed or serviceable hermetic unit, and shall be resiliently mounted to avoid vibration and noise. Compressor shall be provided with antislugging protection, crankcase heater, and time delay on recycling of the compressor. Two internal compressor motor thermal cutouts and a hot gas cutout shall protect the compressor in addition to high- and low-pressure safeties. Compressor shall be rated for operation down to 5 deg. F (-20 deg. C).

e. Condenser fan(s) shall be direct-driven on the shaft of the slow-speed motor, which shall be designed to operate exposed to the weather.

f. Condenser coils shall have a subcooling section.

g. Refrigerant circuit shall include reversing valve, filter dryer, moisture indicator, sight glass, and gauge ports on suction and discharge of compressor.

h. Filter rack shall be provided for filters 1 in.(25 mm) thick and shall filter both outdoor air and return air. See Section 15299 of these Specifications for type of filters and the number of filter changes to be furnished with the equipment.

i. Evaporator fan shall be quiet-type centrifugal blower directly connected to an adjustable-speed motor or belt driven with an adjustable-pitch pulley on the motor.

j. Controls shall include solid-state defrost control, low ambient control, outdoor thermostat and start controls, and low temperature control.

k. Provide auxiliary electric heat as shown on the drawings or as required to provide the total heating capacity specified. Heaters shall be provided with separate power circuit and circuit protection, but controls shall be interlocked with the RTU controls.

2.2 ACCESSORY EQUIPMENT

a. Condenser coil guards shall be provided for all units mounted within reach of the public.

b. Roof mounting frame shall be provided for all units mounted on the roof. Frame shall be approved by the National Roofing Contractors Association. Provide all necessary flashing and counterflashing.

c. Provide "power saver" dampers and controls to provide "free cooling" from 0 to 100% outdoor air (OA) when the outside air humidity and temperature are acceptable. Provide OA, return air, and relief air dampers in a factory-provided enclosure. All air shall be filtered and bird screen shall be installed.

d. Provide a warm-up thermostat to prevent the OA dampers from opening if the return air temperature is below the set point (65 deg. F) (18 deg. C).

e. Provide status panel for mounting in the conditioned space, to indicate Cooling Mode, Heating Mode, Compressor, No Heat, Filter Change Required.

f. Provide factory-trained service person to check out the system, calibrate the controls, and see that the RTU is operating properly. The service person making the settings shall make a written report to the engineer and the owner with all set points listed for future reference.

g. Rooftop units mounted on slabs or other fixed locations shall be provided with adapters for end discharge and return to the unit.

h. Provide thermostat and other controls required to produce the control functions called for.

i. Provide duct enclosures as required for system indicated on the drawings. Enclosure shall mount above the basic roof mounting curb and provide a complete weatherproof system.

j. Provide combination supply and return ceiling supply/return diffuser packages as indicated on the drawings.

2.3 ACCEPTABLE MANUFACTURERS

a. RTU shall be the make and model number shown on the schedule on the drawings, or acceptable equivalents by Carrier, Lennox, Trane, or York.

PART THREE - EXECUTION

3.1 INSTALLATION

a. Install the curb as required by the job conditions and as recommended by the manufacturer, and install proper flashing and counterflashing. See details on the drawings.

b. Set the unit in place, taking care to protect the adjacent roofing, and connect the supply and return ductwork.

c. Make electrical connections, taking care that these do not block access to any part of the equipment requiring service.

d. Have the factory service person check out the unit and make a written report. Place the unit in service.

3.2 BALANCING AND TEST

a. Operate the RTU on heating and cooling and check for proper supply air quantity, noise, and correct operation.

b. Report the airflow, static pressure, voltage and current draw of each item, refrigerant pressure readings, etc., as required by Section 15111 of these Specifications. This system is not complete until these readings have been made, submitted to the engineer, and accepted.

END OF SECTION 15215

SECTION 15216
****
ROOFTOP SINGLE-PACKAGED HEAT PUMP

PART ONE - GENERAL

1.1 The GENERAL and SPECIAL CONDITIONS, Section 15100, are included as a part of this Section as though written in full in this document.

1.2 Scope of the Work shall include the furnishing and complete installation of the equipment covered by this Section, with all auxiliaries, ready for owner's use.

PART TWO - PRODUCTS

2.1 ROOFTOP UNIT

a. Rooftop unit (RTU) shall be single packaged heat pump dual compressor unit, complete with reversing valves, defrost controls, auxiliary heat, and steps of cooling and heating as shown on the drawings.

b. Unit shall be factory-charged and -tested; unit and all accessories shall be UL-labeled and ARI-certified by Standard 240 and 270.

c. Unit casing shall be heavy-gauge galvanized steel or heavy-gauge aluminum with protective coat of baked enamel. Weatherproof access panels shall be provided for access to all parts requiring service.

d. Compressor(s) shall be sealed or serviceable hermetic heavy-duty type, rated for heat pump service, and shall be resiliently mounted to avoid vibration and noise. Compressor shall be provided with antislugging protection, crankcase heater, and time delay on recycling of the compressor. Two internal compressor motor thermal cutouts and a hot gas cutout shall protect each compressor, in addition to high- and low-pressure safeties. Compressor shall be rated for operation down to 5 deg. F (-20 deg. C).

e. Condenser fan(s) shall be direct-driven on the shaft of the slow-speed motor, which shall be designed to operate exposed to the weather.

f. Condenser coils shall have a subcooling section.

g. Refrigerant circuit shall include reversing valve, filter dryer, moisture indicator, sight glass, and gauge ports on suction and discharge of compressor.

h. Filter rack shall be provided for filters 1 in.(25 mm) thick and shall filter both outdoor air and return air. See Section 15299 of these Specifications for type of filters and the number of filter changes to be furnished with the equipment.

i. Evaporator fan shall be quiet-type centrifugal blower with belt drive with an adjustable-pitch pulley on the motor.

j. Controls shall include solid-state defrost control, low ambient control, outdoor thermostat and start controls, and low temperature control.

k. Provide electric heat as shown on the drawings or as required to provide the total heating capacity specified. Heaters shall be factory-served from the single power circuit and circuit protection for the RTU, and controls shall be interlocked with the RTU controls.

2.2 ACCESSORY EQUIPMENT

a. Condenser coil guards shall be provided for all units mounted within reach of the public.

b. Roof mounting frame shall be provided for all units mounted on the roof. Frame shall be approved by the National Roofing Contractors Association. Provide all necessary flashing and counterflashing.

c. Provide "power saver" dampers and controls to provide "free cooling" from 0 to 100% outdoor air (OA) when the outside air humidity and temperature are acceptable. Provide OA, return air, and relief air dampers in a factory-provided enclosure. All air shall be filtered and bird screen shall be installed.

d. Provide a warm-up thermostat to prevent the OA dampers from opening if the return air temperature is below the set point (65 deg. F) (18 deg. C).

e. Provide status panel for mounting in the conditioned space to indicate Cooling Mode, Heating Mode, Compressor, No Heat, Filter Change Required.

f. Provide a remote-switching status panel to control and observe the operation of the system, with signal lights for Cool Mode, Heat Mode, Compressor No. 1, Compressor No. 2, No Heat, and Filter lights.

g. Provide additional controls for the status panel to include system selector switches for OFF-HEAT-AUTO-COOL-EMERGENCY HEAT and AUTO-ON fan switch. This control shall include manually operated, after-hours timer for 0 to 12 hours.

h. Provide night setback controls to program equipment for day-night operation.

i. Provide factory-trained service person to check out the system, calibrate the controls, and see that the RTU is operating properly. The service person making the settings shall make a written report to the engineer and the owner with all set points listed for future reference.

j. Rooftop units mounted on slabs or other fixed locations shall be provided with adapters for end discharge and return to the unit.

k. Provide thermostat and other controls required to produce the control functions called for.

l. Provide duct enclosures as required for system indicated on the drawings. Enclosure shall mount above the basic roof mounting curb and provide a complete weatherproof system.

m. Provide combination supply and return ceiling supply/return diffuser packages, as indicated on the drawings.

n. Provide hot water heating coil and all controls necessary for control of such coil in the status panels and thermostatic controls.

2.3 ACCEPTABLE MANUFACTURERS

a. RTU shall be the make and model number shown on the schedule on the drawings, or acceptable equivalents by Carrier, Lennox, Trane, or York.

PART THREE - EXECUTION

3.1 INSTALLATION

a. Install the curb as required by the job conditions and as recommended by the manufacturer, and install proper flashing and counterflashing. See details on the drawings.

b. Set the unit in place, taking care to protect the adjacent roofing, and connect the supply and return ductwork.

c. Make electrical connections, taking care that these do not block access to any part of the equipment requiring service.

d. Have the factory service person check out the unit and controls specified and make a written report. Place the unit in service.

3.2 BALANCING AND TEST

a. Operate the RTU on heating and cooling and check for proper supply air quantity, noise, and correct operation.

b. Report the airflow, static pressure, voltage and current draw of each item, refrigerant pressure readings, etc., as required by Section 15111 of these Specifications. This system is not complete until these readings have been made, submitted to the engineer, and accepted.

END OF SECTION 15216

SECTION 15217

****

ROOFTOP SINGLE-PACKAGED HEAT PUMP

PART ONE - GENERAL

1.1 The GENERAL and SPECIAL CONDITIONS, Section 15100, are included as a part of this Section as though written in full in this document.

1.2 Scope of the Work shall include the furnishing and complete installation of the equipment covered by this Section, with all auxiliaries, ready for owner's use.

PART TWO - PRODUCTS

2.1 ROOFTOP UNIT

a. Rooftop unit (RTU) shall be single-packaged heat pump dual compressor with independent refrigerant circuits, each complete with reversing valves, and defrost controls. Provide steps of cooling and heating as shown on the drawings.

b. Unit shall be factory-charged and -tested; unit and all accessories shall be UL-labeled and ARI-certified by Standard 240 and 270.

c. Unit casing shall be heavy-gauge galvanized steel or heavy-gauge aluminum with protective coat of baked enamel. Weatherproof access panels shall be provided for access to all parts requiring service.

d. Compressor(s) shall be sealed or serviceable hermetic heavy-duty type, rated for heat pump service, and shall be resiliently mounted to avoid vibration and noise. Compressor shall be provided with antislugging protection, crankcase heater, and time delay on recycling of the compressor. Two internal compressor motor thermal cutouts and a hot gas cutout shall protect each compressor, in addition to high- and low-pressure safeties. Compressor shall be rated for operation down to 5 deg. F (-20 deg. C).

e. Condenser fan(s) shall be direct-driven on the shaft of the slow-speed motor, which shall be designed to operate exposed to the weather.

f. Condenser coils shall have a subcooling section.

g. Refrigerant circuit shall include reversing valve, filter dryer, moisture indicator, sight glass, and gauge ports on suction and discharge of compressor.

h. Filter rack shall be provided for filters 1 in.(25 mm) thick and shall filter both outdoor air and return air. See Section 15299 of these Specifications for type of filters and the number of filter changes to be furnished with the equipment.

i. Evaporator fan shall be quiet-type centrifugal blower with belt drive with an adjustable-pitch pulley on the motor.

j. Controls shall include solid-state defrost control, low ambient control, outdoor thermostat and start controls, and low temperature control.

k. Provide auxiliary heat as shown on the drawings or as required to provide the total heating capacity specified. Heaters shall be factory-installed. Electric heaters shall be served from the single power circuit and circuit protection for the RTU, and controls shall be interlocked with the RTU controls.

2.2 ACCESSORY EQUIPMENT

a. Condenser coil guards shall be provided for all units mounted within reach of the public.

b. Roof mounting frame shall be provided for all units mounted on the roof. Frame shall be approved by the National Roofing Contractors Association. Provide all necessary flashing and counterflashing.

c. Provide "power saver" dampers and controls to provide "free cooling" from 0 to 100% outdoor air (OA) when the outside air humidity and temperature are acceptable. Provide OA, return air, and relief air dampers in a factory-provided enclosure. All air shall be filtered and bird screen shall be installed.

d. Provide a warm-up thermostat to prevent the OA dampers from opening if the return air temperature is below the set point (65 deg. F) (18 deg. C).

e. Provide remote readout panel for mounting in the conditioned space to indicate System On, Heating Inoperative, Condensing Unit Inoperative, Dirty Filter, and provide system switch, manual override timer for after-hours operation, and switch to shut off the condensing unit.

f. Provide a night setback control to operate the system for day-night selection with a 12-hour timed override.

g. Provide factory-trained service person to check out the system, calibrate the controls, and see that the RTU is operating properly. The service person making the settings shall make a written report to the engineer and the owner with all set points listed for future reference.

h. Provide hot water heating coil, and all controls necessary for control of such coil, in the status panels and thermostatic controls.

2.3 ACCEPTABLE MANUFACTURERS

a. RTU shall be the make and model number shown on the schedule on the drawings, or acceptable equivalents by Carrier, Lennox, Trane, or York.

PART THREE - EXECUTION

3.1 INSTALLATION

a. Install the curb as required by the job conditions and as recommended by the manufacturer, and install proper flashing and counterflashing. See details on the drawings.

b. Set the unit in place, taking care to protect the adjacent roofing, and connect the supply and return ductwork.

c. Make electrical connections, taking care that these do not block access to any part of the equipment requiring service.

d. Have the factory service person check out the unit and controls specified and make a written report. Place the unit in service.

3.2 BALANCING AND TEST

a. Operate the RTU on heating and cooling and check for proper supply air quantity, noise, and correct operation.

b. Report the airflow, static pressure, voltage and current draw of each item, refrigerant pressure readings, etc., as required by Section 15111 of these Specifications. This system is not complete until these readings have been made, submitted to the engineer, and accepted.

END OF SECTION 15217

SECTION 15225

****

THROUGH-WALL HEATING AND COOLING UNIT
ELECTRIC COOLING - ELECTRIC HEATING

PART ONE - GENERAL

1.1 The GENERAL and SPECIAL CONDITIONS, Section 15100, are included as a part of this Section as though written in full in this document.

1.2 Scope of the Work shall include the furnishing and complete installation of the equipment covered by this Section, with all auxiliaries, ready for Owner's use.

PART TWO - PRODUCTS

2.1 THROUGH-WALL UNITS

a. The heating and cooling unit shall be a small-capacity packaged unit with electric cooling and electric resistance heat of the capacity shown on the drawings and shall be designed for mounting in an opening through a masonry wall.

b. Unit shall have a heavy-gauge galvanized wall sleeve with a primer coat and a finish coat of baked enamel. All internal dividers shall be galvanized steel. Provide condenser coil guard of expanded galvanized and painted metal.

c. Chassis shall include the hermetic compressor, evaporator and condenser coils, and a double-ended fan motor to drive the evaporator and condenser fans.

d. Heating element shall be nonglow resistance-type heating element with dual overheat protection.

e. Controls shall be front-mounted under an access door and shall consist of an adjustable thermostat, high-medium-low fan with high-medium-low cooling and heating selector switch, and a ventilation damper control.

f. Provide plug and cord of the type required by the code for the voltage and power draw of the unit.

ACCESSORY EQUIPMENT

a. Provide a decorative front for the unit in place of the commercial front.

b. Provide an architectural condenser grille to replace the standard expanded metal grille.

c. Provide a subbase to support the unit for installation in curtain wall construction.

d. Provide remote control accessory for use with a remote thermostat to control units not readily accessible. Provide necessary wall thermostat.

ACCEPTABLE MANUFACTURERS

a. Unit shall be make and model number shown on the drawings, or acceptable equivalents by York (Model T2BU), or Carrier (Model 51SYA).

PART THREE - EXECUTION

3.1 INSTALLATION

a. Install the wall sleeve and subbase (if specified or shown on drawings), as recommended by the manufacturer and as shown on the drawings. Seal all areas around the sleeve with silicone sealer.

b. Install the chassis and connect the power cord.

c. Install the architectural condenser grille if required on drawings.

d. Install the filter and the commercial or the decorative front, as shown on the drawings.

e. Install the remote unit and the wall thermostat, and connect the two as required by electrical codes. Run all wire concealed or provide finish conduit as required.

3.2 TEST

a. Operate the unit and check for proper heating and cooling.

b. Eliminate all unnecessary noise and vibration.

END OF SECTION 15225

SECTION 15249
****
ISOLATION CURBS FOR ROOFTOP UNITS

PART ONE - GENERAL

1.1 The GENERAL and SPECIAL CONDITIONS, Section 15100, are included as a part of this Section as though written in full in this document.

1.2 Scope of the Work shall include the furnishing and complete installation of the equipment covered by this Section, with all auxiliaries, ready for Owner's use.

PART TWO - PRODUCTS

2.1 ISOLATION CURBS

a. Isolators for rooftop equipment shall be designed to reduce noise and vibration transmitted to the structure by inserting a spring-supported separation in the curb with a deflection of up to 1 in.(25 mm). A foam neoprene seal shall seal the unit for air and water leakage.

b. Isolator shall mount on the standard curb with the unit mounted on the isolator. Isolator shall be adjustable to compensate for uneven weight distribution of the rooftop unit. Provide restraints to minimize motion due to wind loads.

2.2 ACCEPTABLE MANUFACTURERS

a. Isolation curbs shall be make and model number shown on the drawings or equivalents by Peabody (Model ASR).

PART THREE - EXECUTION

3.1 INSTALLATION

a. Curb shall be fabricated to fit the standard roof curb and to match the unit base.

b. Adjust the springs as required to support the unit evenly and level on all sides.

c. Install restraints to prevent motion from wind, as directed by the manufacturer.

d. Set the unit in place and check for proper water and wind seal of the entire installation.

END OF SECTION 15249

SECTION 15250

****

AIR-COOLED CONDENSING UNITS

PART ONE - GENERAL

1.1 The GENERAL and SPECIAL CONDITIONS, Section 15100, are included as a part of this Section as though written in full in this document.

1.2 Scope of the Work shall include the furnishing and complete installation of the equipment covered by this Section, with all auxiliaries, ready for owner's use.

PART TWO - PRODUCTS

2.1 AIR-COOLED CONDENSING UNITS

a. Air-cooled condensing unit shall be designed for use with split system having a remote direct-expansion (DX) cooling coil mounted in evaporator fan unit. Capacity shall be as called for on the drawings when matched to the appropriate evaporator coil.

b. Condensing unit (CU) shall consist of high-efficiency hermetic compressor, air-cooled condenser with quiet fan, necessary controls, and refrigeration circuit and valves.

c. Cabinet shall be heavy-gauge galvanized steel with bonding primer and baked-enamel finish coat. The entire cabinet shall be protected from rust.

d. Compressor shall be protected from excessive current and temperatures and shall be provided with a thermostatically controlled crankcase heater to operate only when needed for protection of the compressor. Compressor shall be spring-mounted on rubber isolators. Provide a high-capacity dryer in the system to remove moisture and dirt.

e. Condenser fan shall be directly connected to a weather-protected, quiet, high-efficiency motor. Fan guard shall be provided and shall be protected from rust by a PVC finish.

f. Connections for refrigerant suction and liquid lines shall be extended outside the cabinet and provided with service valves with gauge connections. Power connections shall be made to the connectors located inside the electrical connection box.

g. Standard operating and safety controls shall include high-pressure switch, low-pressure switch, and solid-state timed-off control.

h. All components of the sealed refrigeration circuit shall be warranted by the manufacturer for five years.

2.2 AUXILIARY EQUIPMENT

a. Auxiliary equipment shall consist of refrigerant lines prepared for the unit involved. These lines shall be cleaned, dried, and pressurized at the factory.

b. Low ambient kit to allow operation at outside temperature below 35 deg. F (2 deg. C) shall be provided.

c. Expansion valve shall be provided with the evaporator coil.

d. Provide thermostat to match the requirements of the job. Thermostat shall provide subbase with Heat-Cool-Off and Fan On-Auto switch. See section on controls for other related requirements.

e. Provide polyethylene structural base designed for that service and intended to support the unit and eliminate vibration transmission.

2.3 ACCEPTABLE MANUFACTURERS

a. CU shall be the make and model number shown on the drawings or acceptable equivalents by Lennox (HS13 SEER 10+)(HS16 SEER 8+), Carrier, York, or Trane.

PART THREE - EXECUTION

3.1 INSTALLATION

a. Install the CU on proper foundation as shown on the drawings, and in location that will not restrict the air entry or discharge from the unit.

b. Install refrigerant lines as recommended by the manufacturer, taking care not to lose the refrigerant charge contained in the lines, or allow air to enter the lines or equipment. Locate the lines in such a way as to not obstruct access to the CU or other equipment. Lines located under ground or under concrete shall be installed in a PVC pipe conduit for protection.

c. Provide electrical connections as required by the applicable codes. Provide control wiring required. All power wiring and control wiring shall be in conduit and located so as not to obstruct access to the unit or other equipment.

3.2 TESTING

a. Operate the CU and the system to assure that unit is operating properly and without excessive noise and vibration.

b. Read and record the power draw and the refrigeration suction and liquid pressures as required by Balancing and Test, Section 15111.

END OF SECTION 15250

CHAPTER 5

Air-Moving Equipment

SECTION 15300
****
CENTRAL STATION AIR-HANDLING UNIT

PART ONE - GENERAL

1.1 The GENERAL and SPECIAL CONDITIONS, Section 15100, are included as a part of this Section as though written in full in this document.

1.2 Scope of the Work shall include the furnishing and complete installation of the equipment covered by this Section, with all auxiliaries, ready for owner's use.

PART TWO - PRODUCTS

2.1 CENTRAL STATION AIR-HANDLING UNIT
a. Central station air-handling unit shall be designed for the moving and conditioning of air in air conditioning, heating, and ventilation systems. The equipment shall consist of a heavy-gauge steel casing consisting of mixing box section, filter section, coil section (heating and cooling if specified), and fan section including the fan, fan drive, and motor. The unit shall be equal to that shown on the drawings with capacity not less than that shown.

2.2 CASING
a. Casing shall be fabricated from heavy-gauge steel, braced to produce a rigid assembly composed of prefabricated sections. All metals in the casing shall be given a rust-preventative treatment equal to galvanizing and a finish coat of baked enamel.
b. Casing shall be provided with factory-installed handholes covered with hinged doors and latches to allow the observation of all interior areas and to provide for washing of coil surfaces.
c. The casing shall be insulated internally with rigid insulation to prevent sweating of the casing under normal operating conditions. The air side of the insulation shall be protected against erosion by the air velocity.
d. The casing shall be supported on rubber isolators selected for the weight imposed on each support. All connections to the casing shall be made with flexible connectors in ductwork and piping.

2.3 MIXING BOX SECTION
a. Mixing box section shall allow the mixing of return and outside air generally as shown on the drawings and as required to provide evenly mixed air to the filters.
b. Mixing box shall include opposed blade dampers in both the return and outside air. Outside air damper shall be low-leakage type, allowing not more than 5% leakage with these dampers closed tight and the return dampers full open.

2.4 FILTER SECTION
a. Filter section shall provide racks and access doors for the filters as specified in the section on filters. Where high-efficiency filters are required, this section shall include prefilters and leak-proof doors to seal the filters and prevent leakage.

2.5 COIL SECTION

a. The coil section shall house the heating coil(s) and/or the cooling coil(s) to produce the conditions called for on the drawings.

b. The coils shall be selected for minimum air pressure drop and velocity, but not in excess of that shown on the drawings. Water coils shall be selected to provide a minimum water pressure drop, but not to exceed that shown on the drawings.

c. Coil fin spacing shall be as wide as possible within other limitations.

d. Coil section shall include an insulated drain pan under the cooling coil and downstream of the coil to collect all condensation from the coil.

2.6 FAN SECTION

a. Fan section shall house the centrifugal fan(s) and scroll(s) required to produce the airflow required against the system static pressure existing in the system as shown on the drawings. Submit copy of the fan curve showing the operating point of the fan at design conditions called for. Selection point shall be as recommended by the manufacturer for stable operation.

b. Fan wheel(s) shall be forward-curved, backward-inclined, backward-inclined air foil type, or other acceptable designs as required to best meet the conditions shown with the least amount of power and the lowest noise generation. The fan outlet velocity shall not exceed that shown on the drawings.

c. Fan(s) shall be mounted on a large-diameter shaft which shall operate well below critical speed. Shaft may be enlarged in the center portion for rigidity and reduced in diameter on the ends for bearings and drive pulleys.

d. Fan bearings shall be grease-lubricated with grease fittings extended to accessible locations.

e. Bearing shall be sized and selected for operation for not less than 200,000 hours.

2.7 FAN DRIVE

a. Provide belts and belt guards for connection of the fan to the motor. The driven pulley shall be fixed-diameter and the drive pulley shall be adjustable-pitch to allow variation in the fan speed, unless the motor is provided with modulating speed adjustment such as variable-frequency control equipment or variable-pitch equipment.

b. Belts shall be single or multiple as required for the power rating needed. Belts shall be selected for 150% of the maximum power draw of the fan. Multiple belts shall be "matched" from the same production run.

2.8 FAN MOTOR

a. The fan motor shall be selected for the power required to deliver the supply air quantity against a system static pressure of 50% that shown on the drawings without overloading. Selection of motor shall not consider the overload rating of the motor.

b. Fan motors for air-handling units mounted in the condition space shall be provided with resilient mounting to reduce the motor noise.

2.7 AUXILIARY EQUIPEMENT

a. Provide spring-type isolators for the unit in place of the rubber isolators called for above.

b. Provide man-sized access doors in the casings for access to the cooling and heating coil faces.

c. Provide high-efficiency motor to reduce energy consumption.

d. Coils shall be copper fins on copper tubes.

2.8 ACCEPTABLE MANUFACTURERS

a. Air-handling unit shall be the make and model shown on the drawings or equivalent units by Trane, York, Carrier, or as accepted by the engineer.

PART THREE - EXECUTION

3.1 INSTALLATION

a. Mount the casing on the vibration mounts and assemble the unit as recommended by the manufacturer for an airtight assembly.

b. Connect the outside air, return air, and supply air ducts with fireproof fabric flexible connectors.

c. Make all required pipe connections using flexible connectors to prevent vibration transmission to the pipe.

d. Connect the required drain from the drain pan to the nearest acceptable open site drain. Provide a trap in the line and slope as much as possible.

e. Install the motor and drive; adjust the belts for proper tension. Check all bearings for proper lubrication before starting the unit. Adjust the variable-pitch pulley for the proper fan speed. Install the belt guards, filters, etc. for a complete installation.

3.2 TESTING

a. Operate the unit in the system and make the readings and adjustments required under Balancing and Testing, Section 15111.

END OF SECTION 15300

SECTION 15325

****

GAS-FIRED FURNACE WITH COOLING COIL

PART ONE - GENERAL

1.1 The GENERAL and SPECIAL CONDITIONS, Section 15100, are included as a part of this Section as though written in full in this document.

1.2 Scope of the Work shall include the furnishing and complete installation of the equipment covered by this Section, with all auxiliaries, ready for owner's use.

PART TWO - PRODUCTS

2.1 FURNACE AND COIL

a. Furnace shall be natural (LP) gas-fired, warm-air type, for upflow, downflow, or horizontal arrangement as shown on the drawings. Furnace shall be AGA-approved, UL-labeled, and shall have capacity as shown on the drawings.

b. Furnace cabinet shall be furniture-grade steel with rust-protective coating and with a finish coat of baked enamel.

c. The heat exchanger shall be heavy-gauge cold-rolled or aluminized steel, designed to expand and contract without noise or fatigue. The heat exchanger shall be warranted for 10 years from date of owner's acceptance.

d. The burners shall be of rust-resistant steel and shall be designed to quietly spread the ignition.

e. The blower shall be centrifugal type, rated for "high" static pressure for use with a cooling coil, and shall be belt-driven or directly connected to a changeable-speed motor of capacity not less than that shown on the drawings.

f. Controls shall include the following:

1. 110- to 24-volt transformer.
2. High-limit switch and fan control.
3. 100% shutoff gas valve with separate automatic safety pilot shut-off valve.
4. Electric gas valve and pressure regulator.

g. Provide cooling relay for interlock with the condensing unit.

h. Provide cooling coil casing to match and line up with the furnace, with direct expansion cooling coil of capacity not less than shown on the drawings. Coil shall be provided with refrigerant expansion valve. Provide precharged line sets to connect the coil to the condensing unit.

i. Provide a wall thermostat with a subbase to control the unit on heating or cooling. Switches shall allow selection of Heat-Off-Cool or Fan On-Auto.

2.2 AUXILIARY EQUIPMENT

a. Provide a double-wall, UL-listed vent to the outside, generally as shown on the drawings, with a UL-listed vent cap. Size shall be as recommended by the manufacturer and required by code.

b. Provide vent damper to close the vent when the furnace is not in operation. Vent shall be interlocked to prevent furnace operation until the damper opens.

c. Provide electronic ignition of the pilot. Pilot shall burn only when the furnace is calling for heat.

2.3 ACCEPTABLE MANUFACTURERS

a. Furnace and coil shall be the make and model number shown on the drawings or equivalent models by Carrier, Lennox, York, or Trane.

PART THREE - EXECUTION

3.1 INSTALLATION

a. Install the furnace level and on a non-combustible floor, or provide a mounting accessory designed to protect combustible floor. Provide clearance to combustible materials on all sides and for the vent, as required by the manufacturer and the codes.

b. Provide adequate room in front of the furnace for service and to the sides for pipe and ductwork required.

c. Install the coil casing and coil on the furnace as directed by the manufacturer. Seal all air leaks with sheet metal for a neat and proper installation. Duct tape will not be allowed, either inside or outside the duct. If such tape is used, the system will be rejected. Provide a drain from the coil to an appropriate drain. Provide a trap in the drain.

d. Make the necessary return, outside air, and supply air duct connections to the furnace using non-combustible fabric connectors.

e. Connect gas pipe as required by code. Provide a union and gas cock to allow removal of the furnace. Install the vent pipe as required by the codes.

f. Make electrical connections and provide a shut-off switch adjacent to the furnace. Connect the controls to the thermostat and the condensing unit as recommended by the manufacturer.

3.2 TESTING

a. Operate the furnace and check operation of all controls, including the safety controls. Adjust the fan speed to supply the air quantity called for. Make the readings and adjustments called for under Section 15111, Balancing and Test.

END OF SECTION 15325

SECTION 15327
****
GAS-FIRED FURNACE WITH COOLING COIL

PART ONE - GENERAL

1.1 The GENERAL and SPECIAL CONDITIONS, Section 15100, are included as a part of this Section as though written in full in this document.

1.2 Scope of the Work shall include the furnishing and complete installation of the equipment covered by this Section, with all auxiliaries, ready for owner's use.

PART TWO - PRODUCTS

2.1 FURNACE AND COOLING COIL

a. Furnace shall be natural (LP) gas-fired, warm-air type for upflow, downflow, or horizontal arrangement, as shown on the drawings. Furnace shall be AGA-approved and UL-listed and and shall have capacity as shown on the drawings.

b. Furnace cabinet shall be furniture-grade steel with rust-protective coating and with a finish coat of baked enamel.

c. The heat exchanger shall be heavy-gauge cold-rolled or aluminized steel, designed to expand and contract without noise or fatigue. The heat exchanger shall be warranted for 10 years from date of owner's acceptance.

d. The burners shall be of rust-resistant steel and shall be designed to quietly spread the flame.

e. The blower shall be centrifugal type, rated for "high" static pressure for use with a cooling coil, and shall be belt-driven or direct connected to a changeable-speed motor, with power not less than shown on the drawings.

f. Controls shall include the following:

1. 110- to 24-volt transformer.
2. High-limit switch and fan control.
3. 100% shutoff gas valve with separate automatic safety pilot shutoff valve.
4. Electric gas valve and pressure regulator.
5. Electric ignition of the standing pilot when the thermostat calls for heat.

g. Provide cooling relay for interlock with the condensing unit.

h. Provide cooling coil casing to match and line up with the furnace, with direct-expansion cooling coil of capacity not less than shown on the drawings. Coil shall be provided with refrigerant expansion valve. Provide precharged line sets to connect the coil to the condensing unit.

i. Provide a wall thermostat with a subbase to control the unit on heating or cooling. Switches shall allow selection of Heat-Off-Cool or Fan On-Auto.

2.2 AUXILIARY EQUIPMENT

a. Provide a double-wall UL-listed vent to the outside generally as shown on the drawings, with a UL-listed vent cap. Size shall be as recommended by the manufacturer and required by code.

b. Provide vent damper to close the vent when the furnace is not in operation. Vent shall be interlocked to prevent furnace operation until the damper opens.

2.3 ACCEPTABLE MANUFACTURERS

a. Furnace and coil shall be the make and model number shown on the drawings or equivalent models by Carrier, Lennox, York, or Trane.

PART THREE - EXECUTION

3.1 INSTALLATION

a. Install the furnace level and on a non-combustible floor, or provide a mounting accessory designed to protect combustible floor. Provide clearance to combustible materials on all sides and for the vent, as required by the manufacturer and the codes.

b. Provide adequate room in front of the furnace for service and to the sides for pipe and ductwork required.

c. Install the coil casing and coil on the furnace as directed by the manufacturer. Seal all air leaks with sheet metal for a neat and proper installation. Duct tape will not be allowed either inside or outside the duct. If such tape is used, the system will be rejected. Provide a drain from the coil to an appropriate location. Provide a trap in the drain.

d. Make the necessary return, outside air, and supply air duct connections to the furnace using non-combustible fabric connectors.

e. Connect gas pipe as required by code. Provide a union and gas cock to allow removal of the furnace. Install the vent pipe as required by the codes.

f. Make electrical connections and provide a shut-off switch adjacent to the furnace. Connect the controls to the thermostat and the condensing unit as recommended by the manufacturer.

3.2 TESTING

a. Operate the furnace and check operation of all controls, including electric ignition and the safety controls. Adjust the fan speed to supply the air quantity called for. Make the readings and adjustments called for under Section 15111, Balancing and Test.

END OF SECTION 15327

SECTION 15329

****

GAS-FIRED FURNACE WITH COOLING COIL

PART ONE - GENERAL

1.1 The GENERAL and SPECIAL CONDITIONS, Section 15100, are included as a part of this Section as though written in full in this document.

1.2 Scope of the Work shall include the furnishing and complete installation of the equipment covered by this Section, with all auxiliaries, ready for owner's use.

PART TWO - PRODUCTS

2.1 FURNACE AND COIL

a. Furnace shall be a condensing-type natural gas-fired, warm-air upflow, with 90% (+) annual Fuel Utilization Efficiency, as shown on the drawings. Furnace shall be AGA-approved and UL-listed and shall have capacity as shown on the drawings.

b. Furnace cabinet shall be furniture-grade steel with rust-protective coating and with a finish coat of baked enamel.

c. The primary and secondary (condensing) heat exchanger shall be warranted for 20 years from date of owner's acceptance.

d. The burners shall have spark ignition and shall be part of the totally enclosed combustion system, bringing outside air in under negative pressure from the outside, and expelling the products of combustion under pressure.

e. The fresh air intake and the flue gas vent shall be PVC vents not more then 35 ft. (10 m) in length and with not more than five elbows. Provide AGA-approved vent terminal kit.

f. The blower shall be centrifugal type, rated for "high" static pressure for use with a cooling coil, and shall be directly connected to a changeable-speed motor with power not less than shown on the drawings.

g. Controls shall include the following:

1. 110- to 24-volt transformer.
2. High-limit switch and solid-state fan control.
3. Redundant (two) 100% shutoff gas valves with separate automatic safety pilot shutoff valves.
4. Electric gas valve and pressure regulator.
5. Electric ignition of the standing pilot when the thermostat calls for heat.

h. Provide cooling relay for interlock with the condensing unit.

i. Provide cooling coil casing to match and line up with the furnace, with direct-expansion cooling coil of capacity not less than shown on the drawings. Coil shall be provided with refrigerant expansion valve. Provide precharged line sets to connect the coil to the condensing unit.

j. Provide a wall thermostat with a subbase to control the unit on heating or cooling. Switches shall allow selection of Heat-Off-Cool or Fan On-Auto.

2.2 AUXILIARY EQUIPMENT

a. Provide an intake and exhaust vent to the outside, generally as shown on the drawings, with a UL-listed vent cap. Size shall be as recommended by the manufacturer and required by code.

2.3 ACCEPTABLE MANUFACTURERS

a. Furnace and coil shall be the make and model number shown on the drawings or equivalent model by Carrier (58SX).

PART THREE - EXECUTION

3.1 INSTALLATION

a. Install the furnace level and on a non-combustible floor. Provide clearance to combustible materials on all sides, as required by the manufacturer and the codes.

b. Provide a drain from the condensate drain trap to a point of suitable discharge of the contaminated water.

c. Provide adequate room in front of the furnace for service and to the sides for pipe and ductwork required.

d. Install the coil casing and coil on the furnace as directed by the manufacturer. Seal all air leaks with sheet metal for a neat and proper installation. Duct tape will not be allowed either inside or outside the duct. If such tape is used, the system will be rejected. Provide a drain from the coil to an appropriate location. Provide a trap in the drain.

d. Make the necessary return, outside air, and supply air duct connections to the furnace using non-combustible fabric connectors.

e. Connect gas pipe as required by code. Provide a union and gas cock to allow removal of the furnace. Install the vent pipe as required by the codes.

f. Make electrical connections and provide a shutoff switch adjacent to the furnace. Connect the controls to the thermostat and the condensing unit as recommended by the manufacturer.

3.2 TESTING

a. Operate the furnace and check operation of all controls, including electric ignition and the safety controls. Adjust the fan speed to supply the air quantity called for. Make the readings and adjustments called for under Section 15111, Balancing and Testing.

END OF SECTION 15329

SECTION 15331
****
PULSE-TYPE GAS-FIRED FURNACE WITH COOLING COIL

PART ONE - GENERAL

1.1 The GENERAL and SPECIAL CONDITIONS, Section 15100, are included as a part of this Section as though written in full in this document.

1.2 Scope of the Work shall include the furnishing and complete installation of the equipment covered by this Section, with all auxiliaries, ready for owner's use.

PART TWO - PRODUCTS

2.1 FURNACE

a. Furnace shall be pulse type with 91% minimum Annual Fuel Utilization Efficiency. Furnace shall be natural (LP) gas-fired type with capacity not less than shown on the drawings. Furnace shall be AGA-approved and UL-listed.

b. Furnace cabinet shall be furniture-grade steel with rust-protective coating and with a finish coat of baked enamel.

c. The combustion chamber shall be cast iron, aluminized steel, and stainless steel, designed to burn the gas in pulses. Provide air and gas inlet valves, spark igniter, and flame sensor to verify combustion.

d. The heat exchanger shall be warranted for 10 years from date of owner's acceptance.

e. The blower shall be centrifugal type, rated for "high" static pressure for use with a cooling coil, and shall be belt-driven or directly connected to a changeable-speed motor of capacity not less than that shown on the drawings.

f. Controls shall include the following:

1. 110- to 24-volt transformer.
2. High-limit switch and fan control.
3. 100% shutoff gas valve with separate automatic safety shutoff valve.
4. Electric gas valve and pressure regulator.

g. Provide cooling relay for interlock with the condensing unit.

h. Provide cooling coil casing to match and line up with the furnace, with direct-expansion cooling coil of capacity not less than shown on the drawings. Coil shall be provided with refrigerant expansion valve. Provide precharged line sets to connect the coil to the condensing unit.

i. Provide a wall thermostat with a subbase to control the unit on heating or cooling. Switches shall allow selection of Heat-Off-Cool or Fan-On-Auto.

2.2 AUXILIARY EQUIPMENT

a. Provide 2-in. (50-mm) PVC air intake and vent pipe to the outside. Provide vent screens for both.

b. Provide vent/intake wall terminal kit as furnished by the manufacturer.

2.3 ACCEPTABLE MANUFACTURERS

a. Furnace and coil shall be the make and model number shown on the drawings or equivalent models by Lennox (G14 Series).

PART THREE - EXECUTION

NOTE:
INSTALLATION OF THIS TYPE OF FURNACE IS UNUSUAL. FOLLOW MANUFACTURER'S INSTRUCTIONS CAREFULLY!

3.1 INSTALLATION

a. Install the furnace level and on the isolation mounting pads furnished with the furnace. Install the base insulation pad as recommended by the manufacturer.

b. Install the air intake and vent pipe, and extend the drain from the furnace to an adequate drain point.

c. Provide adequate room in front of the furnace for service and to the sides for pipe and ductwork required.

d. Install the coil casing and coil on the furnace as directed by the manufacturer. Seal all air leaks with sheet metal for a neat and proper installation. Duct tape will not be allowed, either inside or outside the duct. If such tape is used, the system will be rejected. Provide a plastic pipe drain from the coil to an appropriate drain. Provide a trap in the drain.

e. Make the necessary return, outside air, and supply air duct connections to the furnace using noncombustible fabric connectors.

f. Connect gas pipe as required by code. Provide a union, gas cock, and flexible connector to allow removal of the furnace.

g. Make electrical connections using flexible conduit, and provide a shutoff switch adjacent to the furnace. Connect the controls to the thermostat and the condensing unit as recommended by the manufacturer.

h. Support all ductwork, refrigerant pipe, etc., on flexible hangers as recommended by the manufacturer.

3.2 TESTING

a. Operate the furnace and check operation of all controls, including the safety controls. Adjust the fan speed to supply the air quantity called for. Make the readings and adjustments called for under Section 15111, Balancing and Testing.

END OF SECTION 15331

SECTION 15390
****
THROWAWAY FILTERS

PART ONE - GENERAL

1.1 The GENERAL and SPECIAL CONDITIONS, Section 15100, are included as a part of this Section as though written in full in this document.

1.2 Scope of the Work shall include the furnishing and complete installation of the equipment covered by this Section, with all auxiliaries, ready for owner's use.

PART TWO - PRODUCTS

2.1 FILTERS

a. Filters shall be 1 in. (25 mm) thick, throwaway type with dust-holding capacity of 120 to 360 grams per 1000-cfm cell, with ASHRAE atmospheric dust spot efficency of 5 to 15%.

b. Washable- or cleanable-type filters are not acceptable, since the Owner is not in position to apply the necessary adhesives after the filter is cleaned.

c. Replaceable-media-type units, having capacity above, are preferred when provided with aluminum frames and a metal media holder.

PART THREE - EXECUTION

3.1 INSTALLATION

a. Install filters in every air-handling unit at the time of installation for any air-handling system.

b. Install filters in heating-only units only when required on drawings or Specifications for that unit. Heating-only units installed above ceilings, or otherwise inaccessible, shall not be equipped with filters, but shall have filter frames as is standard with the equipment.

c. Installed filters in the units at the time of the installation shall be replaced prior to final review by the engineer or just prior to the owner's occupancy of the building, whichever is earlier. The air quantities read for the Balancing and Testing of the system shall be made with the new filters in place.

d. Two full sets of filters of proper size for every unit shall be left on the job, for the owner's use. These shall be stored in the equipment room in which they will be used, as far as possible, or turned over to the owner where such storage is not possible. These filters are for the owner's use and are not to be used by the contractor in making any filter changes required by the Contract documents during the warranty period.

END OF SECTION 15390

SECTION 15392

****

MEDIUM-EFFICENCY FILTERS

PART ONE - GENERAL

1.1 The GENERAL and SPECIAL CONDITIONS, Section 15100, are included as a part of this Section as though written in full in this document.

1.2 Scope of the Work shall include the furnishing and complete installation of the equipment covered by this Section, with all auxiliaries, ready for owner's use.

PART TWO - PRODUCTS

2.1 FILTERS

a. Filters shall be 1 in. (25 mm), 2 in. (50 mm), or 3 in. (76 mm) thick, as shown on the drawings. Filters shall be disposable type with efficency of 25 to 30% when tested under ASHRAE Test Standard 52-76, with arrestance of 90 to 92% by the same standard.

b. Filter media shall be pleated, nonwoven cotton fabric with a welded wire supporting grid and a support frame of rigid, high-wet-strength beverage board.

c. Holding frame for filters shall be 16-gauge (1.5-mm) galvanized steel with gaskets and spring clips for a positive seal.

d. Pressure loss through the filter at 500 fpm (2.5 m/s) face velocity shall not exceed 0.45 in. of water (130 Pa) for 1 in. (25 mm) thickess, 0.28 in. of water (62 Pa) for 2 in. (50 mm) thickness, and 0.35 in. of water (40 Pa) for 3 in. (76 mm) thickness.

2.2 ACCEPTABLE MANUFACTURERS

a. Filters shall be the size and quantity shown on the drawings, or as recommended by the filter manufacturer for the application. Filters shall be Farr 30/30, Continental Prime "V," or accepted equivalent.

PART THREE - EXECUTION

3.1 INSTALLATION

a. Install filters in every air-handling unit at the time of installation for any air-handling system.

b. Install filters in heating-only units only when required on drawings or specifications for that unit. Heating-only units installed above ceilings, or otherwise inaccessible, shall not be equipped with filters but shall have filter frames as is standard with the equipment.

c. Installed filters in the units at the time of the installation shall be replaced prior to final review by the engineer or just prior to the owner's occupancy of the building, whichever is earlier. The air quantities read for the Balancing and Testing of the system shall be made with the new filters in place.

d. Two full sets of filters of proper size for every unit shall be left on the job for the owner's use. These shall be stored in the equipment room in which they will be used, as far as possible, or turned over to the owner where such storage is not possible. These filters are for the owner's use and are not to be used by the contractor in making any filter changes required by the contract documents during the warranty period.

END OF SECTION 15392

SECTION 15394
****
HIGH-EFFICENCY FILTERS

PART ONE - GENERAL

1.1 The GENERAL and SPECIAL CONDITIONS, Section 15100, are included as a part of this Section as though written in full in this document.

1.2 Scope of the Work shall include the furnishing and complete installation of the equipment covered by this Section, with all auxiliaries, ready for owner's use.

PART TWO - PRODUCTS

2.1 FILTERS

a. Filters shall be 12 in. (30 cm) thick, throwaway type with efficency of 90 to 95 % when tested under the ASHRAE Teat Standard 52-76 and shall have an arrestance of 99% under this standard.

b. Filter media shall be high-density microfine glass fibers laminated to nonwoven synthetic backing.

c. Media support grid shall be welded wire grid bonded to the filter media.

d. Enclosing frame shall be galvanized steel with the filter pack bonded to the frame to prevent air leaks.

2.2 ACCESSORY EQUIPMENT

a. High-efficiency filters shall be provided with medium-efficency prefilters to prolong the life of the main filter. Prefilter shall have approximately 30% efficency.

b. Filter housings shall be provided to match the filters and to provide for sealing the filters in place to prevent air bypass.

c. Side access housings shall be provided with locking gasketed doors on both sides and shall have extruded aluminum rails to hold the main filters and the prefilters.

d. Static pressure loss through the high-efficiency filters shall not exceed 0.65 in. of water (161 Pa) at a face velocity of 500 fpm (2.5 m/s). Final resistance of a dirty filter shall be as high as 1.5 in. of water (372 Pa).

2.3 ACCEPTABLE MANUFACTURERS

a. Filters shall be the size and quantity shown on the drawings, or as recommended by the filter manufacturer for the application. Filters shall be Farr 30/30, Continental Prime "V," or accepted equivalent.

b. Filter housings shall be Farr Model 3P Glide/Pack or acceptable equivalent by Continental.

PART THREE - EXECUTION

3.1 INSTALLATION

a. Install filters in every air-handling unit at the time of installation for any air-handling system.

b. Install with the necessary room and access to allow the doors to be opened without obstruction. Provide access to the area as required to handle the filters in unopened boxes.

c. Install the filter housing as recommended by the manufacturer, and make the necessary duct connections.

d. Install the main filters and the prefilters in the housing.

e. Take care that no more dust is generated in the space than is absolutely necessary so that the filters are not loaded prior to owner's acceptance of the building. Change the prefilters prior to the final review of the engineer.

END OF SECTION 15394

SECTION 15399
****
AIR-HANDLING-UNIT DRAIN

PART ONE - GENERAL

1.1 The GENERAL and SPECIAL CONDITIONS, Section 15100, are included as a part of this Section as though written in full in this document.

1.2 Scope of the Work shall include the furnishing and complete installation of the equipment covered by this Section, with all auxiliaries, ready for owner's use.

PART TWO - PRODUCTS

2.1 PIPE

a. Pipe for drains shall be 1 in. (25 mm) minimum type M copper or Sch 40 PVC, installed with solvent-welded fittings.

PART THREE - EXECUTION

3.1 INSTALLATION

a. Install drains with a slope of no less than 1 in. per 40 feet (1 mm per 1.6 m) and, if possible, 1/4 in. per foot (1 mm/50 mm). Do not allow any sags or low places to pocket.

b. Provide for cleanout of drain at all possible locations.

c. Connections to units shall be with a P trap, of size equal to the drain connection on the unit.

d. Drains over ceilings or other sensitive areas shall be insulated with 1/2-in. (13-mm) foamed plastic insulation, same as runouts.

END OF SECTION 15399

CHAPTER 6

Ventilation Equipment and Fans

SECTION 15400

****

CEILING EXHAUST FAN

PART ONE - GENERAL

1.1 The GENERAL and SPECIAL CONDITIONS, Section 15100, are included as a part of this Section as though written in full in this document.

1.2 Scope of the Work shall include the furnishing and complete installation of the equipment covered by this Section, with all auxiliaries, ready for owner's use.

PART TWO - PRODUCTS

2.1 FANS

a. Exhaust fans shall be designed for mounting in the ceiling of the area served, with a finished appearance on the exposed side and shallow mounting depth.

b. Fan shall be centrifugal type, with motor and fan enclosed in an insulated metal box to reduce noise level.

c. Fan shall be AMCA-certified for the capacity called for on the drawings, at a noise level to match the equipment named and at a noise level normal for the location.

2.2 AUXILIARY EQUIPMENT

a. Fan shall be equipped with a back-draft damper and manual variable electronic speed control with time delay to keep fan in operation after lights in area have been turned off. Length of delayed operation shall depend on length of time lights have been on.

b. Fans located in ceilings with fire ratings shall be provided with a fire damper adapter to provide such rating.

2.3 ACCEPTABLE MANUFACTURERS

a. Ceiling exhaust fans shall be the model shown on the drawings or approved equivalent by Penn, Broan, or Nutone.

b. Controls shall be provided by the fan manufacturer to match the fan requirements equal to Penn Airminder and Penn Electrol.

PART THREE - EXECUTION

3.1 INSTALLATION

a. Mount ceiling exhaust fans as directed by the manufacturer and securely fasten to the ceiling or other proper support to prevent rattling, vibration, and other noise to space served or adjacent spaces.

b. Control the exhaust fan with the lights in the area served, unless noted otherwise.

c. Test the proper operation of the fan and the speed control, as well as the delayed shutoff after varying time of operation.

END OF SECTION 15400

SECTION 15401

****

CENTRIFUGAL CEILING/WALL EXHAUSTER

PART ONE - GENERAL

1.1 The GENERAL and SPECIAL CONDITIONS, Section 15100, are included as a part of this Section as though written in full in this document.

1.2 Scope of the Work shall include the furnishing and complete installation of the equipment covered by this Section, with all auxiliaries, ready for owner's use.

PART TWO - PRODUCTS

2.1 FANS

a. Slow-speed motor and fan shall be mounted in an insulated steel housing, complete with exhaust grille and back-draft damper.

2.2 AUXILIARY EQUIPMENT

a. Roof cap shall be aluminum gravity vent equal to Penn Omega with base.

2.3 ACCEPTABLE MANUFACTURERS

a. Exhaust fans shall be the model shown on the drawings or approved equivalent by Penn, Broan, or Nutone.

PART THREE - EXECUTION

3.1 INSTALLATION

a. Provide duct connections as shown on the drawings. All products shall be installed as indicated on the drawings and according to manufacturer's directions.

END OF SECTION 15401

SECTION 15402
****
EXHAUST FAN/LIGHT COMBINATION

PART ONE - GENERAL

1.1 The GENERAL and SPECIAL CONDITIONS, Section 15100, are included as a part of this Section as though written in full in this document.

1.2 Scope of the Work shall include the furnishing and complete installation of the equipment covered by this Section, with all auxiliaries, ready for owner's use.

PART TWO - PRODUCTS

2.1 FAN AND LIGHT

a. Unit shall be combination light/fan, with plug-in adapter for each. Fan shall be quiet centrifugal type with rubber mounting and back-draft damper.

2.2 ACCEPTABLE MANUFACTURERS

a. Combination unit shall be model shown on the drawings or approved equivalent by Nutone (Model 8672).

2.3 ACCESSORIES

a. Exhaust hood shall be aluminum wall type for wall, hooded vent type for pitched roof, or gravity vent type for low slope roof.

PART THREE - EXECUTION

3.1 INSTALLATION

a. Mount enclosure as directed by manufacturer and install fan complete with ductwork as shown on the drawings. Install light complete with proper size bulb.

b. Connect wall switch for fan and a separate switch for light.

c. Operate the fan and light, and prove correct operation.

END OF SECTION 15402

SECTION 15403

****

CEILING EXHAUST FAN, LIGHT, AND HEATER

PART ONE - GENERAL

1.1 The GENERAL and SPECIAL CONDITIONS, Section 15100, are included as a part of this Section as though written in full in this document.

1.2 Scope of the Work shall include the furnishing and complete installation of the equipment covered by this Section, with all auxiliaries, ready for owner's use.

PART TWO - PRODUCTS

2.1 FIXTURE

a. Unit shall be a combination fan, light, and heater, mounted in one single ceiling fixture as follows:

b. The entire fixture shall bear the UL Label by a testing laboratory.

2.2 FAN

a. Fan shall be centrifugal quiet-type fan for small areas with back-draft damper.

2.3 LIGHT

a. Light shall be incandescent type with reflector and frosted lens.

2.4 HEATER

a. Heater shall be forced-air type with centrifugal fan for quiet operation.

2.5 AUXILIARY EQUIPMENT

a. Provide duct connections for exhaust up to roof or to wall hood, as specified on the drawings. Hoods shall be aluminum and manufactured by the fixture manufacturer.

2.6 ACCEPTABLE MANUFACTURERS

a. Fixture shall be model number shown on the drawings or approved equivalent by Nutone.

PART THREE - EXECUTION

3.1 INSTALLATION

a. Install the fixture as recommended with fittings approved by the manufacturer.

b. Test the operation of the fan, light, and heater to ensure that each is working properly.

c. Eliminate all vibration and noise to fit the application.

END OF SECTION 15403

SECTION 15405

****

CABINET-TYPE CENTRIFUGAL FAN

PART ONE - GENERAL

1.1 The GENERAL and SPECIAL CONDITIONS, Section 15100, are included as a part of this Section as though written in full in this document.

1.2 Scope of the Work shall include the furnishing and complete installation of the equipment covered by this Section, with all auxiliaries, ready for owner's use.

PART TWO - PRODUCTS

2.1 FAN AND CABINET

a. Cabinet fan shall consist of direct-shaft-driven centrifugal fan(s) mounted in an insulated enclosure. Fan and motor shall be isolated by rubber supports to reduce noise. Provide back-draft damper in outlet and white egg-crate removeable inlet grille having 85% free area.

b. Capacity shall be as shown on drawings with Sone level as shown or no greater than model called out. Capacity and sound level shall be AMCA-rated with UL Label.

2.2 AUXILIARY EQUIPMENT

a. Speed control shall be provided by solid-state speed control furnished by the fan manufacturer.

b. Provide duct and grille adapters, as are standard with the cabinet fan.

c. Where ceiling is fire-rated, provide three-hour fire protection between the grille and the duct adapter. The fire damper assembly shall bear the UL Label.

2.3 ACCEPTABLE MANUFACTURERS

a. Cabinet fans shall be the model number shown on the drawings or approved equivalent fan by Penn (Zepher Model), ILG, or Brundage (Series 1800).

PART THREE - EXECUTION

3.1 INSTALLATION

a. Install cabinet fan on vibration isolators as specified under that section.

b. Connect ductwork with flexible connections that are not under any strain and do not have air leaks.

c. This system is not complete until the system has been balanced to provide the correct air quantity and has been tested to demonstrate the correct system performance. See Balancing and Testing, Section 15111.

END OF SECTION 15405

SECTION 15407
****
AIR CURTAIN
(FLY FAN)

PART ONE - GENERAL

1.1 The GENERAL and SPECIAL CONDITIONS, Section 15100, are included as a part of this Section as though written in full in this document.

1.2 Scope of the Work shall include the furnishing and complete installation of the equipment covered by this Section, with all auxiliaries, ready for owner's use.

PART TWO - PRODUCTS

2.1 FAN

a. Fan shall have a long axis to provide air discharge over a narrow slot outlet for air distribution to cover the door width.

b. Fan shall be centrifugal type, directly connected to the motor shafts. Motor shall be permanently lubricated with automatic thermal overload protection with HP as shown on the drawings.

c. Provide no less than four adjustable nozzles (deflector blades) to allow air distribution required. Terminal velocity at the discharge point shall be approximately 2500 ft/min (12.7 m/s) with 800 ft/min (4.1 m/s) terminal velocity at 10-ft (2.5-m) distance. Width of airstream at 10-ft (2.5-m) distance shall be approximately 800 ft/min (4.1 m/s).

d. Provide attractive housing, designed for mounting above the door to be protected. Multiple units may be shown on wide openings. (See drawings.)

2.2 AUXILIARY EQUIPMENT

a. Provide manual switch, as well as door-operated switch to activate the air curtain when the door is opened. Provide an On-Off-Automatic switch to provide for manual or automatic operation. Switch shall be mounted near the door, as directed by the engineer.

b. Door switch shall be coordinated with the door supplier for a neatly concealed installation where possible. Provide motor starter or switch as required by the codes and the installation.

2.3 ACCEPTABLE MANUFACTURERS

a. Air curtains (fly fans) shall be the model number shown on the drawings or approved equivalent by Loren Cook (Gemini Model) or Peerless (Model F81-82C).

PART THREE - EXECUTION

3.1 INSTALLATION

a. Provide the necessary mounting brackets and hardware for a proper vibration-free installation of the air curtain above the door. Secure to the door head or to the overhead structural members, as required.

b. Operate the air curtain and adjust the airstream for the most advantageous distribution and the air quantity to blanket the door.

c. Test out the manual switch and door switches and the On-Off-Auto controls, and ensure proper operation.

END OF SECTION 15407

SECTION 15410
****
IN-LINE CENTRIFUGAL FAN

PART ONE - GENERAL

1.1 The GENERAL and SPECIAL CONDITIONS, Section 15100, are included as a part of this Section as though written in full in this document.

1.2 Scope of the Work shall include the furnishing and complete installation of the equipment covered by this Section, with all auxiliaries, ready for owner's use.

PART TWO - PRODUCTS

2.1 FAN

a. Fan shall be direct-drive centrifugal type, selected for quiet operation and high efficiency at the conditions shown on drawings. Motor power shown on the drawings shall not be exceeded.
b. Fan shall be supported on the motor shaft. Motor shall be out of the airstream and accessible for servicing through access panels.
c. Motor bearings shall be grease-lubricated, with grease fittings brought out to accessible locations.
d. Housing shall be heavy-gauge aluminum, except motor supports shall be steel for strength and rigidity.
e. Cooling shall be provided by breathing tubes.
f. Motors shall be high-efficiency type with internal thermal overload protection.
g. Unit shall bear the AMCA Seal as having been tested under Standard 210.
h. Provide an electrical junction box mounted on the exterior of the unit, with all wiring brought out to this box. Provide explosion-proof fittings for units where required.

2.2 AUXILIARY EQUIPMENT

a. Supports for units shall be standard with the manufacturer for the application: either legs, angles, or casing brackets.
b. Provide vibration isolators for all supports.
c. Provide duct companion flanges for round duct connections to the unit. Provide back-draft dampers of the gravity type for units discharging to the outside or against any pressure from other sources.
d. Controls shall be as specified under the control section for On-Off-Automatic operation. Where units are selected with a reduced air quantity, the fan may be provided with a solid-state electronic speed controller mounted in an accessible location. Fans "over capacity" shall be adjusted to the specified airflow against the actual system static as installed.
e. Units in exposed locations shall be provided with weather protection.

2.3 ACCEPTABLE MANUFACTURERS

a. In-line centrifugal fans shall be the model number shown on the drawings or approved equivalent by Penn (Centrex Model), Ammerman (Model ICD), Cook (Centri-vane), or ILG.

PART THREE - EXECUTION

3.1 INSTALLATION

a. Install unit as recommended by the manufacturer using the proper supports and isolators for quiet operation.

b. Verify proper rotation of the fan.

c. Adjust belts or speed control to provide the design system air quantity at the system static pressure.

d. Measure the current draw and compare to the nameplate current shown.

e. This system is not complete until the system has been balanced to provide the correct air quantity and has been tested to demonstrate the correct system performance. See Balancing and Testing, Section 15111.

END OF SECTION 15410

SECTION 15411

****

VERTICAL-DISCHARGE CENTRIFUGAL FAN

PART ONE - GENERAL

1.1 The GENERAL and SPECIAL CONDITIONS, Section 15100, are included as a part of this Section as though written in full in this document.

1.2 Scope of the Work shall include the furnishing and complete installation of the equipment covered by this Section, with all auxiliaries, ready for owner's use.

PART TWO - PRODUCTS

2.1 FAN

a. Fan shall be centrifugal type, selected at the conditions shown on drawings. Motor power shown on the drawings shall not be exceeded.
b. Fan shall be supported on the motor shaft for directly connected models or on a large-diameter shaft belt connected to the motor with variable-pitch drive pulley. Motor and drive shall be out of the airstream and accessible for servicing without entering the duct system. See drawings for type of drive.
c. Fan and motor bearings shall be grease-lubricated, with grease fittings brought out to accessible locations.
d. Housing shall be heavy-gauge aluminum, except motor supports shall be steel for strength and rigidity.
e. Provide an access panel on the side of the housing for access to the motor in direct-driven unit. Cooling shall be provided by breathing tubes for direct-driven units.
f. Motors shall be high-efficiency type with internal thermal overload protection.
g. Unit shall bear the AMCA Seal as having been tested under Standard 210.
h. Provide an electrical junction box mounted on the exterior of the unit with all wiring brought out to this box. Provide explosion-proof fittings for units where required.

2.2 AUXILIARY EQUIPMENT

a. Unit shall be supported by a curb mounted on the roof. See detail on drawings.
b. Provide duct companion flanges for round duct connections to the unit.
c. Controls shall be as specified under the control section for On-Off-Automatic operation. Where units are selected with a reduced air quantity, the fan may be provided with a solid-state electronic speed controller mounted in an accessible location. Fans "over capacity" shall be adjusted to the specified airflow against the actual system static as installed.

d. Units shall be provided with weather protection for exposed drive belts and motors by weather covers.
e. Units mounted for direct up-blast discharge shall be provided with a discharge head having butterfly-type dampers allowing vertical discharge.
f. Units mounted vertically, requiring intake of outdoor air into the fan for discharge downward, shall be provided with a mushroom-type intake head with bird screen.

2.3 ACCEPTABLE MANUFACTURERS
a. In-line centrifugal fans shall be the model number shown on the drawings or approved equivalent by Penn (Centrex Model), Ammerman (Model ICD), Cook (Centri-vane), or ILG.

PART THREE - EXECUTION

3.1 INSTALLATION
a. Install unit as recommended by the manufacturer.
b. Verify proper rotation of the fan.
c. Adjust belts or speed control to provide the design system air quantity at the system static pressure.
d. Measure the current draw and compare to the nameplate current shown.
e. This system is not complete until the system has been balanced to provide the correct air quantity and has been tested to demonstrate the correct system performance. See Balancing and Testing, Section 15111.

END OF SECTION 15411

SECTION 15412
****
IN-LINE CENTRIFUGAL FAN

PART ONE - GENERAL

1.1 The GENERAL and SPECIAL CONDITIONS, Section 15100, are included as a part of this Section as though written in full in this document.

1.2 Scope of the Work shall include the furnishing and complete installation of the equipment covered by this Section, with all auxiliaries, ready for owner's use.

PART TWO - PRODUCTS

2.1 FAN

a. Fan shall be centrifugal type, selected for quiet operation and good efficiency at the conditions shown on the drawings. Motor power shown on the drawings shall not be exceeded.

b. Fan shall be supported on a large-diameter shaft belt connected to the motor with variable-pitch drive pulley. Motor and drive shall be out of the airstream and accessible for servicing without entering the duct system.

c. Fan and motor bearings shall be grease-lubricated, with grease fittings brought out to accessible locations.

d. Housing shall be heavy-gauge aluminum, except motor supports shall be steel for strength and rigidity.

e. Provide access to belt and driven pulley through tube.

f. Motors shall be high-efficiency type with internal thermal overload protection.

g. Unit shall bear the AMCA Seal as having been tested under Standard 210.

h. Provide an electrical junction box mounted on the exterior of the unit with all wiring brought out to this box. Provide explosion-proof fittings for units where required.

2.2 AUXILIARY EQUIPMENT

a. Supports for units shall be standard with the manufacturer for the application: either legs, angles, or casing brackets.

b. Provide vibration isolators for all supports.

c. Provide duct companion flanges for round duct connections to the unit. Provide back-draft dampers of the gravity type for units discharging to the outside or against any pressure from other sources.

d. Controls shall be as specified under the control section for On-Off-Automatic operation. Where units are selected with a reduced air quantity, the fan may be provided with a solid-state electronic speed controller mounted in an accessible location. Fans "over capacity" shall be adjusted to the specified airflow against the actual system static as installed.

e. Units in exposed locations shall be provided with weather protection for exposed drive belts and motors.

2.3 ACCEPTABLE MANUFACTURERS

a. In-line centrifugal fans shall be the model number shown on the drawings or approved equivalent by Penn (Centrex Model), Ammerman (Model ICB), Cook (Centri-vane), or Powerline (Type BIC).

PART THREE - EXECUTION

3.1 INSTALLATION

a. Install unit as recommended by the manufacturer, using the proper supports and isolators for quiet operation.

b. Verify proper rotation of the fan.

c. Adjust belts or speed control to provide the design system air quantity at the system static pressure.

d. Measure the current draw and compare to the nameplate current shown.

e. This system is not complete until the system has been balanced to provide the correct air quantity and has been tested to demonstrate the correct system performance. See Balancing and Testing, Section 15111.

END OF SECTION 15412

SECTION 15413
****
LARGE-CAPACITY IN-LINE CENTRIFUGAL FAN

PART ONE - GENERAL

1.1 The GENERAL and SPECIAL CONDITIONS, Section 15100, are included as a part of this Section as though written in full in this document.

1.2 Scope of the Work shall include the furnishing and complete installation of the equipment covered by this Section, with all auxiliaries, ready for owner's use.

PART TWO - PRODUCTS

2.1 FAN

a. Fan shall be centrifugal type, selected for quiet operation and good efficiency at the conditions shown on the drawings. Motor power shown on the drawings shall not be exceeded.
b. Fan shall be supported on a large-diameter shaft belt connected to the motor with variable-pitch drive pulley. Motor and drive shall be out of the airstream and accessible for servicing without entering the duct system.
c. Fan and motor bearings shall be grease-lubricated, with grease fittings brought out to accessible locations.
d. Housing shall be heavy-gauge aluminum, except motor supports shall be steel for strength and rigidity.
e. Provide access to belt and driven pulley through tube.
f. Motors shall be high-efficiency type with internal thermal overload protection.
g. Unit shall bear the AMCA Seal as having been tested under Standard 210.
h. Provide an electrical junction box mounted on the exterior of the unit with all wiring brought out to this box. Provide explosion-proof fittings for units where required.

2.2 AUXILIARY EQUIPMENT

a. Supports for units shall be standard with the manufacturer for the application: either legs, angles, or casing brackets.
b. Provide vibration isolators for all supports.
c. Provide duct companion flanges for round duct connections to the unit. Provide back-draft dampers of the gravity type for units discharging to the outside or against any pressure from other sources.

d. Controls shall be as specified under the control section for On-Off-Automatic operation. Where units are selected with a reduced air quantity, the fan may be provided with a solid-state electronic speed controller mounted in an accessible location. Fans "over capacity" shall be adjusted to the specified airflow against the actual system static as installed.

e. Units in exposed locations shall be provided with weather protection for exposed drive belts and motors.

2.3 ACCEPTABLE MANUFACTURERS

a. In-line centrifugal fans shall be the model number shown on the drawings or approved equivalent by Trane (Model Q), New York Blower (Acousta Foil), or Peerless (Centrifan).

PART THREE - EXECUTION

3.1 INSTALLATION

a. Install unit as recommended by the manufacturer, using the proper supports and isolators for quiet operation.

b. Verify proper rotation of the fan.

c. Adjust belts or speed control to provide the design system air quantity at the system static pressure.

d. Measure the current draw and compare to the nameplate current shown.

e. This system is not complete until the system has been balanced to provide the correct air quantity and has been tested to demonstrate the correct system performance. See Balancing and Testing, Section 15111.

END OF SECTION 15413

SECTION 15414

****

WALL EXHAUST FAN
CENTRIFUGAL - DIRECT DRIVE

PART ONE - GENERAL

1.1 The GENERAL and SPECIAL CONDITIONS, Section 15100, are included as a part of this Section as though written in full in this document.

1.2 Scope of the Work shall include the furnishing and complete installation of the equipment covered by this Section, with all auxiliaries, ready for owner's use.

PART TWO - PRODUCTS

2.1 WALL EXHAUST FAN

a. Wall exhaust fan shall be spun aluminum enclosure, designed for mounting directly on a wall and discharging the air away from the wall.

b. The aluminum centrifugal fan shall be mounted on the motor shaft and the motor housing supported on rubber or springs.

c. The capacity of the fan and the noise rating shall be as shown on the drawings. The motor power and speed given shall not be exceeded.

d. The fan shall bear the AMCA Seal as having been tested and meeting Standard 210.

2.2 AUXILLARY EQUIPMENT

a. Provide a bird screen in the fan discharge.

b. Provide a disconnect switch or removable plug located under the fan housing.

c. Provide back-draft damper mounted in the throat of the intake.

d. Provide solid-state motor speed control for single-phase motors as recommended by the manufacturer.

2.3 ACCEPTABLE MANUFACTURERS

a. Exhaust fans shall be the model number shown on the drawings or approved equivalent by Penn (Domex), ILG, or Jenn (CW).

PART THREE - EXECUTION

3.1 INSTALLATION

a. Install the fan securely to the wall.

b. Install ductwork as shown and secure to the wall.

c. Install back-draft dampers and check to make sure they are free to open and close.

d. Connect power and check rotation of fan.

f. This system is not complete until the system has been balanced to provide the correct air quantity and has been tested to demonstrate the correct system performance. See Balancing and Testing, Section 15111.

END OF SECTION 15414

SECTION 15415
****
WALL EXHAUST FAN
CENTRIFUGAL - BELT DRIVE

PART ONE - GENERAL

1.1 The GENERAL and SPECIAL CONDITIONS, Section 15100, are included as a part of this Section as though written in full in this document.

1.2 Scope of the Work shall include the furnishing and complete installation of the equipment covered by this Section, with all auxiliaries, ready for owner's use.

PART TWO - PRODUCTS

2.1 WALL EXHAUST FAN

a. Wall exhaust fan shall be spun aluminum, designed for mounting on a wall and discharging away from the wall.

b. The aluminum centrifugal fan shall be mounted on a large-diameter shaft supported by two ball bearings and connected to the motor by belt drive. The motor drive pulley shall be variable-pitch. The motor shall be supported on adjustable heavy steel brackets, and the entire fan assembly shall be supported on rubber or springs.

c. The capacity of the fan and the noise rating shall be as shown on the drawings. The motor power and speed given shall not be exceeded.

d. The fan shall bear the AMCA Seal as having been tested and meeting Standard 210.

2.2 AUXILIARY EQUIPMENT

a. Provide a bird screen in the fan discharge.
b. Provide a disconnect switch located under the fan housing.
c. Provide back-draft damper mounted in the throat of the curb.
d. Provide automatic belt tensioner.

2.3 ACCEPTABLE MANUFACTURERS

a. Exhaust fans shall be the model number shown on the drawings or approved equivalent by Penn (Domex), ILG, or Acme (PWB).

PART THREE - EXECUTION

3.1 INSTALLATION

a. Install the fan securely to the wall using expansion bolts with metal inserts, through bolts, etc.

b. Install ductwork as shown and secure to the deck or base of the curb.

c. Install back-draft dampers and check to make sure they are free to open and close.

d. Connect power and check rotation of fan.

e. Adjust belt tensioner to properly load the belt(s).

f. This system is not complete until the system has been balanced to provide the correct air quantity and has been tested to demonstrate the correct system performance. See Balancing and Testing, Section 15111.

END OF SECTION 15415

SECTION 15416

****

ROOF EXHAUST FAN

PROPELLER - DIRECT DRIVE

PART ONE - GENERAL

1.1 The GENERAL and SPECIAL CONDITIONS, Section 15100, are included as a part of this Section as though written in full in this document.

1.2 Scope of the Work shall include the furnishing and complete installation of the equipment covered by this Section, with all auxiliaries, ready for owner's use.

PART TWO - PRODUCTS

2.1 ROOF EXHAUST FAN

a. Roof exhaust fan shall be spun aluminum enclosure mounted on an aluminum curb cap.

b. The aluminum propeller fan shall be mounted on the motor shaft and the motor housing supported on rubber or springs.

c. The capacity of the fan and the noise rating shall be as shown on the drawings. The motor power and speed given shall not be exceeded.

d. The fan shall bear the AMCA Seal as having been tested and meeting Standard 210.

2.2 AUXILIARY EQUIPMENT

a. Provide a prefabricated curb of aluminum to match and line up with the fan. Curb shall be sound-attenuating type, if specified, or as required to meet the sound level given on the drawings.

b. Provide a bird screen in the fan discharge.

c. Provide a disconnect switch or removable plug located under the fan housing.

d. Provide back-draft damper mounted in the throat of the curb.

e. Provide solid-state motor speed control for single-phase motors as recommended by the manufacturer.

2.3 ACCEPTABLE MANUFACTURERS

a. Exhaust fans shall be the model number shown on the drawings or approved equivalent models by Jenn (AR).

PART THREE - EXECUTION

3.1 INSTALLATION

a. Install the curb and secure to the roof deck.

b. Install ductwork as shown and secure to the deck or base of the curb.

c. Install back-draft dampers and check to make sure they are free to open and close.

d. Install the fan and secure to the curb with stainless steel screws.

e. Connect power and check rotation of fan.

f. This system is not complete until the system has been balanced to provide the correct air quantity and has been tested to demonstrate the correct system performance. See Balancing and Testing, Section 15111.

END OF SECTION 15416

SECTION 15417

****

ROOF VENTILATOR
PROPELLER - BELT DRIVE

PART ONE - GENERAL

1.1 The GENERAL and SPECIAL CONDITIONS, Section 15100, are included as a part of this Section as though written in full in this document.

1.2 Scope of the Work shall include the furnishing and complete installation of the equipment covered by this Section, with all auxiliaries, ready for owner's use.

PART TWO - PRODUCTS

2.1 ROOF VENTILATOR

a. Roof ventilator shall be heavy-gauge galvanized steel housing, reinforced and structurally supported.

b. The steel propeller fan shall be mounted on a large-diameter shaft supported by two ball bearings and connected to the motor by belt drive. The motor drive pulley shall be variable-pitch. The motor shall be supported on adjustable heavy steel brackets, and the entire fan assembly shall be supported on rubber or springs.

c. The capacity of the fan and the noise rating shall be as shown on the drawings. The motor power and speed given shall not be exceeded.

d. The fan shall bear the AMCA Seal as having been tested and meeting Standard 210.

2.2 AUXILIARY EQUIPMENT

a. Provide a prefabricated curb of aluminum to match and line up with the fan. Curb shall be sound-attenuating type, if specified, or as required to meet the sound level given on the drawings.

b. Provide a bird screen in the fan discharge.

c. Provide a disconnect switch located under the fan housing.

d. Provide back-draft damper mounted in the throat of the curb.

2.3 ACCEPTABLE MANUFACTURERS

a. Exhaust fans shall be the model number shown on the drawings or approved equivalent models by Penn (Airette), ILG, or Ammerman (LMST).

PART THREE - EXECUTION

3.1 INSTALLATION

a. Install the curb and secure to the roof deck.

b. Install ductwork as shown and secure to the deck or base of the curb.

c. Install back-draft dampers and check to make sure they are free to open and close.

d. Install the fan and secure to the curb with stainless steel screws.

e. Connect power and check rotation of fan.

f. This system is not complete until the system has been balanced to provide the correct air quantity and has been tested to demonstrate the correct system performance. See Balancing and Testing, Section 15111.

END OF SECTION 15417

SECTION 15418
****
ROOF SUPPLY VENTILATORS
PROPELLER - BELT DRIVE

PART ONE - GENERAL

1.1 The GENERAL and SPECIAL CONDITIONS, Section 15100, are included as a part of this Section as though written in full in this document.

1.2 Scope of the Work shall include the furnishing and complete installation of the equipment covered by this Section, with all auxiliaries, ready for owner's use.

PART TWO - PRODUCTS

2.1 ROOF VENTILATOR

a. Roof ventilator shall be heavy-gauge galvanized steel housing, reinforced and structurally supported. (Contractor has the option to furnish all aluminum construction.)

b. The propeller fan shall be mounted on a large-diameter shaft supported by two ball bearings and connected to the motor by belt drive. The motor drive pulley shall be variable-pitch. The motor shall be supported on adjustable heavy steel brackets, and the entire fan assembly shall be supported on rubber or springs.

c. The housing shall be designed for air intake with 2-in. (50.8-mm) filters held in the housing and removable without tools. Filters shall be expanded metal mesh and woven wire cloth and shall be cleanable.

d. The capacity of the fan and the noise rating shall be as shown on the drawings. The motor power and speed given shall not be exceeded.

e. The fan shall bear the AMCA Seal as having been tested and meeting Standard 210.

2.2 AUXILIARY EQUIPMENT

a. Provide a prefabricated curb of aluminum to match and line up with the fan. Curb shall be sound-attenuating type, if specified, or as required to meet the sound level given on the drawings.

b. Provide a disconnect switch located under the fan housing.

2.3 ACCEPTABLE MANUFACTURERS

a. Exhaust fans shall be the model number shown on the drawings or approved equivalent by Penn (Airette), ILG, or Ammerman (LMST).

PART THREE - EXECUTION

3.1 INSTALLATION

a. Install the curb and secure to the roof deck.

b. Install ductwork as shown and secure to the deck or base of the curb.

c. Install the fan and secure to the curb with stainless steel screws.

d. Connect power and check rotation of fan.

e. This system is not complete until the system has been balanced to provide the correct air quantity and has been tested to demonstrate the correct system performance. See Balancing and Testing, Section 15111.

END OF SECTION 15418

SECTION 15419
****
ROOF VENTILATOR
PROPELLER - UP-BLAST

PART ONE - GENERAL

1.1 The GENERAL and SPECIAL CONDITIONS, Section 15100, are included as a part of this Section as though written in full in this document.

1.2 Scope of the Work shall include the furnishing and complete installation of the equipment covered by this Section, with all auxiliaries, ready for owner's use.

PART TWO - PRODUCTS

2.1 VENTILATOR

a. Ventilators shall be industrial-grade roof exhausters designed to remove hot and dirty air. Ventilator shall be constructed from heavy-gauge galvanized steel with an epoxy finish or heavy-gauge aluminum.

b. The fan blades shall be mounted on the motor shaft and the motor shall be supported by heavy steel supports. The motor shall be a totally enclosed type designed for this service.

c. The butterfly dampers shall be aluminum, center-hinged, and held closed by magnets and shall be opened by the airstream. The leaving airstream shall repel most of the water when the fan is in operation, and drain gutters shall take away the water when the fan is off and the dampers are closed by gravity.

d. A heavy-gauge wind band shall protect the fan from side currents and direct the air vertically.

e. The entire unit shall be supported on a heavy-gauge reinforced base designed to fit on a roof curb. See detail on drawings.

f. Fan capacity and operating static pressure shall be as shown on the drawings. The fan power and speed shown shall not be exceeded. Where noise criterion is given, it shall not be exceeded, but where noise level is not given, the fan shall be selected for a normal noise level for the application.

g. The fan shall have an AMCA Seal indicating that it has been tested and certified by Standard 300.

2.2 AUXILIARY EQUIPMENT

a. Curb for mounting of fan shall be approved by the roofing contractor if installed on a metal roof. Seals around the curb shall be supplied by the roofer to match the profile of the metal roof.

b. Motor control shall be provided for overload protection and for start/stop.

2.3 ACCEPTABLE MANUFACTURERS

a. Fans shall be the make and model number shown on the drawings or approved equivalent by Acme (UB), Penn (Hi-Ex), or Bayley (SBL).

PART THREE - EXECUTION

3.1 INSTALLATION

a. Install the curb on the roof as required by the roof construction.
b. Install the fan on the curb and secure with stainless steel screws.
c. Connect the fan and check for proper rotation and power draw.
d. Check the butterfly dampers for proper opening and free closing.

END OF SECTION 15419

SECTION 15420

****

ROOF EXHAUST FAN
CENTRIFUGAL - DIRECT DRIVE

PART ONE - GENERAL

1.1 The GENERAL and SPECIAL CONDITIONS, Section 15100, are included as a part of this Section as though written in full in this document.

1.2 Scope of the Work shall include the furnishing and complete installation of the equipment covered by this Section, with all auxiliaries, ready for owner's use.

PART TWO - PRODUCTS

2.1 ROOF EXHAUST FAN

a. Roof exhaust fan shall be spun aluminum enclosure mounted on an aluminum curb cap.

b. The aluminum centrifugal fan shall be mounted on the motor shaft and the motor housing supported on rubber or springs.

c. The capacity of the fan and the noise rating shall be as shown on the drawings. The motor power and speed given shall not be exceeded.

d. The fan shall bear the AMCA Seal as having been tested and meeting Standard 210.

2.2 AUXILIARY EQUIPMENT

a. Provide a prefabricated curb of aluminum to match and line up with the fan. Curb shall be sound-attenuating type, if noted on the drawings, or as required to meet the sound level given on the drawings.

b. Provide a bird screen in the fan discharge.

c. Provide a disconnect switch or removable plug located under the fan housing.

d. Provide back-draft damper mounted in the throat of the curb.

e. Provide solid-state motor speed control for single-phase motors as recommended by the manufacturer.

2.3 ACCEPTABLE MANUFACTURERS

a. Exhaust fans shall be the model number shown on the drawings or approved equivalent models by Penn (Domex), ILG, or Jenn (CR).

PART THREE - EXECUTION

3.1 INSTALLATION

a. Install the curb and secure to the roof deck.

b. Install ductwork as shown and secure to the deck or base of the curb.

c. Install back-draft dampers and check to make sure they are free to open and close.

d. Install the fan and secure to the curb with stainless steel screws.

e. Connect power and check rotation of fan.

f. This system is not complete until the system has been balanced to provide the correct air quantity and has been tested to demonstrate the correct system performance. See Balancing and Testing, Section 15111.

END OF SECTION 15420

SECTION 15421

****

ROOF EXHAUST FAN
CENTRIFUGAL - BELT DRIVE

PART ONE - GENERAL

1.1 The GENERAL and SPECIAL CONDITIONS, Section 15100, are included as a part of this Section as though written in full in this document.

1.2 Scope of the Work shall include the furnishing and complete installation of the equipment covered by this Section, with all auxiliaries, ready for owner's use.

PART TWO - PRODUCTS

2.1 ROOF EXHAUST FAN

a. Roof exhaust fan shall be spun aluminum enclosure mounted on an aluminum curb cap.

b. The aluminum centrifugal fan shall be mounted on a large-diameter shaft supported by two ball bearings and connected to the motor by belt drive. The motor-drive pulley shall be variable-pitch. The motor shall be supported on adjustable heavy steel brackets, and the entire fan assembly shall be supported on rubber or springs.

c. The capacity of the fan and the noise rating shall be as shown on the drawings. The motor power and speed given shall not be exceeded.

d. The fan shall bear the AMCA Seal as having been tested and meeting Standard 210.

2.2 AUXILIARY EQUIPMENT

a. Provide a prefabricated curb of aluminum to match and line up with the fan. Curb shall be sound-attenuating type, if noted on the drawings, or as required to meet the sound level given on the drawings.

b. Provide a bird screen in the fan discharge.

c. Provide a disconnect switch located under the fan housing.

d. Provide back-draft damper mounted in the throat of the curb.

2.3 ACCEPTABLE MANUFACTURERS

a. Exhaust fans shall be the model number shown on the drawings or approved equivalent models by Penn (Domex), ILG, or Jenn (BCR).

PART THREE - EXECUTION

3.1 INSTALLATION

a. Install the curb and secure to the roof deck.

b. Install ductwork as shown and secure to the deck or base of the curb.

c. Install back-draft dampers and check to make sure they are free to open and close.

d. Install the fan and secure to the curb with stainless steel screws.

e. Connect power and check rotation of fan.

f. This system is not complete until the system has been balanced to provide the correct air quantity and has been tested to demonstrate the correct system performance. See Balancing and Testing, Section 15111.

END OF SECTION 15421

SECTION 15423

****

RESTAURANT HOOD EXHAUST FAN
CENTRIFUGAL - BELT DRIVE - UP-DISCHARGE

PART ONE - GENERAL

1.1 The GENERAL and SPECIAL CONDITIONS, Section 15100, are included as a part of this Section as though written in full in this document.

1.2 Scope of the Work shall include the furnishing and complete installation of the equipment covered by this Section, with all auxiliaries, ready for owner's use.

PART TWO - PRODUCTS

2.1 ROOF EXHAUST FAN

a. Roof exhaust fan shall be spun aluminum enclosure, discharging vertically away from the base and mounted on an aluminum curb cap. The base shall provide a drain trough and drain connection to accumulate any liquid and to provide means of removal.

b. The aluminum centrifugal fan shall be mounted on a large-diameter shaft supported by two ball bearings and connected to the motor by belt drive. The motor drive pulley shall be variable-pitch. The motor shall be supported on adjustable heavy steel brackets, and the entire fan assembly shall be supported on rubber or springs. The motor compartment shall be isolated from the fumes being exhausted and shall be cooled by vent tubes connecting to the fresh air.

c. The capacity of the fan and the noise rating shall be as shown on the drawings. The motor power and speed given shall not be exceeded.

d. The fan shall bear the AMCA Seal as having been tested and meeting Standard 210.

2.2 AUXILIARY EQUIPMENT

a. Provide a prefabricated curb of aluminum to match and line up with the fan. Curb shall be smooth metal designed for this service. Provide a hinge adapter to allow the fan to be raised for access to the duct below the fan.

b. Provide shaft seal and heat shield to protect the motor compartment from heat and fumes.

c. Provide a disconnect switch located under the fan housing.

d. Provide a shroud extension to ensure discharge of contaminated air 40 in. (1 m) above the roof.

e. Provide belt tensioner to automatically maintain correct tension on the drive belt(s).

f. Provide mounting pedestal to house a back-draft damper above the roof and to provide access to the damper through the side of the pedestal (not needed on restaurant applications).

g. Provide gravity-operated back-draft damper (not allowed on restaurant hood applications).

2.3 ACCEPTABLE MANUFACTURERS

a. Exhaust fans shall be the model number shown on the drawings or approved equivalent models by Penn (Fumex), ILG, or Jenn (BTD).

PART THREE - EXECUTION

3.1 INSTALLATION

a. Install the curb and secure to the roof deck.

b. Install ductwork as shown and secure to the deck or base of the curb.

c. Install the mounting pedestal with the back-draft dampers and check to make sure they are free to open and close (not allowed on restaurant hood applications).

d. Install the fan and secure to the curb with stainless steel screws.

e. Connect power and check rotation of fan.

f. This system is not complete until the system has been balanced to provide the correct air quantity and has been tested to demonstrate the correct system performance. See Balancing and Testing, Section 15111.

END OF SECTION 15423

SECTION 15425
****
UTILITY VENT SET

PART ONE - GENERAL

1.1 The GENERAL and SPECIAL CONDITIONS, Section 15100, are included as a part of this Section as though written in full in this document.

1.2 Scope of the Work shall include the furnishing and complete installation of the equipment covered by this Section, with all auxiliaries, ready for owner's use.

PART TWO - PRODUCTS

2.1 FAN AND HOUSING

a. Fan shall be single-inlet, end-suction centrifugal type mounted directly on the motor shaft. Fan may be used for supply or exhaust as indicated on the drawings.

b. Fan wheel shall be forward-curved, selected at a point of high efficiency for the air quantity and pressure shown on the drawings. The motor speed and power required shall not exceed that shown on the drawings.

c. Fan wheel and housing shall be constructed of heavy-gauge steel, fabricated by welding, and properly braced for heavy service. All surfaces shall be protected from rust by appropriate coating for the service shown.

d. Fan and housing shall bear the AMCA Seal. All electrical components shall bear the UL Label.

2.2 AUXILIARY EQUIPMENT

a. Fan base shall be supported as shown on the drawings, with vibration isolators as shown or as specified in section on vibration isolators.

b. Duct connections shall be flexible-type duct material appropriate for the service involved and the location and exposure.

c. Outlet shutters formed of aluminum or stainless steel shall be installed on all fans discharging directly to the outdoors, unless otherwise shown.

d. Fans exhausting noxious materials shall be provided with exhaust stacks as shown on the drawings.

e. Fans located outside and drawing in outside air directly shall be provided with intake hoods and shutters to prevent entry of moisture and to prevent draft when not in operation.

f. Fans located in protected areas with open intakes shall be provided with an inlet guard to prevent entry of trash and to protect persons in the area.

g. Fans located outside and exposed to weather shall be provided with a weather cover to protect the motor.

2.3 ACCEPTABLE MANUFACTURERS

a. Fans shall be the model numbers shown on the drawings or approved equivalent by New York Blower (Junior Fans), Bayley (FC), or Brundage (DS).

b. Submit fan curves of proposed units with operating points marked.

PART THREE - EXECUTION

3.1 INSTALLATION

a. Install fan as shown on the drawings and provide duct connections as shown and required. NOTE: Changes in inlet conditions from those shown on the drawings may completely change fan selection. Where possible, the inlet to the fan shall be a straight duct section of several wheel diameters.

b. Mount the fan on vibration supports as shown and specified. Provide duct flexible connections as specified and required for the service.

c. Check the fan wheel for proper rotation and the motor for power draw.

d. Balance and test the system and adjust for the proper airflow. See Balancing and Testing, Section 15111.

END OF SECTION 15425

SECTION 15426

****

UTILITY FAN

PART ONE - GENERAL

1.1 The GENERAL and SPECIAL CONDITIONS, Section 15100, are included as a part of this Section as though written in full in this document.

1.2 Scope of the Work shall include the furnishing and complete installation of the equipment covered by this Section, with all auxiliaries, ready for owner's use.

PART TWO - PRODUCTS

2.1 FAN AND HOUSING

a. Fan shall be single-inlet centrifugal type mounted on large-diameter steel fan shaft supported by two bearings. Bearing shall be self-aligning pillow block, grease-lubricated ball type, selected for 400,000 hours of operation at maximum rating of the fan. Grease fittings shall be provided with tubes for greasing without having to remove the protective housing.

b. Motor shall be high-efficiency, open, drip-proof type, selected for no less than the maximum fan power required. The motor shall be mounted on a heavy steel adjustable baseplate, securely held in position by large steel tensioning bolts.

c. Drive shall consist of adjustable variable-pitch motor pulley and a fixed pulley on the fan shaft. Both pulleys shall be keyed to their shafts, with the keys locked by setscrews

d. Belts shall be selected for 200% of motor rating and of size and number recommended by the belt manufacturer. Where two or more belts are used, they shall be "matched sets," bearing the same lot number.

e. Fan wheel shall be forward-curved, backward-curved, or airfoil design as indicated by the model number shown on the drawings and as required for the airflow and pressure shown. Fan shall be selected at a point near maximum efficiency for lowest noise generation and lowest power requirements.

f. The fan selected shall not exceed the sound power levels shown on the drawings or, if not shown, the normal recommended noise criteria for the application.

g. The construction of the fan shall be the class recommended by the manufacturer for the airflow and pressure.

h. The fan speed and power required shall not exceed that shown on the drawings.

i. Fan wheel and housing shall be constructed of heavy-gauge steel, fabricated by welding, and properly braced for heavy service. All surfaces shall be protected from rust by appropriate coating for the service shown.

j. Fan and housing shall bear the AMCA Seal. All electrical components shall bear the UL Label.

k. Protective coatings, if required for lab hood or other severe duty, shall be applied to all surfaces, inside and out, as specified in chemical corrosion protection.

2.2 AUXILIARY EQUIPMENT

a. Fan base shall be supported as shown on the drawings, with vibration isolators as shown or as specified in the section on vibration isolators.

b. Duct connections shall be flexible-type duct material, appropriate for the service involved and the location and exposure.

c. Outlet shutters formed of aluminum or stainless steel shall be installed on all fans discharging directly to the outdoors, unless otherwise shown.

d. Fans exhausting noxious materials shall be provided with exhaust stacks as shown on the drawings.

e. Fans located outside and drawing in outside air directly shall be provided with intake hoods and shutters to prevent entry of moisture and to prevent draft when not in operation.

f. Fans located in protected areas with open intakes shall be provided with an inlet guard to prevent entry of trash and to protect persons in the area.

g. Fans located outside and exposed to weather shall be provided with a weather cover to protect the motor.

2.3 ACCEPTABLE MANUFACTURERS

a. Utility fans shall be the model numbers shown on the drawings or approved equivalent by Carrier, Trane, New York Blower, American Air Filter, Clarage, ILG, Lau, or Penn.

b. Submit fan curves of proposed units with operating points marked.

PART THREE - EXECUTION

3.1 INSTALLATION

a. Install fan as shown on the drawings and provide duct connections as shown and required. NOTE: Changes in inlet conditions from those shown on the drawings may completely change fan selection. Where possible, the inlet to the fan shall be a straight duct section of several wheel diameters.

b. Mount the fan on vibration supports as shown and specified. Provide duct flexible connections as specified and required for the service.

c. Check the fan wheel for proper rotation and the motor for power draw.

d. Balance and test the system and adjust for the proper airflow. See Balancing and Testing, Section 15111.

e. Verify that location of the disconnect switch does not interfere with the removal of the protective covers or otherwise interfere with the proper servicing of the fan.

END OF SECTION 15426

SECTION 15427
****
WALL FAN

PART ONE - GENERAL

1.1 The GENERAL and SPECIAL CONDITIONS, Section 15100, are included as a part of this Section as though written in full in this document.

1.2 Scope of the Work shall include the furnishing and complete installation of the equipment covered by this Section, with all auxiliaries, ready for owner's use.

PART TWO - PRODUCTS

2.1 FAN

a. Fan shall be propeller type, directly connected to the motor for supply or exhaust, as shown on the drawings.

b. Fan motor shall be supported by the metal guard and the guard supported on rubber to eliminate noise.

c. Fan speed shall not exceed that shown on the drawings, with selection of the fan not exceeding the recommended noise level for the area served.

d. Fan shall be rated for air quantity and static pressure as shown on the drawings. Motor power shall not exceed that listed.

2.2 AUXILIARY EQUIPMENT

a. Fan guard shall be furnished and installed.

b. Furnish and install gravity-type shutters on the discharge of exhaust fans.

c. Supply fans shall be provided with a weatherproof louver and gravity shutters on the intake.

d. Furnish and install electronic-type speed control to adjust the capacity of the fan.

e. Furnish and install switch and pilot light (or combination) to control the fan, if the fan switch is not located in the area served and visible.

f. Fans serving toilet rooms for exhaust shall be provided with a combination light switch and fan controller that will cause the fan to operate for a period of time after the lights are turned off, depending on how long the lights have been on.

PART THREE - EXECUTION

3.1 INSTALLATION

a. Install the fan as shown on the drawings and as recommended by the manufacturer.

b. Provide the shutter dampers and the controls.

c. Test the operation of the fan and adjust the speed to meet the airflow requirements. See Balancing and Testing, Section 15111.

END OF SECTION 15427

SECTION 15428

****

PROPELLER-TYPE WALL FAN

PART ONE - GENERAL

1.1 The GENERAL and SPECIAL CONDITIONS, Section 15100, are included as a part of this Section as though written in full in this document.

1.2 Scope of the Work shall include the furnishing and complete installation of the equipment covered by this Section, with all auxiliaries, ready for owner's use.

PART TWO - PRODUCTS

2.1 FAN

a. Fan shall be slow-speed propeller type for supply (SF) or exhaust (EF) as shown on the drawings. The fan shall be mounted on a large-diameter shaft supported by two pillow-block-type sealed bearings.

b. Fan motor shall be mounted in rubber cradle on an adjustable base.

c. Drive shall consist of one or more belts as required for the power. Where multiple belts are required, they shall be matched from the same production run. The drive pulley shall be adjustable-pitch type. All pulleys shall be keyed to the shaft and the keys held by setscrews.

d. Fan frame shall be heavy steel, welded to rigidly support the fan and motor.

e. Fan capacity and static pressure shall be as called for on the drawings. Motor power shall not exceed that shown on the drawings.

f. Fan shall bear AMCA Seal for air and sound performance, as tested under Standards 210 and 300.

2.2 AUXILIARY EQUIPMENT

a. Fan, motor, and belts shall be protected by a heavy-duty guard with provisions for access to the motor and drive. In gymnasiums and other areas where the fan will be subject to hard treatment, the guard shall be made of a welded frame covered with expanded metal. The back or side must be hinged for access.

b. Shutters and/or louvers shall be provided as specified under respective sections.

2.3 ACCEPTABLE MANUFACTURERS

a. The fan shall be the model number shown on the drawings or an approved equivalent by Penn (Type FB), ILG, Ammerman (Type WFB), or American Coolair (Model CBQ or CBM).

PART THREE - EXECUTION

3.1 INSTALLATION

a. Install fan in a heavy, treated-wood frame secured in wall or metal frame secured to wall. Frame shall be anchored to wall and fan secured to frame.

b. Check motor rotation and adjust the fan drive to the required air quantity under normal operating conditions.

c. Provide maximum possible airflow and check motor power to determine that the motor is not drawing more than rated current.

d. Check operation of shutters or louvers to see that they are free and do not bind open or closed.

END OF SECTION 15428

SECTION 15429
****
CORROSION PROTECTION

PART ONE - GENERAL

1.1 The GENERAL and SPECIAL CONDITIONS, Section 15100, are included as a part of this Section as though written in full in this document.

1.2 Scope of the Work shall include the furnishing and complete installation of the equipment covered by this Section, with all auxiliaries, ready for owner's use.

PART TWO - PRODUCTS

2.1 EQUIPMENT REQUIRING PROTECTION

a. The equipment to be protected from corrosion or chemical attack shall be carefully prepared and coated in accordance with the coating manufacturer's instructions and shall be warranted by the coating manufacturer.

b. The coating shall be a pure, thermosetting phenolic resin, with suitable modifying resins applied in multiple thin coats and oven-baked to a dry film thickness of no less than 0.0005 in. (0.0127 mm). The final finish shall have a high gloss and shall resist chemical attack from those chemicals found in normal use in university teaching-lab hoods including the following:

(SPEC WRITER SHOULD INCLUDE THE NAMES OF CHEMICALS THAT WILL BE USED IN THE HOOD. LIST SHOULD BE OBTAINED FROM THE OWNER.)

c. The coating on equipment exposed to sunlight shall be resistant to attack from the sun.

d. The coating shall be applied to all surfaces, both those exposed to chemical attack (such as air side of fans) and those surfaces not normally exposed except to the environment (such as the outside of fans). Surfaces such as shafts and bearings, that cannot be normally coated, shall be easily replaceable.

2.2 ACCEPTABLE SUPPLIERS

a. The equipment manufacturer shall submit, for consideration, the name and product identification of the coating proposed for use, including a total list of those chemicals for which the coating is considered superior, acceptable, and not acceptable.

b. The manufacturer's submittal shall include a description of the application methods to be used and the type of curing required to ensure the chemical resistance described in 2.1.

2.3 PREPARATION OF THE EQUIPMENT

a. All projecting bolts, cap nuts, and other such attachments in the coated area shall be removed and replaced with connectors that will not break the final film surface when the equipment is assembled.

b. All rough and projecting weld surfaces shall be ground smooth, and all voids shall be filled and smoothed with filler acceptable to the coating manufacturer.

c. All intermittent welds shall be extended to make them continuous, and all welds shall be ground smooth.
d. After fabrication and prior to assembly, the parts shall be thoroughly cleaned to remove all grease and scale as recommended by the coating manufacturer and then shall be coated to the proper thickness and baked. Bolt holes shall be oversized and left open so that all edges shall be coated. Self-tapping screws and captive nuts are not acceptable, as they break the coating.
e. Bolts used for fabrication shall be of material resistant to the chemicals to which the coating is resistant, or they shall be coated after final assembly to the same thickness of the equipment.
f. Prior to assembly, all surfaces shall be checked for coating void by using a charged wet sponge on an appropriate high-voltage tester. All voids of any type shall be recoated and rebaked.

2.4 FINAL ASSEMBLY AND BALANCING OF EQUIPMENT

a. The equipment shall be carefully assembled and balanced, statically and dynamically, at the operating speed to within the tolerances listed below as measured on the bearing housings:

Fan speed, rev/min	All planes (mils)
up to 300	5
300-500	4
500-1000	3
over 1000	2

b. An inspection report shall be submitted for each fan, detailing the type of coating applied, the number of coats used, the method of curing, the dry film thickness, and the final results of the balancing.
c. A label shall be affixed to each fan, stating that a special coating has been applied and listing the person to call (the coating manufacturer) should the coating be damaged and need repairs.
d. A warning label with large letters shall be adhered to the fan, indicating that the equipment has a special coating and should not be drilled, burned, scraped, or otherwise damaged.

PART THREE - EXECUTION

3.1 HANDLING

a. Equipment having special coating shall be handled very carefully to protect the coating from even the smallest scratch or damage. Any such damage shall be repaired at the installer's cost, which may include sending the entire fan back to the factory.
b. Attachment of ductwork, electrical connections, supports, etc., shall be done in such a way as to not damage the coating.

END OF SECTION 15429

CHAPTER 7

Ventilation Equipment and Systems

WALL LOUVERS - INTAKE AND EXHAUST   SECTION 15462   1 Page   7-9

Galvanized steel, shop-fabricated louvers with 4-in. (10-mm) blades with multiple breaks with slope of 45 deg. With bird screen. Frame to recess in the opening.

WALL LOUVERS - INTAKE AND EXHAUST   SECTION 15463   1 Page   7-10

Galvanized steel, shop-fabricated louvers with 4-in. (10-mm) blades with double breaks with slope of 45 deg. With bird screen. Frame provided with outer lip for surface mounting on wall.

LOUVERED PENTHOUSE   SECTION 15465   2 Pages   7-11

Extruded-aluminum, prefabricated penthouse mounted on prefabricated curb for roof mounting.

GOOSENECK INTAKE/RELIEF   SECTION 15469   1 Page   7-13

Galvanized steel, shop fabricated, with bird screen.

DOMESTIC RANGE HOOD   SECTION 15470   1 Page   7-14

Residential kitchen hood, vented type, with fan and filter.

KITCHEN HOOD   SECTION 15475   4 Pages   7-15

Stainless steel canopy-type commercial hood with exhaust and supply ductwork. Fire-extinguishing system included with hood.

KITCHEN HOOD   SECTION 15477   3 Pages   7-19

Canopy-type small commercial hood acceptable to local authority, with exhaust duct, but no supply. Fire-extinguishing system acceptable to local authority.

DISHWASHER HOOD   SECTION 15479   1 Page   7-22

Stainless steel hood for mounting over dishwasher, with stainless steel exhaust duct. No supply.

HEATING AND VENTILATION UNIT   SECTION 15480   2 Pages   7-23

Indirect gas fired, H&V unit with dual gas valves for minimum ventilation supply and valve to provide for space temperature control.

MAKE-UP AIR HEATER   SECTION 15483   3 Pages   7-25

Direct gas fired modulating burner directly in the airstream.

MAKE-UP AIR HEATER   SECTION 15484   3 Pages   7-28

Indirect gas fired, stainless steel heat exchanger for use in 100% outside air.

PLATE-TYPE HEAT RECOVERY UNIT   SECTION 15485   2 Pages   7-31

Embossed aluminum plates for normal operating temperatures with alternate for stainless steel plates with wash-down for kitchen hood use.

ENGINE EXHAUST SYSTEM   SECTION 15490   3 Pages   7-33

Vehicle exhaust for installation in shop ceiling. Retractable stainless steel hose and fittings.

SECTION 15450

****

VENTILATOR

(AIR INTAKE OR EXHAUST)

PART ONE - GENERAL

1.1 The GENERAL and SPECIAL CONDITIONS, Section 15100, are included as a part of this Section as though written in full in this document.

1.2 Scope of the Work shall include the furnishing and complete installation of the equipment covered by this Section, with all auxiliaries, ready for owner's use.

PART TWO - PRODUCTS

2.1 VENTILATOR

a. Ventilator shall be stationary type, with aspirating action as well as gravity movement of air.

b. Neck size shall be as shown on the drawings.

c. Ventilator shall be constructed of no less than 0.025-gauge (0.635-mm) 35-alloy aluminum, unless otherwise noted on the drawings.

2.2 AUXILIARY EQUIPMENT

a. Provide flat flange mounting base of 0.064-gauge (1.6-mm) aluminum for installation into flat built-up roofs or single-ply roofs.

b. Provide slope base or ridge-type base of not less than 0.064-gauge (1.6-mm) aluminum for mounting on slope roofs of all types.

c. Provide base for mounting on a curb of not less than 0.064-gauge (1.6-mm) aluminum where shown on drawings. Curb shall be prefabricated type, not less than 12 in. (30 cm) high, as required to match the roof construction.

d. Provide bird screen of aluminum No. 2 expanded metal.

e. Provide single-leaf butterfly dampers of aluminum, with pull chain and necessary damper accessories such as pulleys, chain clips, etc., to make a complete installation.

f. Provide 160 deg. F (70 deg. C) fusible link to close (open) damper in case of fire.

g. Provide simple ring grille on exposed underside of vents installed in finished areas and at ceiling of finished rooms connected to ventilators.

2.3 ACCEPTABLE MANUFACTURERS

a. Ventilators shall be the model number shown on the drawings or equivalent products by Penn, Breidert, Greenheck.

PART THREE - EXECUTION

3.1 INSTALLATION

a. Install ventilators as shown on the drawings and connect ductwork as shown and required.

b. Install ring grille on finished ceiling, where specified and required.

END OF SECTION 15450

SECTION 15455
****
SHUTTER-TYPE RELIEF DAMPERS

PART ONE - GENERAL

1.1 The GENERAL and SPECIAL CONDITIONS, Section 15100, are included as a part of this Section as though written in full in this document.

1.2 Scope of the Work shall include the furnishing and complete installation of the equipment covered by this Section, with all auxiliaries, ready for owner's use.

PART TWO - PRODUCTS

2.1 RELIEF DAMPERS

a. Dampers shall be pressure-opened and gravity-closed, parallel blade type, to regulate pressure or control back-draft.

b. Frame shall be heavy-gauge extruded-aluminum, designed to mount in ductwork or be installed in outside wall, either with a flange or recessed. See detail on drawings for type of mounting.

c. Blades shall be roll-formed aluminum ribbed blades mounted on ball-bearing pivots with self-lubricating 1/2-in. (12.5-mm) minimum-size shafts. Blades shall be linked together to operate in unison. Linkage shall be attached to each blade with a lubricated pivot and shall be out of the airstream to reduce air friction. Edges of blades on all sides shall be provided with vinyl seal strips to prevent air leakage.

d. Provide aluminum counterbalance for field adjustment of operating pressure.

2.2 AUXILIARY EQUIPMENT

a. Provide motor operator for shutters, if specified on the drawings. The motor shall be by a recognized control manufacturer and of a type that can remain in energized position for long periods of time without problems. Controls for the motor shall be provided, as well as all components for a complete system.

b. Provide an aluminum bird screen on the outside of the shutters, as recommended by the manufacturer.

c. Provide flanges as required by details shown on the drawings. Large shutters shall be installed in sections, with adequate mullions between sections.

PART THREE - EXECUTION

3.1 INSTALLATION

a. Install shutters as recommended by the manufacturer.

b. Caulk completely around the shutter with silicone caulk to prevent entry of water.

c. Test operation of the shutters and adjust the counterbalance to the proper air pressure for operation.

END OF SECTION 15455

SECTION 15456
****
RELIEF DAMPERS

PART ONE - GENERAL

1.1 The GENERAL and SPECIAL CONDITIONS, Section 15100, are included as a part of this Section as though written in full in this document.

1.2 Scope of the Work shall include the furnishing and complete installation of the equipment covered by this Section, with all auxiliaries, ready for owner's use.

PART TWO - PRODUCTS

2.1 DAMPERS

a. Relief dampers shall be motorized-type, opposed blade dampers constructed of heavy-gauge extruded aluminum, operated by a motor under control of the control system.

b. Frame shall be heavy-gauge extruded aluminum, designed to mount in ductwork or be installed in outside wall either with a flange or recessed. See detail on drawings for type of mounting.

c. Blades shall be roll-formed aluminum ribbed blades mounted on ball-bearing pivots with self-lubricating 1/2-in. (12.5-mm) minimum-size shafts. Blades shall be linked together to operate in unison. Linkage shall be attached to each blade with a lubricated pivot and shall be out of the airstream to reduce air friction. Edges of blades on all sides shall be provided with vinyl seal strips to prevent air leakage.

2.2 AUXILIARY EQUIPMENT

a. Provide support brackets for motor operator. The motor shall be by a recognized control manufacturer and of type that can remain in energized position for long periods of time without problems. Controls for the motor shall be provided, as well as all components for a complete system.

b. Provide an aluminum bird screen on the outside of the shutters, as recommended by the manufacturer.

c. Provide flanges as required by details shown on the drawings. Large shutters shall be installed in sections, with adequate mullions between sections.

PART THREE - EXECUTION

3.1 INSTALLATION

a. Install shutters as recommended by the manufacturer.

b. Caulk completely around the shutter with silicone caulk to prevent entry of water.

c. Test operation of the shutters and adjust the counterbalance to the proper air pressure for operation.

SECTION 15460

****

WALL LOUVERS - INTAKE AND EXHAUST

PART ONE - GENERAL

1.1 The GENERAL and SPECIAL CONDITIONS, Section 15100, are included as a part of this Section as though written in full in this document.

1.2 Scope of the Work shall include the furnishing and complete installation of the equipment covered by this Section, with all auxiliaries, ready for owner's use.

PART TWO - PRODUCTS

2.1 a. Wall louvers shall be heavy-gauge extruded-aluminum type with multiple breaks or ridges to prevent water from traveling up the blade. Blade depth shall be 4 in. (10 cm), unless shown otherwise on the drawings.
b. Blade shall slope at 45 deg. and shall be reinforced on the outer and inner edges with a reinforced lip.
c. Louvers shall bear the AMCA Seal as having been rated in accordance with Standard 500 for air performance and moisture penetration.
d. Louvers shall be sized with a maximum pressure loss of 1/10 in. (2.54 mm) at 750 ft/min (3.8 m/s) intake or exhaust.
e. Finish shall be factory-anodized aluminum color, unless specified otherwise. If specific color is required, a sample of the required color will be furnished.
f. Top and side mullions shall be designed for mounting against masonry openings with a caulking lip.

2.2 AUXILIARY EQUIPMENT
a. Louvers shall be provided with an aluminum bird screen in a removable frame on the inside. Insect screen, if shown on the drawings for intake louvers, shall be aluminum in an aluminum frame and shall be mounted on the exterior of the louver.
b. Provide sill extension on louver to provide drip ledge to carry water away from the building.
c. For larger-size louvers, provide mullions to provide rigidity and allow expansion.
d. See drawings for duct connections, dampers, etc., that attach to the louver.

PART THREE - EXECUTION

3.1 INSTALLATION
a. Install the louver in the masonry opening with the sill extension extending from the building 1/4 in. (6.5 mm), and secure to the masonry to prevent unauthorized entry.
b. Caulk the perimeter of the louver at the junction of the wall with silicone caulk, applied as recommended by the manufacturer. Allow room for expansion and contraction without damage to the caulking.

END OF SECTION 15460

SECTION 15461
****
WALL LOUVERS - INTAKE AND EXHAUST

PART ONE - GENERAL

1.1 The GENERAL and SPECIAL CONDITIONS, Section 15100, are included as a part of this Section as though written in full in this document.

1.2 Scope of the Work shall include the furnishing and complete installation of the equipment covered by this Section, with all auxiliaries, ready for owner's use.

PART TWO - PRODUCTS

2.1 WALL LOUVERS

a. Wall louvers shall be heavy-gauge extruded-aluminum type with multiple breaks or ridges to prevent water from traveling up the blade. Blade depth shall be 4 in. (10 cm), unless shown otherwise on the drawings.

b. Blade shall slope at approximately 45 deg. and shall be reinforced on the outer and inner edges with a reinforced lip.

c. Louvers shall bear the AMCA Seal as having been rated in accordance with Standard 500 for air performance and moisture penetration.

d. Louvers shall be sized with a maximum pressure loss of 1/10 in. (2.54 mm) at 750 ft/min (3.8 m/s) intake or exhaust.

e. Finish shall be factory-anodized aluminum color, unless specified otherwise. If specific color is required, a sample of the required color will be furnished.

f. Louver shall be designed for flange mounting against the face of the building.

2.2 AUXILIARY EQUIPMENT

a. Louvers shall be provided with an aluminum bird screen in a removable frame on the inside. Insect screen, if shown on the drawings for intake louvers, shall be aluminum in an aluminum frame and shall be mounted on the exterior of the louver.

b. For larger-size louvers, provide mullions to provide rigidity and allow expansion.

d. See drawings for duct connections, dampers, etc., that attach to the louver.

PART THREE - EXECUTION

3.1 INSTALLATION

a. Install the louver in the masonry opening with the flange extending around the perimeter of the louver and 1/4 in. (6.5 mm) away from the wall to allow caulking. Secure the louver in such a way to prevent entry into the building.

b. Caulk the perimeter of the louver at the junction of the wall with silicone caulk, applied as recommended by the manufacturer. Allow room for expansion and contraction without damage to the caulking.

END OF SECTION 15461

SECTION 15462

****

WALL LOUVERS - INTAKE AND EXHAUST
GALVANIZED STEEL - SHOP FABRICATED

PART ONE - GENERAL

1.1 The GENERAL and SPECIAL CONDITIONS, Section 15100, are included as a part of this Section as though written in full in this document.

1.2 Scope of the Work shall include the furnishing and complete installation of the equipment covered by this Section, with all auxiliaries, ready for owner's use.

PART TWO - PRODUCTS

2.1 WALL LOUVERS

a. Wall louvers shall be heavy-gauge galvanized steel with multiple breaks or ridges to prevent water from traveling up the blade. Blade depth shall be 4 in. (10 cm), unless shown otherwise on the drawings.
b. Blade shall slope at 45 deg. and shall be reinforced on the outer and inner edges with a double break.
c. Louver shall be shop-fabricated in accord with the best practices of SMACNA or other industry standards.
d. Louver shall be designed for recess mounting inside a masonry or metal opening, with edges turned back to provide a rounded caulking edge.

2.2 AUXILIARY EQUIPMENT

a. Louvers shall be provided with a bird screen in a removable frame on the inside. Insect screen, if shown on the drawings for intake louvers, shall be mounted on the exterior of the louver and the bird screen omitted.
b. Provide sill extension on louver to provide drip ledge to carry water away from the building.
c. For larger-size louvers, provide mullions to provide rigidity and allow expansion.
d. See drawings for duct connections, dampers, etc., that attach to the louver.

PART THREE - EXECUTION

3.1 INSTALLATION

a. Install the louver in the masonry opening with the sill extension extending from the building 1/4 in. (6.5 mm), and secure to the masonry to prevent unauthorized entry.
b. Caulk the perimeter of the louver at the junction of the wall with silicone caulk, applied as recommended by the manufacturer. Allow room for expansion and contraction without damage to the caulking.

END OF SECTION 15462

SECTION 15463
****

WALL LOUVERS - INTAKE AND EXHAUST
GALVANIZED STEEL - SHOP FABRICATED

PART ONE - GENERAL

1.1 The GENERAL and SPECIAL CONDITIONS, Section 15100, are included as a part of this Section as though written in full in this document.

1.2 Scope of the Work shall include the furnishing and complete installation of the equipment covered by this Section, with all auxiliaries, ready for owner's use.

PART TWO - PRODUCTS

2.1 WALL LOUVERS

a. Wall louvers shall be heavy-gauge galvanized steel with multiple breaks or ridges to prevent water from traveling up the blade. Blade depth shall be 4 in. (10 cm), unless shown otherwise on the drawings.

b. Blade shall slope at 45 deg. and shall be reinforced on the outer and inner edges with a double break.

c. Top and side mullions shall be designed for flange mounting against masonry or metal opening with a caulking lip.

d. Louver shall be shop-fabricated in accord with the best practices of SMACNA or other industry standards.

2.2 AUXILIARY EQUIPMENT

a. Louvers shall be provided with a galvanized aluminum bird screen in a removable frame on the inside. Insect screen, if shown on the drawings for intake louvers, shall be aluminum in an aluminum frame and shall be mounted on the exterior of the louver.

b. Provide sill extension on louver to provide drip ledge to carry water away from the building.

c. For larger-size louvers, provide mullions to provide rigidity and allow expansion.

d. See drawings for duct connections, dampers, etc., that attach to the louver.

PART THREE - EXECUTION

3.1 a. Install the louver in the masonry opening with the sill extension extending from the building 1/4 in. (6.5 mm), and secure to the masonry to prevent unauthorized entry.

b. Caulk the perimeter of the louver at the junction of the wall with silicone caulk, applied as recommended by the manufacturer. Allow room for expansion and contraction without damage to the caulk.

c. Paint the entire louver with color selected by owner. Clean all grease and prime with proper primer before applying final coat.

END OF SECTION 15463

SECTION 15465
****
LOUVERED PENTHOUSE

PART ONE - GENERAL

1.1 The GENERAL and SPECIAL CONDITIONS, Section 15100, are included as a part of this Section as though written in full in this document.

1.2 Scope of the Work shall include the furnishing and complete installation of the equipment covered by this Section, with all auxiliaries, ready for owner's use.

PART TWO - PRODUCTS

2.1 LOUVERED PENTHOUSE

a. Penthouse shall be fabricated by welding section of extruded aluminum louvers with a double break or hook to prevent water from traveling up the blade. Blade width shall be no less than 4 in. (100 mm) and thickness shall be no less than 0.081 in. (2 mm). Spacing shall be 2 in. (50 mm) apart.

b. Frame shall be heavy-gauge aluminum structural members to support and brace the louvers. Provide concealed mullions as required.

c. Top shall be formed from 0.081-in. (2-mm) aluminum sheet with crown for rigidity. Corners shall be mitered and welded. Insulate underside with 1-in.(25-mm) thickness of fiberglass insulation with vapor jacket to prevent condensation and reduce noise.

d. Base shall be structural aluminum member to sit over a prefabricated curb and support the entire unit, as well as any dampers required in the throat. Curb shall provide a lip over the curb and shall be anchored to the curb.

e. Where dampers are to be installed in the throat of the penthouse, provide a section of louvers hinged as a door of no less that 24 in. x 24 in. (60 x 60 cm) for access to the dampers and operators. Door shall be provided with a hasp and padlock to prevent unauthorized entry to the building.

f. Provide an aluminum bird screen in aluminum frame inside the louvers, including the door, if provided.

2.2 AUXILIARY EQUIPMENT

a. Provide a prefabricated curb to match and line up with the penthouse. Curb shall be aluminum construction with the slope as required to match the roof.

2.3 ACCEPTABLE MANUFACTURERS

a. Penthouse and curb shall be size and model number shown on the drawings or equivalents by Carnes (Model P330), Arrow, Louvers & Dampers (PL300).

PART THREE - EXECUTION

3.1 a. Install curb and anchor to roof structural members, with adequate connectors to prevent uplift and side movement.

b. Install penthouse on curb and secure to curb with stainless steel screws spaced on 12-in. (30-cm) centers.

c. If penthouse is provided with access door, provide quality padlock and deliver keys to owner at close of job.

END OF SECTION 15465

SECTION 15469

****

GOOSENECK INTAKE/RELIEF

PART ONE - GENERAL

1.1 The GENERAL and SPECIAL CONDITIONS, Section 15100, are included as a part of this Section as though written in full in this document.

1.2 Scope of the Work shall include the furnishing and complete installation of the equipment covered by this Section, with all auxiliaries, ready for owner's use.

PART TWO - PRODUCTS

2.1 a. Gooseneck shall be shop-fabricated from heavy-gauge galvanized steel, as recommended by SMACNA, or other industry standards, and as detailed on the drawings.

b. Provide a curb cap of welded galvanized steel. Clean, prime, and coat with zinc rich paint to protect exposed metal from rust.

c. Provide a galvanized bird screen over the open end.

PART THREE - EXECUTION

3.1 INSTALLATION

a. Install curb of no less than 12 in. (30 cm), and secure to roof.

b. Install gooseneck and connect ductwork as required.

END OF SECTION 15469

SECTION 15470

****

DOMESTIC RANGE HOOD

PART ONE - GENERAL

1.1 The GENERAL and SPECIAL CONDITIONS, Section 15100, are included as a part of this Section as though written in full in this document.

1.2 Scope of the Work shall include the furnishing and complete installation of the equipment covered by this Section, with all auxiliaries, ready for owner's use.

PART TWO - PRODUCTS

2.1 RANGE HOOD

a. Hood shall be attractive in style, designed for residential use, and complete with a quiet-type centrifugal fan with a solid-state speed control and incandescent light. Color shall be selected from five standard colors.

b. Hood size shall be generally as shown on the drawings, unless dimensions are given on the plans.

c. Hood fan shall be two-speed and provided with a back-draft damper. The vent from the hood shall be carried to the outside, unless clearly shown otherwise (well-ventilated attic), and terminated in a roof or wall cap as shown. Vent terminating on flat, or very low pitch roof, shall be a gravity ventilator with a flashing cap. See section on ventilators. Vents terminating on pitched roofs may use low roof cap, as is standard with the hood manufacturer.

2.2 AUXILIARY EQUIPMENT

a. Provide cleanable filter for the exhaust hood.

b. Provide duct connections, as are standard with the unit.

2.3 ACCEPTABLE MANUFACTURERS

a. Hood shall be make and model number shown on the drawings or equivalent hoods by Nutone (V-26 Series) or Broan (7000 Series).

PART THREE - EXECUTION

3.1 INSTALLATION

a. Hood shall be mounted as shown on the drawings and as recommended by the manufacturer.

b. Connect the ductwork and carry to terminal point as shown on the drawings, using sheet metal duct of the size noted or required by the fan.

c. Connect power and verify proper operation of the fan.

END OF SECTION 15470

SECTION 15475

****

KITCHEN HOOD

COMMERCIAL TYPE

PART ONE - GENERAL

1.1 The GENERAL and SPECIAL CONDITIONS, Section 15100, are included as a part of this Section as though written in full in this document.

1.2 Scope of the Work shall include the furnishing and complete installation of the equipment covered by this Section, with all auxiliaries, ready for owner's use.

PART TWO - PRODUCTS

2.1 KITCHEN HOOD

a. Hood shall be a complete system, designed for removal of fumes from a cooking area and with tempered air supplied to replace the exhausted air. System shall consist of canopy no less than (2 ft.-0 in.) (60.6 cm) deep and extending 6 in. (15 cm) beyond the equipment covered by the canopy in all directions, except where the equipment is against a wall. See drawings for size.

b. The air quantity exhausted shall provide a velocity through the face area of the hood of 80 ft/min. (25 m/min) for wall canopy hoods and 100 ft/min (30 m/min) for island hoods, unless greater velocity is required by the local authorities.

c. The supply air quantity delivered at the hood face shall be 80% of the exhaust air and shall be delivered at low velocity, to be drawn in evenly across the face of the hood.

d. The make-up air shall be filtered and tempered to 60 deg. F (15 deg. C) by:

(Spec writer should select one of the following:)

( ) Heat recovery from internal duct transfer within the hood.

( ) Heat recovery by plate-type, air-to-air heat exchanger, as allowed by local authority. See Section 15488 for Specifications.

( ) Heat recovery (by air to water to air) runaround cycle using coils.

( ) Indirectly fired gas furnace with modulating controls. See Section 15606 for Specifications.

( ) Direct natural gas fired heater with modulating controls. See Section 15698 for Specifications.

( ) Steam coil, served by medium- or high-pressure steam, and face and bypass controls. See Section 15667 for Specifications.

( ) Electric resistance duct heater with step controls. See Section 15640 for Specifications.

( ) Gas-fired heating and ventilation unit. See Section 15485 for Specifications.

e. The hood and all associated equipment shall be furnished and installed by the hood manufacturer or a qualified representative. Complete drawings of the hood(s), exhaust duct(s), appliances, and the fire-suppression system shall be submitted to the local authority for review and approval. All components and installation shall meet the requirements of NFPA-96.

2.2 EXHAUST

a. Exhaust shall be provided by a roof-mounted centrifugal fan mounted at least 18 in. (46 cm) above the roof and discharging at least 40 in. (102 cm) above the roof in an upward discharge. See Section 15453 for Specifications.

2.3 SUPPLY AIR

a. Supply air shall be provided by centrifugal-type fan located on the roof and drawing in fresh air through 2-in. (50-mm) throwaway filters, protected by a weatherproof hood. Supply air fan may be a part of the tempering system specified above.
b. Fresh tempered air shall be supplied to the hood face opening through double deflection registers, located around the open edge of the hood. Registers shall be curved-blade type of aluminum that can be outward, down, or into the hood.

2.4 GREASE EXTRACTORS

a. Grease extractors bearing the UL Label shall be installed as required by the label.

2.5 EXHAUST DUCT

a. Exhaust ductwork shall be installed as required by NFPA-96 and shall be UL-labeled or shall be 0.054-in. (1.4-mm) carbon steel or 0.043-in. (1.1-mm) stainless steel.
b. Exhaust ducts shall be installed with 18-in. (46-cm) clearance from combustible material or as approved by the local authority.

2.6 LIGHTS

a. Vapor-proof fluorescent lighting fixtures shall be provided in each hood, as required for good lighting on the working surface. The fixtures shall be UL-listed for this service and approved by the local authority.

2.7 CONSTRUCTION

a. All exposed surfaces shall be constructed of 18-gauge (1.3-mm) type 304 stainless steel with No. 3 finish.
b. All unexposed surfaces shall be 18-gauge (1.3-mm) galvanized steel.
c. All joints shall be liquid-tight, continuously welded, and polished to match the original No. 3 finish.
d. Hood shall be fabricated to comply with NFPA-96 and with NSF and shall bear the NSF Seal. Hood shall be braced for rigidity as required for the size.

2.8 SUPPORTS

a. Hood manufacturer shall design the supports for the hood and furnish the members needed to adequately align and support the hood from the structural members provided by others.

b. The hood shall be mounted 6 1/2 ft. (2 m) above the floor for schools, and at 7 ft. (2.1 m) for commercial applications.

2.9 FIRE-EXTINGUISHING SYSTEM

a. Provide a fixed-pipe fire-extinguishing system to cover those items of equipment under the hood that might be a source of ignition of grease, the grease removal equipment, and the hood with associated exhaust ductwork. The system shall fully comply with NFPA-96 and be accepted and approved by the local authorities.

(CHOUSE ONE OF THE FOLLOWING:)

( ) b. The system shall be a dry chemical extinguishing system, as required by NFPA-17.

( ) b. The system shall be a carbon dioxide extinguishing system, as required by NFPA-12.

2.10 CONTROLS

a. The supply fan(s), exhaust fan(s), and lights in the hood(s) shall be controlled from a single, prewired control panel, located on the hood or located on the wall at a convenient place acceptable to the owner's representative.

b. Controls mounted on the hood shall be mounted on the bottom of the hood so that the switches are not more than 6 ft. 9 in. (2.1 m) above the floor for noncommercial installations and 7 ft. 2 in. (2.2 m) for commercial installations.

c. Motor control equipment shall be located in a locked control cabinet, located on the end of the hood or remotely located from the hood at a convenient location acceptable to the owner's representative. Provide pilot light on the face of the control cabinet, visible from the general area, to indicate which fans are in operation and at what speed.

d. Motor control panel shall be provided with numbered terminal strips to match similar strips in the hood and in the equipment located on the roof. The manufacturer shall provide a wiring diagram, permanently installed inside the control cabinet door, indicating all wiring by terminal number and color. Three copies of this diagram shall be provided to the owner by letter and additional copies made available to the electrician doing the installation. Wiring between components shall be color-coded as shown on the control diagram furnished by the manufacturer.

e. Provide HI-LOW-OFF switches to control the fans for selected speeds. Provide a 10-second time delay relay in the circuit to delay the start of the fans when switched from HI to LOW.

2.11 ACCEPTABLE MANUFACTURERS

a. Hood(s) shall be manufactured and installed by the company shown on the drawings or equivalent products by Sun-Air, Greenheck, Air Systems, Ventilation Equipment.

PART THREE - EXECUTION

3.1 APPROVALS

a. The hood manufacturer, or a qualified representative, shall submit all necessary drawings to the local authorities as required to secure their approval of the equipment to be installed.

b. After installation is complete, the installer shall secure the acceptance of the local authorities for the installation.

c. The manufacturer shall submit copies of preinstallation and final acceptance letters to the owner, through proper channels.

3.2 WORKMANSHIP

a. The final welding and grinding/polishing of the field welds shall match those done in the shop, so that no distinction is visible between welds.

3.3 INSTALLATION

a. The hood(s) shall be hung level and true, at the location shown on the drawings.

b. Install all components such as grease extractors, fans, heat exchangers, heater, etc., as shown on the drawings and as required to make a complete installation.

3.4 WIRING

a. Provide all field wiring between terminals in hood, controls, and rooftop equipment for complete installation.

b. All wiring shall be color-coded as shown on the manufacturer's drawings.

3.5 OPERATION AND TESTING

a. Place all components in operation and check for proper fan rotation, motor power draw, ignition for heaters, high limits, etc.

b. This system is not complete until the system has been balanced to provide the correct air quantity and has been tested to demonstrate the correct system performance. See Balancing and Testing, Section 15111.

END OF SECTION 15475

SECTION 15477

****

KITCHEN HOOD
SMALL COMMERCIAL TYPE

PART ONE - GENERAL

1.1 The GENERAL and SPECIAL CONDITIONS, Section 15100, are included as a part of this Section as though written in full in this document.

1.2 Scope of the Work shall include the furnishing and complete installation of the equipment covered by this Section, with all auxiliaries, ready for owner's use.

PART TWO - PRODUCTS

2.1 KITCHEN HOOD

a. Hood shall be a complete system, designed for removal of fumes from a cooking area and with tempered air supplied to replace the exhausted air. System shall consist of canopy no less than 2 ft. (60 cm) deep and extending 6 in. (15 cm) beyond the equipment covered by the canopy in all directions, except where the equipment is against a wall. See drawings for size.

b. The air quantity exhausted shall provide a velocity through the face area of the hood of 80 ft/min. (25 m/min) for wall canopy hoods and 100 ft/min (30 m/min) for island hoods, unless greater velocity is required by the local authorities.

c. The hood and all associated equipment shall be furnished and installed by the hood manufacturer or a qualified representative. Complete drawings of the hood(s), exhaust ducts(s), appliances, and the fire-suppression system shall be submitted to the local authority for review and approval. All components and installation shall meet the requirements of NFPA-96.

2.2 EXHAUST

a. Exhaust shall be provided by a roof-mounted centrifugal fan, mounted at least 18 in. (46 cm) above the roof and discharging at least 40 in. (102 cm) above the roof in an upward discharge. See Section 15423 for Specifications.

2.3 GREASE EXTRACTORS

a. Grease extractors bearing the UL label shall be installed as required by the label.

2.4 EXHAUST DUCT

a. Exhaust ductwork shall be installed as required by NFPA-96 and shall be UL-labeled or shall be 0.054-in. (1.4-mm) carbon steel or 0.043-in. (1.1-mm) stainless steel.

b. Exhaust ducts shall be installed with 18-in. (46-cm) clearance from combustible material or as approved by the local authority.

2.5 LIGHTS

a. Vapor-proof fluorescent lighting fixtures shall be provided in each hood, as required for good lighting on the working surface. The fixtures shall be UL-listed for this service and approved by the local authority.

2.6 CONSTRUCTION

a. All exposed surfaces shall be constructed of 18-gauge (1.3-mm) type 304 stainless steel with No. 3 finish.

b. All unexposed surfaces shall be 18-gauge (1.3-mm) galvanized steel.

c. All joints shall be liquid-tight, continuously welded, and polished to match the original No. 3 finish.

d. Hood shall be fabricated to comply with NFPA-96 and with NSF and shall bear the NSF Seal. Hood shall be braced for rigidity as required for the size.

2.7 SUPPORTS

a. Hood manufacturer shall design the supports for the hood and furnish the members needed to adequately align and support the hood from the structural members provided by others.

b. The hood shall be mounted 6 1/2 ft. (2 m) above the floor for schools and at 7 ft. 0 in. (2.1 m) for commercial applications.

2.8 FIRE EXTINGUISHING SYSTEM

a. Provide a fixed-pipe fire extinguishing system to cover those items of equipment under the hood that might be a source of ignition of grease, the grease removal equipment, and the hood with associated exhaust ductwork. The system shall fully comply with NFPA-96 and be accepted and approved by the local authorities.

b. The system shall be a dry chemical extinguishing system or accepted equivalent as required by NFPA-17.

2.9 CONTROLS

a. The exhaust fan(s) and lights in the hood(s) shall be controlled from a single, prewired control panel, located on the hood or located on the wall at a convenient place acceptable to the owner's representative.

b. Controls mounted on the hood shall be mounted on the bottom of the hood so that the switches are not more than 6 ft 9 in. (2.1 m) above the floor for noncommercial installations and 7 ft. 2 in. (2.2 m) for commercial installations.

c. Motor control equipment shall be located in a locked control cabinet, located on the end of the hood or remotely located from the hood at a convenient location acceptable to the owner's representative. Provide pilot light on the face of the control cabinet, visible from the general area, to indicate which fans are in operation and at what speed.

d. Motor control panel shall be provided with numbered terminal strips to match similar strips in the hood and in the equipment located on the roof. The manufacturer shall provide a wiring diagram, permanently installed inside the control cabinet door, indicating all wiring by terminal number and color. Three copies of this diagram shall be provided to the owner by letter and additional copies made available to the electrician doing the installation. Wiring between components shall be color-coded as shown on the control diagram furnished by the manufacturer.

e. Provide HI-LOW-OFF switches to control the fans for selected speeds. Provide a 10-second time delay relay in the circuit to delay the start of the fans when switched from HI to LOW.

2.10 ACCEPTABLE MANUFACTURERS

a. Hood(s) shall be manufactured and installed by the company shown on the drawings or equivalent products by Sun-Air, Greenheck, Air Systems, Ventilation Equipment.

PART THREE - EXECUTION

3.1 APPROVALS

a. The hood manufacturer, or a qualified representative, shall submit all necessary drawings to the local authorities as required to secure their approval of the equipment to be installed.

b. After installation is complete, the installer shall secure the acceptance of the local authorities for the installation.

c. The manufacturer shall submit copies of preinstallation and final acceptance letters to the owner, through proper channels.

3.2 WORKMANSHIP

a. The final welding and grinding/polishing of the field welds shall match those done in the shop, so that no distinction is visible between welds.

3.3 INSTALLATION

a. The hood(s) shall be hung level and true, at the location shown on the drawings.

b. Install all components such as grease extractors, fans, etc., as shown on the drawings and as required to make a complete installation.

3.4 WIRING

a. Provide all field wiring between terminals in hood, controls, and rooftop equipment for complete installation.

b. All wiring shall be color-coded as shown on the manufacturer's drawings.

3.5 OPERATION AND TESTING

a. Place all components in operation and check for proper fan rotation, motor power draw, ignition for heaters, high limits, etc.

b. This system is not complete until the system has been balanced to provide the correct air quantity and has been tested to demonstrate the correct system performance. See Balancing and Testing, Section 15111.

END OF SECTION 15477

SECTION 15479

****

DISHWASHER HOOD

PART ONE - GENERAL

1.1 The GENERAL and SPECIAL CONDITIONS, Section 15100, are included as a part of this Section as though written in full in this document.

1.2 Scope of the Work shall include the furnishing and complete installation of the equipment covered by this Section, with all auxiliaries, ready for owner's use.

PART TWO - PRODUCTS

2.1 DISHWASHER HOOD

a. Dishwasher hood shall be provided as shown on the drawings. Hood shall be constructed of no less than 20-gauge (0.965-mm) type 304 stainless steel with No. 3 finish. All corners shall be welded and polished.

b. Hood shall be no less than 2 ft. (60 cm) deep with sloping top, unless hood is carried above the ceiling. Bottom of hood shall be mounted at 6 1/2 ft. (2 m) for noncommercial and 7 ft. (2.1 m) for commercial kitchens.

c. Provide a drip lip inside hood skirt on all sides to catch condensation. Provide watertight aluminum or stainless steel ductwork up to the fan on the roof.

d. Exhaust fan shall have capacity and be located as shown on the drawings. Provide a back-draft damper. See Section 15432 for Specifications.

2.2 ACCEPTABLE MANUFACTURERS

a. Hood(s) shall be manufactured by the company shown on the drawings or equivalents by Sun-Air, Greenheck, Air Systems, Ventilation Equipment.

PART THREE - EXECUTION

3.1 INSTALLATION

a. Install the hood level and at height required. Support shall be from the overhead structural members.

b. Connect the exhaust ductwork so that any water running back down the duct will collect in the hood drain trough.

c. Install exhaust fan with wall-mounted switch and pilot light, located as directed by the owner's representative.

d. Check for proper fan rotation and check power draw against the nameplate.

3.2 TESTING

a. This system is not complete until the system has been balanced to provide the correct air quantity and has been tested to demonstrate the correct system performance. See Balancing and Testing, Section 15111.

END OF SECTION 15479

SECTION 15480

****

HEATING AND VENTILATION UNIT
INDIRECT GAS FIRED

PART ONE - GENERAL

1.1 The GENERAL and SPECIAL CONDITIONS, Section 15100, are included as a part of this Section as though written in full in this document.

1.2 Scope of the Work shall include the furnishing and complete installation of the equipment covered by this Section, with all auxiliaries, ready for owner's use.

PART TWO - PRODUCTS

2.1 HEATING SECTION

a. Heating and ventilation unit shall be indirectly natural (propane) gas-fired and curb-mounted on the roof in weatherproof enclosure.
b. Heat exchanger shall be type 321 stainless steel, guaranteed non-pro rata for 10 years, for service on outside air.
c. There shall be two burners, one controlled by a modulating self-contained (electric) gas valve to act as a low limit with controlling element in the discharge to maintain 60 deg. F (16 deg. C) leaving air minimum (adjustable).
d. There shall be second burner controlled by a modulating gas valve with controlling element in the space to be served.
e. There shall be a main gas valve ahead of burner gas valves to shut off both burners in case of pilot failure or as a high limit. In SUMMER position, this valve shall be closed.
f. Provide standing pilot with automatic reignition for both heaters, with pilot safety on both pilots to shut the main gas valve.

2.2 FAN SECTION

a. Fan shall be centrifugal type, direct-driven or belt-driven, to deliver the air quantity called for on the drawings against the estimated static pressure shown. Fan motor shall be two-speed type as called for on drawings.
b. Fan shall be provided with throwaway filters, 1 in. (25 mm) thick, with weatherproof canopy to filter outside air. Provide another bank of similar filters for the return air.

2.3 CABINET AND HOUSING

a. All components mounted on the roof shall be enclosed in a heavy-gauge galvanized cabinet with a final coat of factory-applied baked-enamel paint.

2.4 MOUNTING

a. All rooftop components shall be mounted on curb built into the roof. Curb may be factory-built or may be job-built. Job-built curb shall have an 18-gauge (1.3-mm) galvanized steel cover with corners turned and soldered. See detail on drawings. All gas pipe, conduit, etc., shall come up to the roof in separate curbs. See Section 15795 for Specifications and details on drawings.

2.5 CONTROLS

a. Controls shall provide for operation of the unit as a space heater, as well as a ventilation unit, and shall control the exhaust fan in the area along with the H&V unit.

b. In the HEATING position, the outside air dampers shall be closed, the return air damper shall be open, the exhaust fan shall be off, and the air shall be supplied to the space. The space thermostat shall control the second heater, and the fan shall operate in low speed.

c. In the LOW VENTILATION position, the return air dampers shall close, the outside air dampers shall open, the supply air shall open, and the supply and exhaust fan shall operate on low speed. The furnace controlled by the space thermostat shall be energized.

d. In the HIGH VENTILATION position, the operation shall be the same as in low ventilation, except both fans shall operate on high speed and both furnaces shall be energized. Provide a 10-second time delay in the controls to prevent immediate starting of the fans in high speed on a switch from LOW to HIGH.

e. In the SUMMER position, the heat shall be off and both fans shall operate on high speed. The outside air dampers shall be open and the return air damper closed.

2.6 ACCEPTABLE MANUFACTURERS

a. The H&V unit shall be the model number shown on the drawings or equivalent by Reznor, Jackson Church, Lennox.

PART THREE - EXECUTION

3.1 INSTALLATION

a. Install all rooftop equipment on the curb as reviewed and accepted. Provide roof curbs for other services such as power, control wiring, gas pipe, etc., and connect all services.

b. Provide ductwork and grilles, registers, dampers, etc., as shown on the drawings.

c. Operate the system, checking for proper fan rotation and power draw against nameplate ratings. Check ignition of the gas furnaces, high limits, pilot safeties, etc.

3.2 a. This system is not complete until the system has been balanced to provide the correct air quantity and has been tested to demonstrate the correct system performance. See Balancing and Testing, Section 15111.

END OF SECTION 15480

SECTION 15483

****

MAKE-UP AIR HEATER
DIRECT GAS FIRED

PART ONE - GENERAL

1.1 The GENERAL and SPECIAL CONDITIONS, Section 15100, are included as a part of this Section as though written in full in this document.

1.2 Scope of the Work shall include the furnishing and complete installation of the equipment covered by this Section, with all auxiliaries, ready for owner's use.

PART TWO - PRODUCTS

2.1 GAS BURNER

a. The gas burner shall be suitable for burning natural gas, propane, or propane air mix, with a modulating turndown ratio of 30:1.

b. The burner assembly and fuel piping arrangement shall include, but is not limited to:

1. Electric modulating main gas valve (proportioning).
2. Main and pilot gas regulators.
3. Main and pilot electric gas valves.
4. Main and pilot hand gas valves.
5. All units shall be provided with double-block and bleed valves, as well as high- and low-pressure gas switches.

c. Pilot shall be arranged so that the flame from the pilot lights the burner with instantaneous ignition.

d. Pilot assembly shall include a flame rod and pilot to be automatically ignited by the spark rod through a 6000-volt ignition transformer.

e. The burner capacity shall be as shown on the drawings. Units with lesser capacity will not be considered. Units with capacity in great excess of that shown must be limited to that shown.

f. Heaters shall be shop-assembled flame-tested, wired completely for operation, and shipped bearing a UL Label.

2.2 CASING

a. The entire unit shall be factory-assembled (except for accessory hoods, etc.) in a heavy-gauge casing with structural reinforcing, as required for a rigid package ready to be supported as shown on the drawings.

b. Roof-mounted units shall be designed for mounting on a roof curb furnished by the unit manufacturer.

c. Provide access to the burner and fan assemblies.

d. Provide observation ports for observation of the main burner and the pilot without opening the unit.

e. Casing shall be prime-coated, inside and out, with a final coating of heat-resistant enamel.

2.3 FANS AND DRIVES

a. Unit shall deliver the air quantity shown on the drawings against the static pressure shown.

b. Fan(s) shall be centrifugal type, mounted on a heavy steel shaft supported by two pillow-block bearings, with grease fittings brought out to accessible location.

c. Drive shall be by belt(s) (multiple belts shall be matched from the same production run) rated for 150% of the motor brake power, driven by a variable-pitch pulley on the motor. Provide belt guard to meet OSHA requirements.

d. Motor shall be a high-efficiency type, mounted on an adjustable base located out of the airstream.

2.4 CONTROLS

a. Controls system shall include, but is not limited to, the following:

1. Main fan airflow switch.
2. High air temperature switch.
3. End switch on burner, modulating motor to ensure low-fire start.
4. Electronic flame safety control.
5. Red warning light and audible alarm to indicate flame failure. Provide silencers for alarm, leaving light on.
6. White light to indicate when the main gas valve is open.
7. Premix airflow switch (if premixed air is used).
8. Modulating temperature controller with sensor in discharge air to regulate outlet air temperature. Controller shall be factory-mounted and -wired.
9. Automatic two-position damper on discharge of unit. Damper is to be wired so that unit fan cannot start until the damper is open and the damper shall automatically close when the unit is deenergized.

b. All necessary disconnect switches, starters, relays, fuse blocks shall be furnished and mounted in a control cabinet with a locking door mounted on the unit. The flame failure controls shall be mounted in this control cabinet.

2.5 WORK PLATFORM

a. Unit shall provide a work ledge mounted to the side to provide access to fan motor and drive, gas manifold, controls, etc., with guard rails as required by OSHA for units mounted indoors and elevated.

b. Units mounted on the roof shall have the above items enclosed in locked doors, accessible from the roof.

2.6 INSURANCE UNDERWRITERS

a. The units shall be manufactured to standards required by Factory Mutual (FM), Factory Insurance Association (FIA), or other insurance carriers, as shown on the drawings.

2.7 ACCEPTABLE MANUFACTURERS

a. Units shall be the model number shown on the drawings or equivalent products by Dravo (Directflow L-1 and L-2 Series), Applied Air Systems (DF Series), Hastings (Model LB), Trane (Torrovent).

PART THREE - EXECUTION

3.1 INSTALLATION

a. Provide necessary structural supports for indoor units and roof curb for units mounted on the roof.

b. Set the unit in place and connect all ductwork as shown.

c. Connect gas piping of the proper pressure required by the unit. Provide gas regulator in the supply if pressure is above the rating of the gas regulator on the unit.

d. Connect the electrical service and operate the unit to check out controls.

e. Operate the fan to check rotation, and check actual readings against nameplate readings.

f. This system is not complete until the system has been balanced to provide the correct air quantity and has been tested to demonstrate the correct system performance. See Balancing and Testing, Section 15111.

END OF SECTION 15483

SECTION 15484

****

MAKE-UP AIR HEATER
INDIRECT GAS FIRED

PART ONE - GENERAL

1.1 The GENERAL and SPECIAL CONDITIONS, Section 15100, are included as a part of this Section as though written in full in this document.

1.2 Scope of the Work shall include the furnishing and complete installation of the equipment covered by this Section, with all auxiliaries, ready for owner's use.

PART TWO - PRODUCTS

2.1 HEATING SECTION

a. Heating section shall consist of type 321 stainless steel heat exchanger, with a 10-year nonprorated guarantee, for 100% outside air. Capacity of the heater shall be as shown on the drawings. Units with less capacity will not be acceptable. Units with capacity in great excess of that shown shall be reduced in output to more nearly match that required.

b. Burners shall be individually removable, of aluminized steel with stainless steel ribbons and cross-lighters and pilot assembly. Burners shall be mounted on slide-out tray.

c. Provide a modulating nonelectric gas valve to maintain a set (adjustable) low limit for the leaving air.

d. Entire unit shall be AGA- and/or CGA-certified.

2.2 POWER VENTOR

a. Provide a power ventor built into the unit to remove the combustion gases from each furnace. Interlock with the burner controls.

2.3 FAN AND DRIVE

a. Blower section shall be provided with one or more forward-curved, double-width, double-inlet centrifugal fans mounted on heavy steel shaft with self-aligning, permanently lubricated ball bearings.

b. Two-speed motor (if shown on the drawings) shall be mounted on an adjustable base and provided with a variable-pitch pulley.

c. Drive shall be one or more belts designed for 150% of the motor rating and provided with belt guards where exposed. Where more than one belt is required, the belts shall be matched from the same production run.

d. Fans shall deliver the air quantity as specified on the drawings against the static pressure shown.

2.4 CABINETS

a. All equipment shall be designed for roof mounting on a prefabricated curb furnished by the manufacturer and shall be housed in aluminized steel cabinets with baked-enamel finish.
b. Provide outside air intake hood with bird screen.
c. Provide 1-in. (25-mm) throwaway (cleanable) filters.
d. Provide outside air shutoff damper with two-position motor, linkage, and end switch.

2.5 CONTROLS

a. Provide a remote control station and interlock with any other controls serving related equipment such as exhaust fan(s).
b. Provide ON-OFF-AUTO switch so that in the ON position the outside air damper opens and the blower operates continuously.
c. Provide a HEAT-VENT switch to select heat or ventilation only.
d. Provide HI-LOW-AUTO motor speed selector, with the AUTO position tied into the speed selector of associated equipment.
e. Remote control station shall contain white indicating light to show when the unit is energized and the blower is operating and a yellow light to indicate when the main gas valve is energized. A green light shall indicate high speed and a blue light low speed, where two-speed operation is required.
f. Control sequence shall be as follows:
   1. With switch in the HEAT position the fan will start, the outside air damper will open, and the blower proving relay will close. The main gas valve will open when the ventor relay is proved, the pilot relay is proved, and the high limit allows operation.
   2. On a call for space heat, the room thermostat provides full heat by permitting operation of the bypass gas valve. If the thermostat is satisfied, the nonelectric gas valve maintains a constant leaving air temperature.
   3. Spark ignition operates continuously until the flame is established.
   4. In the VENT position, the outside air damper will open and the fan will operate in high or low speed, as selected. The main gas valve shall be closed.

2.6 ACCESSORIES

a. Provide fan delay relay to delay fan operation until unit has warm air available.
b. Provide rubber-in-shear vibration isolators.
c. Provide fused disconnect switch for all power to unit.
d. Provide electronic modulation with remote temperature selector in place of modulating gas valve.
e. Provide high-pressure gas regulator where service pressure is above standard.

2.7 ACCEPTABLE MAUNFACTURERS

a. Unit shall be the model number shown on the drawings or equivalent by Hastings (RMU), Applied Air Systems (IF Series), Dravo (Paraflow).

PART THREE - EXECUTION

3.1 INSTALLATION

a. Install curb on roof as required for proper installation.

b. Set the unit in place and connect all ductwork as shown on the drawings.

c. Provide curb for electrical and gas connections to the unit and install disconnect switch and gas cutoff valve.

d. Operate the unit to check for proper fan rotation, and compare the actual readings against the nameplate readings.

e. Operate the make-up air unit and check operation of the heater safeties, gas valves, etc.

f. This system is not complete until the system has been balanced to provide the correct air quantity and has been tested to demonstrate the correct system performance. See Balancing and Testing, Section 15111.

END OF SECTION 15484

SECTION 15485

****

PLATE-TYPE HEAT RECOVERY UNIT

PART ONE - GENERAL

1.1 The GENERAL and SPECIAL CONDITIONS, Section 15100, are included as a part of this Section as though written in full in this document.

1.2 Scope of the Work shall include the furnishing and complete installation of the equipment covered by this Section, with all auxiliaries, ready for owner's use.

PART TWO - PRODUCTS

2.1 HEAT RECOVERY UNIT

a. Heat recovery unit shall be a fixed-plate air-to-air heat exchanger with ratings of air quantity and pressure drop as shown on the drawings.

(SPEC WRITER SHOULD SELECT ONE OF THE FOLLOWING:)

(Standard ventilation system with temperature below 120 deg. F (49 deg. C)).

( ) b. Plates shall be constructed of 0.021-in. (0.53-mm) embossed, corrugated aluminum plates sealed with epoxy bedding material.

( ) c. Casing shall be of 0.050-in. (1.27-mm) aluminum.

(Higher-temperature heat exchanger with better corrosion resistance, for use on kitchen hood exhaust with water/detergent washdown.)

( ) b. Plates shall be constructed of 0.025-in. (0.64-mm) type 316 stainless steel with ceramic seal suitable for temperature up to 1200 deg. F (659 deg. C).

( ) c. Casing shall be of 16-gauge (1.6-mm) type 304 stainless steel.

d. Airflow and temperatures through the intake side and the exhaust side shall be as shown on the drawings. Efficiency shall be no less than that shown for the unit called for, but in no case less than 70%.

e. Air pressure drop through the intake side and the exhaust side shall not be greater than that shown on the drawings.

2.2 ACCESSORIES

a. Provide face and bypass dampers to allow reduction of heat recovery when not wanted.

b. Provide exhaust discharge hood on air leaving the unit. See drawings.

c. Provide intake hood with bird screen for air entering the unit. See drawings.

e. Provide a factory-assembled, programmed water wash control cabinet to automatically feed detergent to the water. Unit shall be fully piped and wired for solenoid valves and pumps; 24-hour time clock with weekend skip function; freeze protection (drain down); programmed sequence controller; fan interlock circuit; detergent injection system with priming bypass; water shutoff valve and shock arrester; pressure gauge on water inlet; detergent suction and return lines with check valve and foot valve; and main power switch with lights to indicate functions.

f. Provide firestat, with alarm bell, to energize the wash spray in case of fire.

g. Differential pressure switch.

h. Ten-hour (minimum) spring or battery reserve for time clock.

i. Stainless steel cabinet for washdown control.

2.3 ACCEPTABLE MAUNFACTURERS

a. The units shall be make and model numbers shown on the drawings or equivalent by Temp-X-Changer.

PART THREE - EXECUTION

3.1 INSTALLATION

a. Mount the washdown control cabinet in location convenient to the unit and as approved by the owner's representative.

b. Connect hot city water to the unit with a water pressure regulator and a 100-mesh strainer and a backflow preventer as required by code.

c. Provide drain from the control cabinet, as needed for draining lines.

d. Provide supply lines from each valve in the control cabinet to the appropriate connections on the heat exchanger unit.

e. Provide drain(s) from heat exchanger to a point of acceptable discharge for the waste water which contains soap and grease. (Connection to the sanitary system leading to the grease trap is recommended.) Comply with Plumbing Code for these connections.

END OF SECTION 15485

SECTION 15490

****

ENGINE EXHAUST SYSTEM
OVERHEAD

PART ONE - GENERAL

1.1 The GENERAL and SPECIAL CONDITIONS, Section 15100, are included as a part of this Section as though written in full in this document.

1.2 Scope of the Work shall include the furnishing and complete installation of the equipment covered by this Section, with all auxiliaries, ready for owner's use.

PART TWO - PRODUCTS

2.1 ENGINE EXHAUST SYSTEM

a. Engine exhaust system shall consist of exhaust fan with discharge duct to outside and exhaust duct with flexible duct mounted overhead to provide connection to vehicle exhaust, all as shown on the drawings.

2.2 FAN

a. Exhaust fan shall be nonferrous, nonoverloading type designed for this service and having an acid-resisting coating. Fan shall be utility type with high static capacity.
b. Fan wheel shall be centrifugal, nonsparking type, mounted on a heavy steel shaft supported by two grease-lubricated, pillow-block ball bearings.
c. Motor shall be open, drip-proof, high-efficiency type with adjustable-pitch pulley.
d. Belts shall be selected at 150% of motor rating. Multiple belts shall be matched from the same production run.
e. Belt guard shall be OSHA-approved type. Where units are exposed to weather, the motor and bearings shall be protected by a steel cover.

2.3 DUCTWORK

a. Exhaust duct shall be suspended overhead and as out of the way as possible. Duct shall be lockseam galvanized steel (stainless steel), as required to meet SMACNA standards for medium-pressure duct. Fittings shall be five-piece elbows or long-radius, smooth die-cast type; fittings for side inlets shall be tapered-boot type designed for this service.
b. Flexible tubing shall be stainless steel with stainless steel flanges and adapters. Flexible tubing shall be sized as shown on the drawings.

c. Roller duct, designed to support the flexible duct inside it, shall be provided to totally conceal the flex. See the drawings for length and location. Roller duct shall be no less than 3 in. (75 mm) larger than the flexible duct it is to contain.

2.4 RECEPTACLES
a. Downstream end of flexible duct shall be provided with a guide ring of heavy-gauge galvanized steel, designed to guide the flex duct and to serve as an anchor point for cable.
b. Inlet side of flex shall be provided with a tapered cone adapter of 20-gauge (0.965-mm) stainless steel with protected edge and with spring clip for attachment to exhaust pipe. Provide exhaust analyzer opening for testing.
c. Receptacle on end of roller duct shall be designed to allow passage of the flex duct and provide for retrieval of the duct. The receptacle shall be provided with a self-closing damper to reduce air intake when not in use. See detail on drawings.

2.5 WINCH
a. Winch fabricated from heavy steel shall be provided for each connection for retrieval of flex duct. Winch shall have capacity for 125 ft (38 m) of 3/8-in. (9.5-mm) aircraft-type cable and shall be provided with a 10-in. (25-cm) wheel with a rubberized hand grip. Winch shall be provided with a spring-actuated safety ratchet.

2.6 CABLE, PULLEYS
a. Cable for retrieving flex shall be galvanized 1/8-in. (3.2-mm) aircraft type with individual wires parallel with the rope axis.
b. Pulleys shall be large-diameter, designed for this service, and shall be cadmium-plated.

2.7 ACCESSORIES
a. See drawings for louvers and discharge ductwork.

2.8 ACCEPTABLE MANUFACTURERS
a. The units shall be make and model numbers shown on the drawings or equivalent by Car-Mon, or National.

PART THREE - EXECUTION

INSTALLATION
a. Install the ductwork complete and generally as shown on the drawings, following manufacturer's recommendations and details.
b. Clean the ductwork of all trash and connect the duct to the fan with approved flex connectors.

c. When system is connected, check the fan for rotation and the motor for proper current draw. Adjust drive pulley for proper airflow at each inlet and make sure the system is working properly.
d. Test each winch at each flexible duct and prove the smooth and proper mechanical function.

END OF SECTION 15490

SECTION 15491

****

ENGINE EXHAUST SYSTEM
UNDERFLOOR

PART ONE - GENERAL

1.1 The GENERAL and SPECIAL CONDITIONS, Section 15100, are included as a part of this Section as though written in full in this document.

1.2 Scope of the Work shall include the furnishing and complete installation of the equipment covered by this Section, with all auxiliaries, ready for owner's use.

PART TWO - PRODUCTS

2.1 a. Engine exhaust system shall consist of exhaust fan with discharge duct to outside, underfloor exhaust duct, and flexible duct located in the underfloor to provide connection to vehicle exhaust, all as shown on the drawings.

2.2 FAN

a. Exhaust fan shall be nonferrous, nonoverloading type designed for this service and having an acid-resisting coating. Fan shall be utility type with high static capacity.
b. Fan wheel shall be centrifugal, nonsparking type, mounted on a heavy steel shaft supported by two grease-lubricated, pillow-block ball bearings.
c. Motor shall be open, drip-proof, high-efficiency type with adjustable-pitch pulley (unless located in a hazardous area).
d. Belts shall be selected at 150% of motor rating. Multiple belts shall be matched from the same production run.
e. Belt guard shall be OSHA-approved type. Where units are exposed to weather, the motor and bearings shall be protected by a steel cover.

2.3 DUCTWORK

a. Exhaust duct from the fan to the underfloor duct shall be lockseam stainless steel, as required to meet SMACNA standards for medium-pressure duct. Fittings shall be five-piece elbows or long-radius, smooth die-cast type; fittings for side inlets shall be tapered-boot type designed for low loss in this service.
b. Underfloor duct shall be Sch 40 PVC pipe and fittings, vitrified clay, or 18-gauge (1.27-mm) lockseam stainless steel. All metal fittings under the floor shall be stainless steel as above.
c. Flexible tubing shall be stainless steel with stainless steel flanges and adapters. Flexible tubing shall be sized as shown on the drawings.

2.4 RECEPTACLES

a. Downstream end of flexible duct shall be provided with a guide ring of heavy-gauge galvanized steel, designed to guide the flex duct back into the underfloor duct. Where PVC duct is used under the floor, special adapters shall be provided, designed to protect the duct on return of the tubing.

b. Inlet side of flex shall be provided with a tailpipe adapter of 20-gauge (0.965-mm) stainless steel with protected edge and with spring clip for attachment to exhaust pipe. Provide exhaust analyzer opening. Adapter shall be designed so that it will not leave the floor receptacle and enter the duct.

c. Receptacle mounted in the concrete floor shall be fabricated from 1/4-in. (6.4-mm) boiler plate or heavy cast aluminum. Body of the receptacle shall be no less than 12-gauge (2.5-mm) black steel with protective coating. Saddle on the underfloor duct and duct connecting the saddle to the receptacle shall be no less than 16-gauge (1.5-mm) black steel, with protective coating or 20-gauge (0.965-mm) stainless steel. Body of the receptacle, connecting duct, and the saddle shall be fully set or encased in concrete. See details on drawings.

2.5 ACCESSORIES

a. See drawings for louvers and discharge ductwork.

b. Provide a connection from the lowest point of the underfloor duct to a sump equipped with a drain or sump pump. (Hydraulic lift pit is a possibility.)

2.6 ACCEPTABLE MANUFACTURERS

a. The units shall be make and model numbers shown on the drawings or equivalent by Car-Mon or National.

PART THREE - EXECUTION

3.1 INSTALLATION

a. The ditch for the underfloor duct shall be graded for slope to the lowest point and all rocks and trash removed from the ditch for a smooth, even bed.

b. The underfloor duct shall be installed and all joints and connections made up tightly to prevent entry of groundwater. Backfill carefully around the duct and compact the dirt very tightly to prevent settlement. The saddles and connecting duct up to the receptacles shall be installed and left clear for the concrete to be poured around them.

c. The receptacles shall be set in place level with the floor and poured in place with the floor for a finished, smooth job.

d. The duct connecting to the fan shall be installed and connected to the fan with approved flex connectors.

e. The discharge duct shall be connected to the fan with flex connectors and carried to the discharge louver as shown on the drawings.

f. Operate the fan and check for proper rotation and current draw. Check all of the floor receptacles to see that they are drawing properly and that the underfloor duct is open and clean.

g. Just before the owner is to start using the system, install the flexible tubes in each receptacle, check for proper airflow on each tube, and make the system ready for owner's use.

END OF SECTION 15491

SECTION 15495

****

FIRE-EXTINGUISHING SYSTEM

DRY CHEMICAL

PART ONE - GENERAL

1.1 The GENERAL and SPECIAL CONDITIONS, Section 15100, are included as a part of this Section as though written in full in this document.

1.2 Scope of the Work shall include the furnishing and complete installation of the equipment covered by this Section, with all auxiliaries, ready for owner's use.

PART TWO - PRODUCTS

2.1 FIRE-EXTINGUISHING SYSTEM

a. System shall be a preengineered high-pressure, dry chemical fire-extinguishing system for the prevention of the hazard, as shown on the drawings or as specified.

b. The system shall be furnished and installed by a qualified contractor meeting all the requirements of NFPA No. 17 and No. 96, the local authorities having jurisdiction, and the owner's fire insurance underwriter. The system shall be listed with Underwriters' Laboratories and shall be installed as required by that listing.

c. The system shall include:

1. Automatic fire detection and system actuation by fixed-temperature mechanical release.
2. Remote manual mechanical actuation, or remote manual electrical actuation, or remote manual pneumatic actuation.
3. Remote manual operation shall operate entirely independently of automatic release.
4. Integral manual actuation at the control head.
5. Shutdown of electrical equipment, fans, heaters, etc.

d. The extinguishing agent shall be sodium bicarbonate (monoammonium phosphate), as approved for this service.

e. Cylinders shall be of capacity required and shall conform to Department of Transportation regulations.

f. Cylinder valve assembly shall have a properly designed valve with a fusible alloy plug (275 deg. F) (135 deg. C.) and a waterproof pressure gauge to indicate that the system is operable.

g. Control shall include:

1. Automatic release by fusible element in the hazard area.

2. Remote manual release by pull cable or electrical manual release, located near the normal exit from the space.

3. Integral manual release at the cylinder(s) not requiring electrical power.

h. Distribution system from the cylinder(s) to the nozzles in the hazard shall be through fixed pipes, sized as required for the hazard and as approved by Underwriters' Laboratories, the insurance underwriter, and the local authority. Pipe and fittings shall be galvanized schedule 40 steel pipe, and nozzles shall be of noncorrosive materials or furnished with a protective coating.

2.2 AUXILIARY EQUIPMENT

a. Accessories required with the system shall include the following:

1. Electric power shutoff switch to interrupt power to all electrical equipment in the hazard area in the event of operation of the fire-extinguishing system. Switch shall be UL-listed.
2. Fuel gas shutoff valve to interrupt the flow of fuel gas in the hazard area in the event of operation of the fire-extinguishing system. Shutoff valve shall be UL-listed.
3. Alarm, of distinctive sound and tone, to sound in the event of the operation of the system, and a red indicator light on the panel. Provide a "silencing" switch that shall leave the light on while stopping the alarm horn.
4. Power indicator lamp shall be provided for all systems that use electrical power. Green lamp shall indicate that power is available.

2.3 ACCEPTABLE MANUFACTURERS

a. System shall be equal in construction and performance to Chemetron, Ansul, or Kidde or as shown on the drawings.

PART THREE - EXECUTION

3.1 INSTALLATION

a. Installation shall be performed by a contractor qualified and licensed for this type of work. The installing contractor shall obtain all necessary permits and secure all required approvals for this system. Electrical portions of the system shall be installed by those qualified and licensed for that part of the work.

b. After final acceptance, all certificates shall be turned over to the owner, by letter, through the proper channels.

c. The owner shall be instructed in the operation and maintenance of the system by the installing contractor.

d. The installing contractor shall include with the installation contract a semiannual inspection of the system (starting with the owner's actual use of the system) for the first year. These inspections shall include written reports on the condition of the system and its readiness for operation.

END OF SECTION 15495

SECTION 15496

****

FIRE-EXTINGUISHING SYSTEM

CARBON DIOXIDE

PART ONE - GENERAL

1.1 The GENERAL and SPECIAL CONDITIONS, Section 15100, are included as a part of this Section as though written in full in this document.

1.2 Scope of the Work shall include the furnishing and complete installation of the equipment covered by this Section, with all auxiliaries, ready for owner's use.

PART TWO - PRODUCTS

2.1 FIRE EXTINGUISHING SYSTEM

a. System shall be an engineered high-pressure, carbon dioxide fire-extinguishing system for the prevention of the hazard, as shown on the drawings or as specified.

b. The system shall be furnished and installed by a qualified contractor meeting all the requirements of NFPA No. 17 and No. 96, the local authorities having jurisdiction, and the owner's fire insurance underwriter. The system shall be listed with Underwriters' Laboratories and shall be installed as required by that listing.

c. The system shall include:

1. Automatic fire detection and system actuation by fixed-temperature mechanical release.
2. Remote manual mechanical actuation, or remote manual electrical actuation, or remote manual pneumatic actuation.
3. Remote manual operation shall operate entirely independently of automatic release.
4. Integral manual actuation at the control head.
5. Shutdown of electrical equipment, fans, heaters, etc.

d. The extinguishing agent shall be carbon dioxide.

e. Cylinders shall be of capacity required and shall conform to Department of Transportation regulations.

f. Cylinder valve assembly shall have a properly designed valve and a waterproof pressure gauge to indicate that the system is operable.

g. Control shall include:

1. Automatic release by fusible element in the hazard area.
2. Remote manual release by pull cable or electrical manual release, located near the normal exit from the space.
3. Integral manual release at the cylinder(s) not requiring electrical power.

h. Distribution system from the cylinder(s) to the nozzles in the hazard shall be through fixed pipes, sized as required for the hazard and as approved by Underwriters' Laboratories, the insurance underwriter, and the local authority. Pipe and fittings shall be galvanized extra-heavy steel, and nozzles shall be of noncorrosive materials or furnished with a protective coating.

2.2 AUXILIARY EQUIPMENT

a. Accessories required with the system shall include the following:

1. Electric power shutoff switch to interrupt power to all electrical equipment in the hazard area in the event of operation of the fire extinguishing system. Switch shall be UL-listed.
2. Fuel gas shutoff valve to interrupt the flow of fuel gas in the hazard area in the event of operation of the fire-extinguishing system. Shutoff valve shall be UL-listed.
3. Alarm, of distinctive sound and tone, to sound in the event of the operation of the system, and a red indicator light on the panel to indicate that the system has discharged.
4. Power indicator lamp shall be provided for all systems that use electrical power. Green lamp shall indicate that power is available.

2.3 ACCEPTABLE MANUFACTURERS

a. System shall be equal in construction and performance to Chemetron, Ansul, or Kidde as shown on the drawings.

PART THREE - EXECUTION

3.1 INSTALLATION

a. Installation shall be performed by a contractor qualified and licensed for this type of work. The installing contractor shall obtain all necessary permits and secure all required approvals for this system. Electrical portions of the system shall be installed by those qualified and licensed for that part of the work.

b. After final acceptance, all certificates shall be turned over to the owner, by letter, through the proper channels.

c. The owner shall be instructed in the operation and maintenance of the system by the installing contractor.

d. The installing contractor shall include with the installation contract a semiannual inspection of the system (starting with the owner's actual use of the system) for the first year. These inspections shall include written reports on the condition of the system and its readiness for operation.

END OF SECTION 15496

SECTION 15497

****

FIRE-SUPPRESSION SYSTEM

HALON 1301

PART ONE - GENERAL

1.1 The GENERAL and SPECIAL CONDITIONS, Section 15100, are included as a part of this Section as though written in full in this document.

1.2 Scope of the Work shall include the furnishing and complete installation of the equipment covered by this Section, with all auxiliaries, ready for owner's use.

PART TWO - PRODUCTS

2.1 FIRE-SUPPRESSION SYSTEM

a. System shall be an engineered Halon 1301 fire-suppression system for the prevention of the hazard, as shown on the drawings or as specified.

b. The system shall be designed, furnished, and installed by a qualified contractor meeting all the requirements of NFPA No. 12A, the local authorities having jurisdiction, and the owner's fire insurance underwriter. The system shall be listed with Underwriters' Laboratories and shall be installed as required by that listing.

c. The system shall include:

1. Automatic fire detection and system actuation by fixed-temperature mechanical release.
2. Remote manual mechanical actuation, or remote manual electrical actuation, or remote manual pneumatic actuation.
3. Remote manual operation shall operate entirely independently of automatic release.
4. Integral manual actuation at the control head.
5. Shutdown of electrical equipment, fans, heaters, etc.
6. Closing of all openings into the room, including automatic door closers, dampers in supply and return ductwork, and any other openings that might dilute or allow escape of the Halon.
7. Installation of an exhaust fan with ductwork, dampers, controls, etc., to purge the Halon from the space after operation of the system.

d. The extinguishing agent shall be Halon 1301, delivered into the space to provide a 6% concentration in all parts of the hazard area with a residual of 3% five minutes after discharge.

e. Cylinders shall be of capacity required and shall conform to Department of Transportation regulations.

f. Cylinder valve assembly shall have a properly designed valve and a waterproof pressure gauge to indicate that the system is operable.
g. Control shall include:

Specification writer should select one or more of the following:

( )1a. Automatic release by thermostat in the hazard area.
( )1b. Automatic release by rate of rise detector(s).
( )1c. Automatic release by smoke detector or ionization detector.

2. Remote manual release by cable or electrical release, located on the front of the control panel and covered by a switch guard.

3. Manual spring-return abort switch to override the predischarge alarm and delay the release. The abort switch shall give a trouble signal if activated.

4. Integral manual release at the cylinder(s) not requiring electrical power.

h. Distribution system from the cylinder(s) to the nozzles in the hazard shall be through fixed pipes, sized as required for the hazard and as approved by Underwriters' Laboratories, NFPA, the local jurisdiction, and the insurance underwriter. Pipe and fittings shall be galvanized steel, and nozzles shall be of noncorrosive materials or furnished with a protective coating.

2.2 AUXILIARY EQUIPMENT

a. Accessories required with the system shall include the following:
 1. Class "B" control panel, as listed by UL for releasing service, shall supervise the system for the total number of detectors shown on the drawings or required by the authorities. The panel shall include an adjustable predischarge timer (0 to 60 seconds) to allow predischarge alarm and delay. The panel shall include an adjustable timer to deenergize the system after discharge (0 to 60 seconds).
 2. The control panel shall provide an SPDT relay for each zone and other necessary relays for trouble signals and equipment shutdown. The panel shall include 24-hour standby battery of the "gell cell" type with battery charger. Provide a battery test switch to test the battery under full load.
 3. Alarm, of distinctive sound and tone, to sound in the event of the operation of the system, and a red indicator light on the panel to indicate that the system has discharged. The alarm shall be actuated by all means of tripping the system.

PART THREE - EXECUTION

3.1 INSTALLATION

a. Installation shall be performed by a contractor qualified and licensed for this type of work. The installing contractor shall obtain all necessary permits and secure all required approvals for this system. Electrical portions of the system shall be performed by those qualified and licensed for that part of the work. After final acceptance, all certificates shall be turned over to the owner, by letter, through the proper channels.

b. This contractor shall install, or have installed, under his or her direction and in his or her contract, the necessary systems required to shut down the ventilation system(s), close the doors to the space(s), shut off all openings into the space, and provide the fan for purging the area after system operation, including the necessary electrical services and controls for the above. Only those items shown on the drawings in other sections, and identified for these functions, shall be omitted from this contractor's responsibility.

c. The installing contractor shall perform an acceptance test on the system to be witnessed by the owner, or an authorized representative, and the officials having jurisdiction. Halon 122 shall be used for the test, and strip charts (recording at least four points, two in the occupied space, one under the floor, and one above the ceiling) shall be recorded and must show that the minimum concentration specified has been maintained for the time specified. During the test period, the owner's representative shall observe the shutdown of the equipment, the closing of all openings, and the complete operation of the system. After the test, the installing contractor shall transmit the recording charts to the owner, through proper channels, for the owner's records. After the test, the system shall be returned to service, with the test cylinders removed and the final cylinders installed and puff-tested.

d. Prior to the test (if possible), but before turning the system over to the owner, the installing contractor shall instruct the owner's operating personnel in the operation and maintenance of the system.

e. The installing contractor shall include, with the installation contract, a semiannual inspection of the system (starting with the date of owner's actual use of the system) for the first year. These inspections shall include written reports on the condition of the system and its readiness for operation.

END OF SECTION 15497

CHAPTER 8

Air Distribution

SECTION 15500

****

DUCTWORK

PART ONE - GENERAL

1.1 The GENERAL and SPECIAL CONDITIONS, Section 15100, are included as a part of this Section as though written in full in this document.

1.2 Scope of the Work shall include the furnishing and complete installation of the equipment covered by this Section, with all auxiliaries, ready for owner's use.

PART TWO - PRODUCTS

Note to Specification writer: Drawings must be very explicit on pressure clasification and details of fittings to be used by the contractor. The design criteria must be communicated to the bidder for costing and the installation must be made on the same basis as the design if the system is to perform as intended.

2.1 SHEET METAL - RECTANGULAR

a. Ductwork, unless otherwise noted, shall be galvanized sheet metal (weight of zink coating shall be   ) and shall be built as required by HVAC Duct Construction Standards, Metal and Flexible, First Edition, January 1985 and diagrammatically shown on the drawings.
b. Ductwork 18 in. (36 cm) width and over shall be cross-broken, or ribbed and stiffened, so that it will not "breathe," rattle, vibrate, or sag.
c. Curved elbows shall have a throat radius equal to the duct width. Provide splitter vane(s) in radius elbows where indicated on drawings.
d. Square elbows shall have double-thickness turning vanes, unless single-thickness vanes are clearly indicated on the drawings.
e. Transitions in ductwork shall be made with a slope not exceeding 1 to 5, preferably 1 to 7.
f. Supply duct splits shall be provided with splitter damper and adjustable locking quadrant. Splitter blade shall be 1.5 times the smaller split width.
g. Supply duct takeoffs shall include an adjustable air-turning device equal to Carnes No. 1250 Variturn Model 2, 3, or 4.
h. Supply duct takeoffs shall include an adjustable air-turning device equal to Carnes No. 1250 Variturn Model 2, 3, or 4.

2.2 SHEET METAL - ROUND

a. Round sheet metal duct shall be constructed as required in HVAC Duct Construction Standards as published by SMACNA and refered to above. Low-pressure duct may be shop-fabricated, using good practice. Medium- and high-pressure duct shall be a manufactured product by a firm regularly engaged in such work and with a catalog listing construction, weight, Specifications, and pressure losses. Flat oval shall be same as medium-pressure duct above.
b. Round duct insulated internally shall be a product of a manufacturer engaged in such production and shall have an internal perforated liner equal to United Acousti K-27.

c. Flat oval duct insulated internally shall be same as "b" above. Flat oval duct may be used in place of round or rectangular duct, provided the insulation thickness specifrk, and diffuser neck shall be rectangular to match duct.

2.3 FIBROUS GLASS DUCT - RECTANGULAR

a. Rectangular duct may be constructed of standard-duty duct board complying with UL Standard No. 181. Fiberglass duct shall be used only as allowed by local codes and NFPA 90A for interior, low-temperature service for heating and air conditioning only, where static pressure will not exceed (-/+) 2 in. of water (-/+ 500 Pa) and where velocity will not exceed 2400 fpm (12 m/s). It shall not be used for kitchen hoods, fume hood exhaust, or under the floor slab, or exposed to moisture or to temperatures in excess of 250 deg. F (120 deg. C).

b. Ducts shall be constructed in accordance with SMACNA Fibrous Glass Duct Construction Standards, with all reinforcing recommended by that Manual, both inside and outside. A COPY OF THIS MANUAL SHALL BE MAINTAINED AT THE JOB SITE FOR REFERENCE BY THE OWNER, ARCHITECT, AND ENGINEER. There shall be no deviation from details shown in this Manual.

c. Joints shall be secured with staples and with metal tape. Tape shall be kept warm and shall be burnished down as recommended. In cold weather, the tape shall be burnished with a heated iron made for this purpose.

d. Duct shall have an aluminum outer jacket of medium-gauge aluminum or shall be foil with scrim reinforcing. SUBMIT SAMPLE OF DUCT MATERIAL, MADE UP INTO AN ELBOW WITH TURNING VANES, for acceptance by the engineer.

2.4 AUXILIARY EQUIPMENT

(a.) Duct reinforcing, hangers, and other ferrous metals used in the duct system may be black steel.

(a.) Duct reinforcing, hangers, and other ferrous metal used in the duct system shall be black steal with protective coat of paint or equal protection.

(a.) Duct reinforcing, hangers, and other ferrous metals used in the duct system shall be galvanized as required for the ductwork.

2.5 ACCEPTABLE MANUFACTURERS

a. Duct splitter dampers - Young Regulator Co., Barber Coleman, Carnes, Hart and Cooley.

b. Spiral duct - Young Regulator Co., United Sheetmetal Div.

c. Turning devices - Carnes, Barber Colman, Hart and Cooley.

PART THREE - EXECUTION

3.1 INSTALLATION

a. Install all ductwork generallay as shown on the drawings and as required by SMACNA manual.

b. Low-pressure ductwork and fittings shall be made tight for minimum air leakage. Large or noisy leaks will not be accepted. Duct tape shall not be used to seal joints, to make transitions, or for any other reason except on the outside of wrapped insulation. If duct tape is used on sheet metal, the job will be rejected!

c. Medium- and high-pressure ductwork shall have all joints and laps sealed with mastic equal to Hardcast DT-5300 to ensure a completely airtight duct system.

END OF SECTION 15500

SECTION 15502

****

UNDERFLOOR DUCTWORK

PART ONE - GENERAL

1.1 The GENERAL and SPECIAL CONDITIONS, Section 15100, are included as a part of this Section as though written in full in this document.

1.2 Scope of the Work shall include the furnishing and complete installation of the equipment covered by this Section, with all auxiliaries, ready for owner's use.

PART TWO - PRODUCTS

2.1 DUCTWORK

a. Ductwork noted and/or detailed as located under a concrete floor slab shall be of stainless steel of weight specified in SMACNA Manual for pressure class required.

b. Fittings shall be five-section fabricated or factory-formed type, of material same as pipe.

PART THREE - EXECUTION

3.1 INSTALLATION

a. Joints in pipe and at fittings shall be sealed using Hardcast DT-5300 sealer or equal.

b. Encase underfloor duct in concrete, as shown in detail on drawing. Encase underfloor in clean construction sand and locate above the vapor barrier.

c. Slope the underfloor duct to a drain pit as shown on drawings and provide sump pump for drain pit.

END OF SECTION 15502

SECTION 15504

****

WEATHERPROOFING OF DUCTWORK

PART ONE - GENERAL

1.1 The GENERAL and SPECIAL CONDITIONS, Section 15100, are included as a part of this Section as though written in full in this document.

1.2 Scope of the Work shall include the furnishing and complete installation of the equipment covered by this Section, with all auxiliaries, ready for owner's use.

PART TWO - PRODUCTS

2.1 DUCTWORK

a. All ductwork exposed to weather shall have all joints, laps, edges, etc., sealed and coated with duct sealer equal to Hardcast DT-5300 and applied with FTO-20 adhesive or approved equivalent.

PART THREE - EXECUTION

3.1 INSTALLATION

a. Clean all dirt and dust from duct before applying sealer.

b. Duct shall be dry when sealer is applied.

c. Apply sealer as shown on manufacturer's instructions.

END OF SECTION 15504

SECTION 15505

****

CURBS FOR ROOF PENETRATIONS

PART ONE - GENERAL

1.1 The GENERAL and SPECIAL CONDITIONS, Section 15100, are included as a part of this Section as though written in full in this document.

1.2 Scope of the Work shall include the furnishing and complete installation of the equipment covered by this Section, with all auxiliaries, ready for owner's use.

PART TWO - PRODUCTS

2.1 ROOF PENETRATIONS

a. All services (ducts, pipes, supports, etc.) passing through the roof shall be provided with a curb or other means of weatherproofing the passage. See details on the drawings. Pitch cups are not acceptable.

b. Ducts and other large items shall be provided with a curb, complete with cant and flat base, for sealing watertight into the roofing material. Flashing collar shall be properly sealed to the duct, as is standard with the industry standards, and shall be so arranged that water cannot penetrate the seal, even when driven by high wind. Provide counter flashing for the roofing as required by the roofer.

c. Round penetrations shall be provided with a curb for large penetrations, with a PVC curb cap and a stainless steel drawband to provide a watertight seal.

d. Pipe, conduit, and other small penetrations shall be provided with a curb or a tall flashing cone of heavy-gauge galvanized metal. Provide a PVC cap and stainless steel drawband.

2.2 ACCEPTABLE MANUFACTURERS

a. Curbs and cones shall be as shown on the details on the drawings or approved equivalents by The Pate Company.

PART THREE - EXECUTION

3.1 INSTALLATION

a. The curb base or the cone flashing shall be set into the roofing material as required for a complete weatherproof seal.

b. For flat or low-slope roofs, the roofing material shall match the existing roofing products, unless installed at the time of roof application. Caution: asphalt and pitch are not compatible! Asphalt should be used for asphalt roofs and pitch for pitch roofs. If in doubt, consult a qualified roofer. "Bull" is not acceptable on either type of roof.

c. Sloped roofs with shingles or metal covers shall have the curb or cone built into the roofing by a qualified roofer.

d. Install the cap around the penetrating material and secure to the curb or cone as directed by the manufacturer or as shown on the drawings. Secure the upper edge of the cap for a permanent weatherproof seal.

e. Allow for expansion and contraction on all installations.

END OF SECTION 15505

SECTION 15510

****

INSULATION OF SHEET METAL DUCTWORK

PART ONE - GENERAL

1.1 The GENERAL and SPECIAL CONDITIONS, Section 15100, are included as a part of this Section as though written in full in this document.

1.2 Scope of the Work shall include the furnishing and complete installation of the equipment covered by this Section, with all auxiliaries, ready for owner's use.

1.3 All rectangular sheet metal supply duct shall be insulated internally, unless noted otherwise.

1.4 All round low-pressure supply duct shall be insulated externally, except as noted. Round supply duct is usually limited in scope to short and small runs.

1.5 All round or round-oval medium-pressure supply and/or return duct shall be internally lined with a perforated liner.

1.6 Return duct shall be insulated internally for sound control.

1.7 Exhaust duct on inlet of fan shall be insulated internally. Duct on discharge side of exhaust fan is not insulated.

PART TWO - PRODUCTS

2.1 INTERNAL INSULATION
a. Internal insulation of the duct shall be duct liner with K factor of 2.6 at 100 deg. F (38 deg. C) and acceptable up to velocity of 4000 fpm.

2.2 EXTERNAL INSULATION
a. Insulation on the exterior of the duct shall be duct wrap with a vapor-proof jacket of aluminum foil or foil-scrim-kraft paper having UL label. Density of wrap shall be 1 lb./ft^3 (16 kg/m^3).

2.3 INSULATION THICKNESS
a. Thickness of insulation shall be 1 in. (25.4 mm) thick for supply and outside air duct protected inside the building; 1/2 in. (13 mm) thick for return, ventilation, and exhaust inside the building; and 2 in. (51 mm) thick for all supply and return duct located in an attic or exposed to outside air conditions, such as equipment rooms.
b. All duct wrap shall be 2 in. (51 mm) thick.

2.4 All insulation shall be UL-labeled for fire and smoke ratings, and all accessories shall also be labeled.

2.5 ACCEPTABLE MANUFACTURERS
a. Insulation shall be the product of a recognized manufacturer of such materials, with complete catalog listing R values, recommended thicknesses, and installation details.

b. Insulation shall be the standard product of a manufacturer with a product line equal to Certain Teed.

PART THREE - EXECUTION

3.1 APPLICATION OF DUCT LINER INSULATION

a. All portions of duct designated to receive duct liner shall be completely covered with liner. Transverse joints shall be neatly butted and there shall be no interruptions or gaps.
b. Duct liner shall be cut to ensure overlapped and compressed longitudinal corner joints.
c. For velocities to 2000 fpm (10 m/s), fasteners shall start within 3 in. (75 mm) of the upstream transverse edges of the liner and 3 in. from the longitudinal joints and shall be spaced at a maximum of 12 in.(30 cm) on center (oc) around the perimeter of the duct (except that they may be a maximum of 12 in. from a corner break). Elsewhere, they shall be a maximum of 18 in. (46 cm) oc, except that they shall be placed not more than 6 in. (15 cm) from a longitudinal joint of the liner or 12 in. from a corner break.
d. For velocities from 2000 to 4000 fpm (10 to 20 m/s), fasteners shall start within 3 in. of the upstream transverse edges of the liner and 3 in. from the longitudinal joints and shall be spaced at a maximum of 6 in. oc around the perimeter of the duct, except that they may be a maximum of 6 in. from a corner break. Elsewhere, they shall be a maximum of 16 in. o.c., except that they shall be placed no more than 6 in. from a longitudinal joint of the liner or 12 in. from a corner break.
e. In addition to the adhesive edge coating of transverse joints, any longitudinal joints shall be similarly coated with adhesive.

3.2 DUCT WRAP

a. Duct wrap shall be installed in a neat and competent manner, with all edges neatly covered with approved metallic duct tape to vapor-proof the entire duct. Laps and joints shall be secured with insulation staples and then covered with approved tape.

3.3 QUALITY OF EXECUTION

a. All work shall be done by workers who are thoroughly familiar with such applications.

3.4 MANUFACTURERS' RECOMMENDATIONS

a. Materials and methods of insulation installation shall be as recommended by the manufacturer of the insulation.
b. Insulation exposed to weather shall be protected as called for in Section 15253 or as instructed by the engineer if that Section is not included in the Specifications.

END OF SECTION 15510

SECTION 15515

****

INSULATED FLEXIBLE DUCTWORK
LOW PRESSURE

PART ONE - GENERAL

1.1 The GENERAL and SPECIAL CONDITIONS, Section 15100, are included as a part of this Section as though written in full in this document.

1.2 Scope of the Work shall include the furnishing and complete installation of the equipment covered by this Section, with all auxiliaries, ready for owner's use.

PART TWO - PRODUCTS

2.1 FLEXIBLE DUCTS

a. Flexible air ducts shall have impervious inner core with wire reinforcement. The inner duct shall be covered with 1.5-in. (38-mm) fiberglass insulation with a polyethylene vapor-proof jacket.

b. Flexible air duct shall be UL-181-listed, Class 1, and shall meet all local codes.

c. Adjustable stainless steel or nylon straps shall be used to secure duct fittings and equipment. A second band shall be applied over the jacket to maintain the vapor barrier.

d. Fittings to connect the flex duct to the trunk duct shall be designed to twist into the trunk duct and shall have a butterfly-type volume damper. Fitting shall extend into the duct to provide some air scoop device for airflow.

2.2 ACCEPTABLE MANUFACTURERS

a. Duct shall be the product of an established manufacturer of such products and equivalent to Certain Teed (Model G-25), Wiremold (WGC), P.P.G. (Gloss Flex).

PART THREE - EXECUTION

3.1 INSTALLATION

a. Flexible duct shall be limited in length to that necessary to make connections between trunk ducts and terminal units. This shall normally not exceed 5 ft. (1.52 m).

b. All duct shall be fully stretched out to reduce resistance.

c. Connections to fittings or terminals shall be made with stainless steel bands. The inner liner shall be clamped tight with the band, then the insulation and jacket pulled up tight against the duct or terminal and secured with a second band or strap. Installation shall be as recommended by the duct manufacturer. If in doubt, verify the installation with the engineer.

d. Support the flexible duct with adequate hangers to relieve strain on any fitting. Unnecessary bends, sags, twists, etc., will not be allowed!

END OF SECTION 15515

SECTION 15516
****
INSULATED FLEXIBLE DUCTWORK
MEDIUM PRESSURE

PART ONE - GENERAL

1.1 The GENERAL and SPECIAL CONDITIONS, Section 15100, are included as a part of this Section as though written in full in this document.

1.2 Scope of the Work shall include the furnishing and complete installation of the equipment covered by this Section, with all auxiliaries, ready for owner's use.

PART TWO - PRODUCTS

2.1 FLEXIBLE DUCTS

a. Flexible air ducts shall have an inner core of two layers of polyester with an encapsulated steel wire helix, with 1.5-in. (38-mm) fiberglass insulation with a vapor-proof jacket of metalized mylar. Provide a reinforced hanger strip designed to support the duct.

b. Flexible air duct shall be UL-181-listed, Class 1, and shall meet all local codes. Pressure rating shall be for 1.25 in. (32 mm) negative to 6 in. (150 mm) positive.

c. Adjustable stainless steel or nylon straps shall be used to secure duct fittings and equipment. A second band shall be applied over the jacket to maintain the vapor barrier.

d. Fittings to connect the flex duct to the trunk duct shall be designed to twist into the trunk duct and shall have a butterfly-type volume damper. Fitting shall extend into the duct to provide some turn device for airflow.

2.2 ACCEPTABLE MANUFACTURERS

a. Ducts shall be the product of an established manufacturer of such products and equivalent to Certain Teed (Model Centraflex 25) or Wiremold (Type WK).

PART THREE - EXECUTION

3.1 INSTALLATION

a. Flexible duct shall be limited in length to that necessary to make connections between trunk ducts and terminal units. This shall normally not exceed 5 ft. (1.5 m).

b. All duct shall be fully stretched out to reduce resistance.

c. Connections to fittings or terminals shall be made with stainless steel or nylon bands, designed for that service. The inner liner shall be clamped tight with the band, then the insulation and jacket pulled up tight against the duct or terminal. Install a second band around the outside of the jacket. Installation shall be as recommended by the duct manufacturer. If in doubt, verify the installation with the engineer.

d. Support the flexible duct by the hanger strip with support wires on 30-in. (760-mm) intervals to relieve strain on any fitting. Unnecessary sags, twists, turns, etc. will not be allowed.

END OF SECTION 15516

SECTION 15525

****

GRILLES, REGISTERS, AND DIFFUSERS

PART ONE - GENERAL

1.1 The GENERAL and SPECIAL CONDITIONS, Section 15100, are included as a part of this Section as though written in full in this document.

1.2 Scope of the Work shall include the furnishing and complete installation of the equipment covered by this Section, with all auxiliaries, ready for owner's use.

PART TWO - PRODUCTS

2.1 GRILLES, REGISTERS, AND DAMPERS

a. Grilles, registers, and diffusers shall be style and type listed below. Select for proper throw and drop for air quantity and size shown on drawings. Maximum noise shall be 32 dB ("a" scale).

b. Not all of the types listed below are required on every job. Refer to drawings for type and size.

2.2 ACCEPTABLE MANUFACTURERS

a. One or more manufacturers' names are given below as a means of identification of type of product. Other manufacturers' products are acceptable, as approved by the engineer.

2.3 REVERSIBLE CORE REGISTERS (WALL OR SILL)

a. Extruded-aluminum core with blades on 1/4-in. (6-mm) spacing shall be mounted in an extruded-aluminum frame with borders 1 1/4 in. (32 mm) or 3/4 in. (19 mm) wide.

b. Core shall be removable and reversible to provide for 5 or 15 deg. of spread.

c. Provide concealed rear blades for deflection.

d. Provide extruded-aluminum opposed blade (OB) dampers.

2.4 LINEAR REGISTERS (WALL, SILL, OR FLOOR)

a. Extruded-aluminum core with 1/8-in. (3-mm) bars, or 1/4-in. (6-mm) or 1/2-in. (13-mm) spacing, with 0 or 15-deg. deflection.

b. Frame shall be heavy-gauge extruded aluminum with border for surface mounting, recess mounting, or casting in concrete. Frame width shall be 3/4 in. (19 mm), 1 1/8 in. (30 mm), or 1/4 in. (6 mm).

c. Accessory equipment shall include:

1. Hinged access door
2. Blank off plates
3. Screens
4. Mitered corners
5. OB damper
6. Direction changer
7. Straightening vanes

d. Linear registers shall be Carnes CD30, 35, 40, 45, 50, or 55, as shown on drawings, with accessories as noted. Selections shall be by manufacturers' charts, with correction factors as required.

2.5 CEILING RETURN AND EXHAUST REGISTERS

a. Extruded-aluminum natural color OB damper, maximum velocity of 700 fpm, maximum pressure drop of 0.1 in. sp.

b. Egg-crate type - 1/2 x 1/2 x 1/2 in. (13 x 13 x 13 mm) blades in extruded-aluminum frame for surface mounting, 90% free area. Carnes 6291AN or equivalent.

c. Angle blade type - extruded-aluminum blades on 3/4-in. (19-mm) spacing in extruded frame for surface mounting, 70% free area. Carnes 6831A or equivalent.

d. Straight blade type - extruded-aluminum blades on 1/2-in. (13-mm) spacing in extruded-aluminum frame for surface mounting, 70% free area. Carnes 6595AH (V) or equivalent.

e. Heavy-duty (gymnasium - industrial) - steel bars 0.12-in. (3-mm) thick on 0.66-in. (17-mm) spacing, welded steel frame with blade reinforcing on 6-in. (15-cm) centers, 70% free area. Carnes 6180H (V) or equivalent.

2.6 DOOR LOUVERS

a. Inverted "V" type with telescoping frame for mounting in door panels. Color to be natural aluminum (paintable). Carnes 930 or equivalent.

2.7 AUXILIARY EQUIPMENT

a. Plaster frames shall be provided where grilles, registers, or diffusers are installed in plaster surfaces or where removal is required for access to filters, dampers, etc.

b. Opposed blade (OB) dampers shall be installed on all outlets or intakes, except where a single unit is connected to a fan having a variable-speed drive.

c. Fusible link attachment for OB dampers shall be provided where shown on drawings. Link shall not affect operations of damper.

PART THREE - EXECUTION

3.1 INSTALLATION

a. Install all grilles, registers, and louvers flush with surface and level or straight with other similar items.

b. Insulate the back of all ceiling diffusers, to cover all exposed metal, with 1/4-in. (6-mm) thickness of self-adhesive insulating tape.

c. Drop from supply duct shall be internally insulated down to diffuser. If necessary, cut insulation back to clear operation of the OB damper and reinsulate only the exposed portion with external duct insulation.

d. Support the register from the duct at the proper level to hold it snug against the ceiling.

e. Door louvers and toe-space RA grilles shall be furnished by the mechanical contractor and installed by the general contractor.

END OF SECTION 15525

SECTION 15528

****

GRILLES AND REGISTERS

PART ONE - GENERAL

1.1 The GENERAL and SPECIAL CONDITIONS, Section 15100, are included as a part of this Section as though written in full in this document.

1.2 Scope of the Work shall include the furnishing and complete installation of the equipment covered by this Section, with all auxiliaries, ready for owner's use.

PART TWO - PRODUCTS

2.1 GRILLES AND REGISTERS
Grilles and registers shown shall be generally equal to the model number shown on the drawings or approved equivalent.

PART THREE - EXECUTION

3.1 INSTALLATION
Install grilles and registers in neat and competent manner, as recommended by the manufacturer.

END OF SECTION 15528

SECTION 15570
****
DOOR LOUVERS

PART ONE - GENERAL

1.1 The GENERAL and SPECIAL CONDITIONS, Section 15100, are included as a part of this Section as though written in full in this document.

1.2 Scope of the Work shall include the furnishing and complete installation of the equipment covered by this Section, with all auxiliaries, ready for owner's use.

PART TWO - PRODUCTS

2.1 DOOR LOUVERS
a. Door louvers shall be steel or extruded aluminum with inverted "V" blades and mounted in a telescoping frame designed for a finished appearance from either side. Free area of louver shall be no less than 70%, and louver shall have adequate vertical bracing, so as not to rattle or work loose in the slamming of the door.
b. Door louver shall be factory-coated in either a neutral beige color or a mill finish of aluminum. Louver and frame shall be suitable for final painting or for use as supplied.
c. See schedule of size shown on the drawings.

2.2 ACCEPTABLE MANUFACTURERS
a. Acceptable manufacturers: Louvers shall be Barber Coleman Model GDV or equivalent.

PART THREE - EXECUTION

3.1 INSTALLATION
a. Door louvers shall be furnished by the mechanical contractor but installed by the general contractor.

END OF SECTION 15570

SECTION 15583

****

BELOW-WINDOW SUPPLY AND RETURN

PART ONE - GENERAL

1.1 The GENERAL and SPECIAL CONDITIONS, Section 15100, are included as a part of this Section as though written in full in this document.

1.2 Scope of the Work shall include the furnishing and complete installation of the equipment covered by this Section, with all auxiliaries, ready for owner's use.

PART TWO - PRODUCTS

2.1 SUPPLY AND RETURN

a. Furnish a finished furniture metal cover, located below windows, for supply and return of slab-mounted A/C units.

b. Cover shown is a convector cover, modified with center baffle and double-deflection supply outlet, as specified for side wall grilles. Bottom is to be left open for return.

c. Cover shall be complete with metal back next to the wall, with cutouts for supply and return duct connections. Install on insulated separator between supply and return.

d. Supply portion shall be fully insulated internally with 1/2-in. (13-mm) foamed plastic insulation, applied with 100% adhesive, to prevent condensation.

2.2 ACCEPTABLE MANUFACTURERS

a. Convector cover shall be Sterling Radiator Co. or approved equivalents.

PART THREE - EXECUTION

3.1 INSTALLATION

a. Provide a 1/4 x 1-in. (6 x 25-mm) strip of foamed plastic adhesive-backed tape around the perimeter of the cover to allow for unevenness of the wall.

b. Secure the cover to the wall with concealed fasteners, such as toggle bolts, and pull up tight.

c. Adjust air supply as required for good distribution.

END OF SECTION 15583

SECTION 15585

****

VARIABLE AIR VOLUME TERMINAL UNITS

PART ONE - GENERAL

1.1 The GENERAL and SPECIAL CONDITIONS, Section 15100, are included as a part of this Section as though written in full in this document.

1.2 Scope of the Work shall include the furnishing and complete installation of the equipment covered by this Section, with all auxiliaries, ready for owner's use.

PART TWO - PRODUCTS

2.1 CASING

a. Casing of control and terminal units shall be heavy-gauge galvanized steel, with all surfaces painted with protective paint. All interior surfaces shall be acoustically and thermally lined with fiberglass insulation, with surface protected from erosion.

2.2 CONTROLS

( ) a. Controls shall be pneumatic, operating from the control air system.

( ) a. System-powered controls requiring additional static pressure above nominal system pressure shall be provided with large-diameter operators. System static shall be adequate, as shown, to provide pressure required to operate the control operator.

2.3 OUTLETS - SUPPLY

a. Air distribution outlet shall be a linear-slot-type diffuser, nominally 2 or 4 ft. long (60 or 120 cm), with one- or two-way blow. Diffusers shall be sized for low noise level at full volume with throw to match the job requirements.

b. Each outlet shall be equipped with an integral fire damper with a fusible link. Damper shall meet UL requirements for mounting in a 1-hr rated ceiling system.

2.4 RETURNS

a. Return slots of 2 or 4 ft. (60 or 120 cm) in length shall match and line up with the supply diffusers and shall be insulated above the ceiling to reduce noise transmission.

b. Every return slot shall have a fire damper with a fusible link, meeting UL requirements for a 1-hr rated ceiling assembly.

c. Every return slot shall be duct-connected to the return system.

2.5 THERMOSTATS

a. Thermostat controlling the unit shall be wall-mounted or mounted in the return air slot with concealed adjustment lever. Pneumatic thermostats shall be nonbleed type.

2.6 APPROVALS REQUIRED

a. Units and all accessories shall be approved for locations shown on the drawings by the Southern Standard Building Code. All wiring, tubing, etc., shall comply with the above.

2.7 ACCEPTABLE MANUFACTURERS

a. Terminals shall be equal in performance to the model number shown on the drawings. Trane or equivalent.

PART THREE - EXECUTION

3.1 INSTALLATION

a. All units shall be supported separately from the structure above, but shall match and line up with the ceiling in elevation and in location. Units with fire dampers shall be supported as required by testing agencies.

b. Duct connections shall be made with flexible duct. Inner lining of flexible duct shall be secured to the unit, and duct jacket shall be sealed to maintain vapor barrier jacket. Provide additional clamp on exterior jacket to make a proper seal.

3.2 BALANCING AND TESTING

a. Units shall be returned to required CFM for proper operation after balance and test contractor finishes making readings at full CFM. See Balancing and Testing Section 15111.

3.3 ACCESS

a. Provide access to every unit where access is required to reach any moving parts, such as control operator.

END OF SECTION 15585

SECTION 15588

****

VARIABLE AIR VOLUME CONTROL UNITS AND TERMINAL UNITS WITH SLOT DIFFUSERS

PART ONE - GENERAL

1.1 The GENERAL and SPECIAL CONDITIONS, Section 15100, are included as a part of this Section as though written in full in this document.

1.2 Scope of the Work shall include the furnishing and complete installation of the equipment covered by this Section, with all auxiliaries, ready for owner's use.

PART TWO - PRODUCTS

2.1 CONTROL AND TERMINAL UNITS

a. Casing of control and terminal units and slot diffusers shall be heavy-gauge galvanized steel with all surfaces protected from rust. All interior surfaces shall be acoustically and thermally lined with fiberglass insulation, with surface protected from erosion.

b. Controls shall be pneumatic, operating from the control air system, or 24-volt electric/electronic controls, all as manufactured by the box manufacturer, or controls may be installed by the control subcontractor. Controls operating from supply air side may be used only if the static pressure of the supply air does not have to be increased above that required for the system.

c. Controls shall include a device to limit the maximum supply air to no more than 10% above that shown on the drawings and a minimum-position setting of approximately 10% of the maximum. This control shall compensate for variations in supply air pressure.

d. Thermostats controlling the boxes shall be wall-mounted. Pneumatic thermostats shall be nonbleed type.

e. Units and all accessories shall comply with the Standard Building Code for locations as shown on the drawings. All wiring, tubing, etc., shall be in compliance with the code.

f. Air distribution outlets shall be linear-type diffusers, designed to maintain air pattern down to 10% of design air. Slots shall be either 2 or 4 ft (60 or 120 cm) in length to match the ceiling grid shown on the drawings. Individual slots shall not be more than 1/2-in. (13 mm) in width, and all slots shall be the same width. Direction of blow shall be one way or two way, as shown on the drawings.

g. Noise level shall not exceed 32 dB ("a" scale) at design air quantity.

h. Return air slot diffusers shall match the supply air outlets in size, type, and appearance and shall be insulated above the ceiling to reduce noise transmission.

i. A 1-hr rated fire damper bearing the UL label shall be provided in every supply and return opening through the ceiling.

2.2 ACCEPTABLE MANUFACTURERS

a. Terminals shall be equal in performance to the model number shown on the drawings. Trane (VCCC or VTCC) or equivalents.

PART THREE - EXECUTION

3.1 INSTALLATION

a. All control boxes and terminal units shall be supported from the structure but shall match and line up with the ceiling in elevation and in location. All units with fire dampers shall be supported as required by the UL test of the damper.

b. Duct connections shall be made with insulated flexible duct. The inner liner shall be secured to the unit connection by an adjustable strap tightened properly. The outer jacket shall be secured by adjustable strap as well, so that both inner and outer liners are secured. Duct tape shall not be required or used.

3.2 BALANCING AND TESTING

a. Set the units to the design CFM, and balancing and testing shall verify the correct air quantity. See Balancing and Testing, Section 15111.

3.3 ACCESS

a. Provide access to every unit that has any moving or adjustable parts, such as control operator, etc.

END OF SECTION 15588

SECTION 15595
****
FIRE DAMPERS AND FIRE DOORS

PART ONE - GENERAL

1.1 The GENERAL and SPECIAL CONDITIONS, Section 15100, are included as a part of this Section as though written in full in this document.

1.2 Scope of the Work shall include the furnishing and complete installation of the equipment covered by this Section, with all auxiliaries, ready for owner's use.

PART TWO - PRODUCTS

2.1 FIRE DAMPERS
a. Furnish fire dampers and fire doors as shown on plans and/or as required to meet the fire regulations as outlined in NFPA No. 90-A.

2.2 DAMPERS
a. Dampers shall be interlocking blade type with UL label, with open area equal to the duct area.

2.3 DOORS
a. Fire doors shall be single or double, as shown on the drawings and as required to match the fire wall rating.

2.4 ACCESS DUCT
a. Access door shall be provided for maintenance of each fire damper or fire door. Duct access shall be double wall type with hinged door. Access through ceiling, if required, shall be provided to match the ceiling.

PART THREE - EXECUTION

3.1 INSTALLATION
a. Dampers shall be installed with a 14-gauge sleeve secured to the fire wall, ceiling, or floor and ductwork connected so that damper will not be damaged, should the duct fall.
b. See SMACNA Low Velocity Duct Standards for installation details, or see details on drawings.
c. Provide duct access door that will allow resetting of damper and replacement of the link. Access shall be 12 x 12-in. (31 x 31-cm) minimum, except for smaller ducts, where access shall be as large as practical.
d. Ceiling access shall be coordinated with the ceiling.

END OF SECTION 15595

CHAPTER 9

Heating Sources

SECTION 15600

****

GAS-FIRED UNIT HEATER

PROPELLER TYPE

PART ONE - GENERAL

1.1 The GENERAL and SPECIAL CONDITIONS, Section 15100, are included as a part of this Section as though written in full in this document.

1.2 Scope of the Work shall include the furnishing and complete installation of the equipment covered by this Section, with all auxiliaries, ready for owner's use.

PART TWO - PRODUCTS

2.1 UNIT HEATER

a. Unit heater shall be natural gas-fired with propeller-type fan, with output capacity as shown on the drawings.

b. Heat exchanger shall be aluminized steel (stainless steel), designed to minimize stress and noise from expansion and contraction.

c. Gas burners shall be designed for use on natural or LP gas (with adapter kit). Burners shall be aluminized steel with stainless steel inserts, or all stainless steel, to prevent rusting. Burners shall have fixed-type air adjustment and crossover device and shall be supported on welded brackets.

d. Gas valve shall provide for control of pilot and main valve and shall shut the gas valve in case of pilot failure. Electric pilot ignition shall be provided to light the pilot when the thermostat calls for heat or if the pilot should be extinguished. The pilot shall be extinguished when the thermostat is satisfied. Standing pilot will not be acceptable.

e. Propeller-type fan with motor assembly shall be mounted on resilient rubber mounts to minimize vibration and noise. Motors shall be totally enclosed sleeve-bearing type with automatic reset built-in thermal overload protection.

f. Controls shall consist of wall-mounted thermostat (may be mounted below heater on conduit if shown on drawings) with low-voltage contacts to cycle the gas valve. Fan switch in the heater shall cause the fan to operate so as to avoid cold drafts.

Controls shall include:

( ) 1. Controls shall provide for two-stage thermostat with two stages of gas firing to allow hi-low operation.

( ) 1. Controls shall provide for modulation of the gas valve from 40 to 100% of capacity to allow a more uniform leaving air temperature.

( ) 2. Provide a summer-winter switch for operation of the fan during the summer for air circulation.

( ) 3. Fan switch shall be mounted on the base of the wall thermostat.

( ) 3. Fan switch shall be pull-chain type mounted on the base of the heater.

2.2 ACCESSORIES REQUIRED

a. Provide heavy-gauge vertical louvers, in addition to the horizontal louvers, for four-way control of the discharge air.

b. Provide UL-listed type B vent and vent cap for the heater.

c. Provide a vent exhauster to remove the products of combustion. Exhauster is to interlock with burner to prevent gas valve opening unless the venter is operating.

2.3 ACCEPTABLE MANUFACTURERS

a. Unit heater shall be the product of a manufacturer having a full line of such equipment, equal to model number shown on the drawings or equivalents by Trane (GPN), Reznor (XL), Hastings (Model F), Sterling (Model B or CF).

PART THREE - EXECUTION

3.1 INSTALLATION

a. Install proper hangers from structural members to support the heater as recommended by the manufacturer. Provide clearance from combustible materials, as required by the codes and the manufacturer's installation instructions.

b. Make connection as required by the applicable codes as follows:

1. Connect gas as required. Provide a gas cock in the gas line remote from the heater, in addition to the cock at the heater.
2. Connect vent and provide vent cap.
3. Make electrical connections for power and controls.

c. Place the heater in operation and check for proper burner and control operation.

END OF SECTION 15600

SECTION 15601

****

GAS-FIRED UNIT HEATER
CENTRIFUGAL BLOWER TYPE

PART ONE - GENERAL

1.1 The GENERAL and SPECIAL CONDITIONS, Section 15100, are included as a part of this Section as though written in full in this document.

1.2 Scope of the Work shall include the furnishing and complete installation of the equipment covered by this Section, with all auxiliaries, ready for owner's use.

PART TWO - PRODUCTS

2.1 UNIT HEATER

a. Unit heater shall be natural gas-fired with centrifugal-type fan, with output capacity as shown on the drawings.

b. Heat exchanger shall be (type 304 stainless steel) (aluminized steel), designed to minimize stress and noise from expansion and contraction. Heat exchanger shall carry a 10-year nonprorated warranty.

c. Gas burners shall be designed for use on natural or LP gas (with adapter kit). Burners shall be aluminized steel with stainless steel inserts, or all stainless steel, to prevent rusting. Burners shall have fixed-type air adjustment, and crossover device and shall be supported on welded brackets.

d. Gas valve shall provide for control of pilot and main valve and shall shut the gas valve in case of pilot failure. Electric pilot ignition shall be provided to light the pilot when the thermostat calls for heat or if the pilot should be extinguished. The pilot shall be extinguished when the thermostat is satisfied. Continuous standing pilot will not be acceptable.

e. Centrifugal-type fan with motor assembly shall be mounted on resilient rubber mounts to minimize vibration and noise. Motors shall be totally enclosed sleeve-bearing type with automatic reset built-in thermal overload protection. Provide belts and adjustable drive pulley unless fan wheel is directly mounted on motor shaft.

2.2 CONTROLS

a. Controls shall consist of the following:
(Specification writer should select the following as applicable.)

( ) 1. Controls shall consist of wall-mounted thermostat with low-voltage contacts to cycle the gas valve. Fan switch in the unit shall cause the fan to operate so as to avoid cold drafts.

( ) 1. Controls shall consist of thermostat mounted on the unit heater. Fan switch in the unit shall cause the fan to operate to avoid cold drafts.

( ) 2. Controls shall provide for two-stage thermostat with two stages of gas firing, to allow hi-low operation.

( ) 2. Controls shall provide for modulation of the gas valve from 40 to 100% of capacity to allow a more uniform leaving air temperature.

( ) 3. Provide a summer-winter switch for operation of the fan during the summer for air circulation. Fan switch shall be mounted on the base of the wall thermostat.

( ) 3. Provide a summer-winter switch for operation of the fan during the summer for air circulation. The fan switch shall be pull-chain type mounted on the bottom of the heater.

2.3 ACCESSORIES REQUIRED

a. Provide heavy-gauge vertical and horizontal louvers for four-way control of the discharge air.

b. Provide UL-listed type B vent and vent cap for the heater.

c. Provide a vent exhauster to remove the products of combustion. Exhauster is to interlock with burner to prevent gas valve opening unless the venter is in operation.

d. Provide cabinet for throwaway filters of 1-in. thickness.

2.4 ACCEPTABLE MANUFACTURERS

a. Unit heater shall be the product of a manufacturer having a full line of such equipment, equal to model number shown on the drawings or equivalents by Trane (HBA), Reznor (XLB), Hastings (Model B), Sterling (CB).

PART THREE - EXECUTION

3.1 INSTALLATION

a. Install proper hangers from structural members to support the heater on rubber isolators as recommended by the manufacturer. Provide clearance from combustible materials, as required by the codes and the manufacturer's installation instructions.

b. Make connection as required by the applicable codes as follows:

1. Connect gas as required. Provide a gas cock in the gas line, remote from the heater, as well as the cock at the heater.
2. Connect vent and provide vent cap.
3. Make electrical connections for power and controls.
4. Provide any duct connections shown on the drawings.

c. Place the heater in operation and check for proper burner, fan, and control operation.

END OF SECTION 15601

SECTION 15602

****

GAS-FIRED RADIANT UNIT HEATERS

PART ONE - GENERAL

1.1 The GENERAL and SPECIAL CONDITIONS, Section 15100, are included as a part of this Section as though written in full in this document.

1.2 Scope of the Work shall include the furnishing and complete installation of the equipment covered by this Section, with all auxiliaries, ready for owner's use.

PART TWO - PRODUCTS

2.1 HEATER

a. Heater shall be natural (LP) gas-fired, operating on the principle of radiant heat without fans or blowers, with capacity as shown on the drawings.

b. Heat exchanger shall be designed for operation up to 1140 deg. F (612 deg. C) with a coating of highly emissive materials on the outside. Provide vent connection for the flue gases.

c. For low-bay mounting, the heater shall be provided with a series of reflectors to spread the heat evenly in a downward direction.

d. For high-bay installations, the heater shall be provided with a series of smooth, bright curved reflectors to concentrate the heat beneath the heater.

2.2 PARTIAL AREA HEATING

a. For partial area heating, the heater shall be provided with reflectors spreading the heat in the area to be heated from the side facing the heated area and shall be provided with bright reflectors on the far side to reflect the heat downward.

2.3 CONTROLS

( ) a. Controls for the heater shall be self-generating type with a wall thermostat located as shown on the drawings.

( ) a. Controls shall be 24-volt type with electric ignition of the pilot when the thermostat calls for heat. The pilot shall not be lit when heat is not required.

b. The controls shall monitor the pilot flame; should the pilot not be lit, the main gas valve shall not open. Heater shall bear the AGA approval.

2.4 ACCEPTABLE MANUFACTURERS

a. Heater shall be the make and model number shown on the drawings or equivalent by Panelblock (Model CR).

PART THREE - EXECUTION

3.1 INSTALLATION

a. Provide adequate and proper mounting system to support the heater from the structural system of the area.

b. Locate the heater as required to provide clearance from all combustible materials above, beside, and under the heater.

c. Provide UL-listed vent pipe and cap for the flue gases.

d. Provide gas connections as directed by the codes. Provide gas cock remote from the heater(s), as well as the cock at the heater.

e. Provide electrical connections as required.

f. Check the operation of the heaters and test the safety devices for proper operation. Leave the heaters in proper operating condition.

END OF SECTION 15602

SECTION 15603

****

ELECTRIC UNIT HEATER
PROPELLER TYPE
ENCLOSED ELEMENT

PART ONE - GENERAL

1.1 The GENERAL and SPECIAL CONDITIONS, Section 15100, are included as a part of this Section as though written in full in this document.

1.2 Scope of the Work shall include the furnishing and complete installation of the equipment covered by this Section, with all auxiliaries, ready for owner's use.

PART TWO - PRODUCTS

2.1 UNIT HEATER

a. Unit heater shall consist of an enclosed, extended-surface heating element with propeller-type fan, with capacity as shown on the drawings. The entire unit and controls shall be UL-labeled. Heater shall be suitable for mounting with vertical or horizontal air discharge.

b. Heating element shall be sheathed type with extended fins, rated at the voltage available at the job site.

c. Fan shall be quiet propeller type, mounted on a totally enclosed motor, with internal overload protection. The motor shall be supported by the fan guard, which shall be mounted on rubber isolators.

d. Cabinet shall be heavy-gauge metal, with rust preventive finish. Provide horizontal louvers on the discharge for direction of the air.

2.2 CONTROLS

a. Controls shall consist of a wall-mounted low-voltage thermostat, contactor(s) for the heating element, switch to operate the fan when the element is hot, and high-limit cutout for the heating element, all as standard with the manufacturer.

2.3 ACCESSORIES REQUIRED

a. Accessories required are as follows:

1. Vertical louvers on discharge to allow four-way control of discharge air.
2. Two-stage thermostat and controls to allow operation at 50 and 100% of capacity.

(Specification writer should select the following as applicable.)

( ) 3. Fan switch to allow summer operation of the fan. Locate switch in thermostat subbase.

( ) 3. Fan switch shall be provided by a pull-chain on the bottom of the heater.

4. Provide internal relay for day-night operation, as controlled by a remote timer.

2.4 ACCEPTABLE MANUFACTURERS

a. Electric unit heater shall be the make and model number shown on the drawings or equivalent products by Trane (UHEC), Federal Pacific (USA), Brasch (BTUL).

PART THREE - EXECUTION

3.1 INSTALLATION

a. Install electric unit heater on support brackets furnished with the unit. Secure to structural support as required.

b. Locate the heater so that clearance from combustible materials, as recommended by the manufacturer, is maintained.

c. Make electrical connections for power and controls as required by the code.

d. Check the heater for proper operation, including safety controls.

END OF SECTION 15603

SECTION 15605

****

GAS-FIRED DUCT FURNACE
STAGED GAS VALVE

PART ONE - GENERAL

1.1 The GENERAL and SPECIAL CONDITIONS, Section 15100, are included as a part of this Section as though written in full in this document.

1.2 Scope of the Work shall include the furnishing and complete installation of the equipment covered by this Section, with all auxiliaries, ready for owner's use.

PART TWO - PRODUCTS

2.1 DUCT FURNACE

a. Duct furnace shall be natural gas-fired with output capacity as shown on the drawings, with AGA and UL listings.

b. Heat exchanger shall be type 321 stainless steel, designed to minimize stress and noise from expansion and contraction. Heat exchanger shall carry a 10-year nonprorated warranty against failure.

c. Gas burners shall be designed for use on natural or LP gas (with adapter kit). Burners shall be aluminized steel with stainless steel inserts, or all stainless steel, to prevent rusting. Burners shall have fixed-type air adjustment and crossover device and shall be supported on welded brackets.

d. Main gas valve shall provide for control of pilot and main valve and shall shut the gas valve in case of pilot failure. Electric pilot ignition shall be provided to light the pilot when the thermostat calls for heat or if the pilot should be extinguished. The pilot shall be extinguished when the space thermostat is satisfied. Standing pilot will not be acceptable. Provide a high-limit control in the furnace to close the gas valve, should the discharge temperature be excessive.

e. Main gas valve shall be two-position to allow operation at 100 and approximately 50% of full capacity.

2.2 CONTROLS

a. Controls shall consist of thermostat mounted below the unit, supported on conduit as shown, with low-voltage contacts to cycle the gas valve. Supply air for the duct furnace shall be furnished by other parts of the system. The main gas valve shall be interlocked with the system blower to ensure air movement prior to opening of the gas valve.

2.3 ACCESSORIES REQUIRED

a. Provide UL-listed type B vent and vent cap for the heater.

b. Provide a vent exhauster to remove the products of combustion. Exhauster is to interlock with burner to prevent gas valve opening unless the venter and the blower are in operation. Power venter must be provided if the air entering the heater is 20 deg. F (-8 deg. C) at any time.

c. Provide internal (external) air by-pass as required to limit air temperature rise to that recommended by the manufacturer and to limit the static pressure loss through the furnace to the maximum allowable, as shown on the drawings.

d. Furnaces located downstream of the cooling coil(s) shall be provided with drain pans and drain connections to dispose of the condensate. The furnaces mounted downstream shall be AGA certified for such mounting.

2.4 ACCEPTABLE MANUFACTURERS

a. Duct furnace shall be the product of a manufacturer having a full line of such equipment, equal to model number shown on the drawings or equivalents by Trane (HBA), Reznor (EEDU), Hastings (Model SD).

PART THREE - EXECUTION

3.1 INSTALLATION

a. Install proper hangers from structural members to support the heater as recommended by the manufacturer. Provide clearance from combustible materials, as required by the codes and the manufacturer's installation instructions.

b. Make connection as required by the applicable codes as follows:

1. Connect gas as required. Provide a gas cock in the gas line, remote from the heater.
2. Connect vent and provide vent cap.
3. Make electrical connections for power and controls.
4. Provide any duct connections shown on the drawings.

c. Provide access for inspection of the heater on entering and leaving sides of the ductwork.

3.3 Place the heater in operation and check for proper burner and control operation. Measure and report maximum temperature rise. See Balancing and Testing, Section 15111.

END OF SECTION 15605

SECTION 15607

****

GAS-FIRED DUCT FURNACE
SEPARATED COMBUSTION

PART ONE - GENERAL

1.1 The GENERAL and SPECIAL CONDITIONS, Section 15100, are included as a part of this Section as though written in full in this document.

1.2 Scope of the Work shall include the furnishing and complete installation of the equipment covered by this Section, with all auxiliaries, ready for owner's use.

PART TWO - PRODUCTS

2.1 DUCT FURNACE

a. Duct furnace shall be natural gas-fired with combustion air brought in from the outside to separate the furnace from the space. Furnace output capacity shall be as shown on the drawings. Furnace shall be AGA- and UL-listed.

b. Heat exchanger shall be type 321 stainless steel, designed to prevent noise and stress from expansion and contraction. Heat exchanger shall be warranted for 10 years nonprorated. Burners shall be stainless steel. Access to burners and controls shall be provided.

c. Controls shall include intermittent spark ignition, with 100% shutoff between heating cycles and if pilot fails to light. Provide lockout with manual reset from thermostat.

d. Gas controls shall provide electronic modulation with ductstat and remote temperature selection to modulate the output from 50 to 100% of capacity. Remote set point shall be provided and coordinated with controls.

2.2 VENT

a. Power venter shall be provided to draw in outdoor air and to expel the products of combustion to the outside, even under negative building conditions. Provide proving device to ensure venter operation prior to opening of the gas valve. Control thermostat shall cause the fan to start, and the venter shall then open the gas valve after the draft is established.

b. Vent terminal assembly shall be provided by the manufacturer for the vent location shown on the drawings. Vent shall provide for preheating of the outside air used for combustion and shall require only one hole in wall or roof. Furnish vent pipe and combustion air pipe as required and as recommended by the manufacturer.

2.3 AUXILIARY EQUIPMENT

a. Auxiliary equipment shall be furnished as follows:

1. Condensate drain pan and drain flange.
2. Fan control to delay operation of the supply fan until the heater provides warm air and to delay shutoff until all heat is dissipated.

PART THREE - EXECUTION

3.1 INSTALLATION

a. Provide adequate supports for the furnace from structural members, as recommended by the manufacturer.
b. Install furnace to allow proper clearance to combustible materials and to allow servicing of the unit as recommended by the manufacturer.
c. Provide duct connections as shown on the drawings, with duct access doors to allow access to the entering and leaving sides of the heat exchanger for inspection.
d. Connect the gas and power as required. Install gas cock remote from the furnace, in addition to the cock at the unit.
e. Connect the vent terminal and the outside and vent pipes, as required by the codes and recommended by the manufacturer.
f. Place the furnace in operation and check all controls for proper function and all safety controls.
g. Measure the air temperature rise and report as required under Section 15111, Balancing and Testing.

END OF SECTION 15607

SECTION 15608

****

ROOFTOP HEATING UNIT

PART ONE - GENERAL

1.1 The GENERAL and SPECIAL CONDITIONS, Section 15100, are included as a part of this Section as though written in full in this document.

1.2 Scope of the Work shall include the furnishing and complete installation of the equipment covered by this Section, with all auxiliaries, ready for owner's use.

PART TWO - PRODUCTS

2.1 ROOFTOP HEATING UNIT

a. Rooftop heating unit (RHU) shall be natural (LP) gas-fired with output capacity as shown on the drawings, with AGA and UL listings.

b. Heat exchanger shall be aluminized steel, designed to minimize stress and noise from expansion and contraction.

c. Gas burners shall be designed for use on natural or LP gas (with adapter kit). Burners shall be aluminized steel with stainless steel inserts, or all stainless steel, to prevent rusting. Burners shall have fixed-type air adjustment and crossover device and shall be supported on welded brackets.

d. Main gas valve shall provide for control of pilot and main valve and shall shut the gas valve in case of pilot failure. Electric pilot ignition shall be provided to light the pilot when the thermostat calls for heat or if the pilot should be extinguished. The pilot shall be extinguished when the space thermostat is satisfied. Standing pilot will not be acceptable. Provide a high-limit control in the furnace to close the gas valve should the discharge temperature be excessive.

e. Main gas valve shall be two-stage to allow operation at 100% and approximately 50% of full capacity.

f. Controls shall consist of wall-mounted thermostat with low-voltage contacts to cycle the gas valve. The main gas valve shall be interlocked with the blower to ensure air movement prior to opening of the gas valve.

g. Cabinet shall be heavy-gauge steel with protective coating, such as galvanizing, and finished with a baked-enamel finish suitable for outdoor mounting. Cabinet shall be weatherproof and watertight, designed for curb mounting, and having bottom return and supply openings within the curb area.

h. Blower section shall consist of centrifugal blower, belt-driven by motor with adjustable-pitch drive pulley. Blower shall deliver the air quantity shown on the drawings at the static pressure shown.

2.2 ACCESSORIES REQUIRED

a. Provide roof mounting curb as required to match the unit and for mounting the heater on the roof.

b. Provide a vent exhauster to remove the products of combustion. Exhauster is to interlock with burner to prevent gas valve opening unless the venter and the blower are in operation. Power venter must be provided if the air entering the heater is 20 deg. F (-8 deg. C) at any time.

c. Provide internal (external) air bypass as required to limit air temperature rise to that recommended by the manufacturer and to limit the static pressure loss through the furnace to the maximum allowable as shown on the drawings.

2.3 ACCEPTABLE MANUFACTURERS

a. Duct furnace shall be the product of a manufacturer having a full line of such equipment, equal to model number shown on the drawings or equivalents by Reznor (RPAK), Hastings, or Lennox.

PART THREE - EXECUTION

3.1 INSTALLATION

a. Install roof curb as required by the Roofing Manufacturers Association and as recommended by the manufacturer. Provide clearance from combustible materials, as required by the codes and the manufacturer's installation instructions.

b. Make connection as required by the applicable codes as follows:

1. Connect gas as required. Provide a gas cock in the gas line remote from the heater.
2. Make electrical connections for power and controls.
3. Provide any duct connections shown on the drawings.

c. Place the heater in operation and check for proper burner and control operation. Measure and report maximum temperature rise. See Balancing and Testing, Section 15111.

END OF SECTION 15608

SECTION 15610

****

GAS-FIRED WARM AIR FURNACE

PART ONE - GENERAL

1.1 The GENERAL and SPECIAL CONDITIONS, Section 15100, are included as a part of this Section as though written in full in this document.

1.2 Scope of the Work shall include the furnishing and complete installation of the equipment covered by this Section, with all auxiliaries, ready for owner's use.

PART TWO - PRODUCTS

2.1 GAS FURNACE

a. Gas-fired furnace shall be upflow, downflow, or horizontal type, with output capacity as shown on the drawings. Furnace shall be AGA and UL-labeled.

b. Furnace cabinet shall be heavy-gauge furniture steel with rust-resistant coating, with final coating of factory-baked enamel.

c. Heat exchanger shall be heavy-gauge cold-rolled steel or aluminized steel, designed to provide for expansion without noise or fatigue failure. Efficiency of the furnace shall be no less that 75%. The heat exchanger shall carry a 10-year nonprorated warranty against burnout.

d. Burners shall be rust-resistant steel, designed for quiet ignition and even lighting from burner to burner.

e. Blower shall be centrifugal type for quiet operation, having capacity to move the air quantity shown on the drawings against the static pressure shown. Drive shall be adjustable-speed motor or belt drive, with variable-pitch drive pulley on the motor.

f. The furnace shall include the following standard auxiliaries:

1. 24-volt control transformer.
2. Fan control and high-limit control.
3. Cooling relay for control as evaporator fan.
4. Electric gas valve.
5. Gas pressure regulator.
6. Provide for 100% shutoff in case of pilot failure.
7. Thermostat to match the unit with Heat-Auto-Cool and Fan On-Auto-Off switch. See section on controls.

2.2 AUXILIARY EQUIPMENT

a. Provide controls for ignition of the pilot only when the furnace is calling for heat. Provide solid-state spark igniter for the pilot.

b. Provide cooling coil to match and line up with the furnace and with capacity to match the condensing unit. Cooling capacity shall be no less than shown on the drawings. Provide expansion valve and precharged line sets to match the cooling system capacity.

2.3 ACCEPTABLE MANUFACTURERS

a. Furnace shall be the make and model shown on the drawings or equivalent units by York (P2UG), Carrier, Trane, Lennox.

PART THREE - EXECUTION

3.1 INSTALLATION

a. Install the furnace as shown on the drawings and as recommended by the manufacturer for the type of installation required. Installation shall be plumb and true, with clearance from combustible materials as recommended by the manufacturer and UL.

b. Connect power, gas, and vent as required by the local codes, and secure their acceptance of the installation.

c. Operate the furnace and test the safety controls, fan relay, overtemperature cutout, etc., for proper setting and operation.

d. Install filters in the unit to filter all air entering the furnace (both sides on occasions), as required under the section on filters.

e. Install the cooling coil and make sure that all duct connections are airtight. Adhesive-backed tape shall not be used, either inside or outside of the ductwork, to cover or seal any openings or leaks. Install the expansion valve and make the refrigerant connections as required and as recommended by the manufacturer. Test the refrigerant system for leaks and place in operation.

f. Connect all controls and verify that they are functioning properly on heating and cooling.

g. Adjust the furnace blower to match the air quantity called for on the drawings. Record the actual system data as required by Section 15111, Balancing and Testing, and report to the engineer as required under that Section.

END OF SECTION 15610

SECTION 15615

****

ELECTRIC RADIANT HEATER
CEILING-SUSPENDED TYPE

PART ONE - GENERAL

1.1 The GENERAL and SPECIAL CONDITIONS, Section 15100, are included as a part of this Section as though written in full in this document.

1.2 Scope of the Work shall include the furnishing and complete installation of the equipment covered by this Section, with all auxiliaries, ready for owner's use.

PART TWO - PRODUCTS

2.1 ELECTRIC HEATER

a. Electric radiant heater shall be supported near the ceiling and designed to emit radiant heat generated by electric heating elements and to direct this heat as desired. Capacity shall be as shown on the drawings. Heater shall be UL-listed for this service.

b. Heating elements shall consist of helically coiled nickle-chromium alloy resistance wire, completely embedded in, and surrounded by, magnesium oxide and enclosed in a corrosion-resistant steel sheath. Elements shall terminate in a gasketed weather-sealed junction box. Element operating temperature shall be no less than 1450 deg. F (788 deg. C).

c. Enclosure shall be constructed of no less than 16-gauge (1.5-mm) highly polished aluminum housing and reflector.

d. Provide heavy-gauge wire screen to protect the elements and the reflectors.

2.2 Accessory equipment shall include:

a. Input controller/percentage timer to vary input from 0 to 100% based on an on-off cycle.

b. Control transformer, pilot light, and pilot duty control circuit, with disconnect switch for larger units.

2.3 ACCEPTABLE MANUFACTURERS

a. Acceptable manufacturers shall include the model number shown on the drawings or equivalents by Markel (Series 3500), Federal Pacific (IR), Chromalox (RDO).

PART THREE - EXECUTION

3.1 INSTALLATION

a. Provide adequate supports for heaters from structural members above.

b. Locate heater to provide clearance from combustible materials above, to the sides, and below the heaters, as recommended by the manufacturer.

c. Provide proper electrical connections and controls as shown on the drawings and as required by code.

d. Place the heater in operation and check for proper functioning.

END OF SECTION 15615

SECTION 15616

****

GAS-FIRED INFRARED HEATER
CEILING-MOUNTED

PART ONE - GENERAL

1.1 The GENERAL and SPECIAL CONDITIONS, Section 15100, are included as a part of this Section as though written in full in this document.

1.2 Scope of the Work shall include the furnishing and complete installation of the equipment covered by this Section, with all auxiliaries, ready for owner's use.

PART TWO - PRODUCTS

2.1 INFRARED HEATERS

a. Infrared heaters shall consist of a radiating surface heated by combustion of gas to produce high-temperature radiant energy and to direct that energy to the desired area. Capacity shall be as shown on the drawings. Heater shall be AGA- and UL-labeled.

b. Radiating surface shall be a porous ceramic material, operating at a temperature of no less than 1800 deg. F (982 deg. C). Ceramic material shall be shock-resistant against breakage, even if sprayed with water when in operation. The ceramic element shall be warranted for 10 years. A secondary radiating surface of rods or screens shall be provided to protect the ceramic and to reradiate. The secondary surface shall be warranted for 10 years.

c. Burner shall provide for proper mixture of gas and air and shall be designed for natural or LP gas or propane-air mixture. Heaters shall be designed for operation without a vent connection to the outside.

d. Controls shall provide for direct ignition of the main burner through a solid-state direct spark ignition system. Loss of gas or power shall cause a 100% shutdown of the burner. Provide a space thermostat to control the on-off operation of the heater(s) as shown on the drawings.

e. Housing shall be heavy-gauge aluminum to support the burner and radiant surfaces and to direct the heat where desired.

2.2 ACCEPTABLE MANUFACTURERS

a. Acceptable manufacturers shall include the make and model number shown on the drawings or equivalents by Detroit Re-Verber-Ray (Model DR), Ray-Tec (Model RT), Barber Solarflo (Model DH).

PART THREE - EXECUTION

3.1 INSTALLATION

a. Provide adequate support for the heaters from the structural members above and suspend the heater as recommended by the manufacturer.

b. Locate the heater to provide necessary clearance from combustible materials above, below, and to the sides of the heater, as recommended by the manufacturer. Install the heater at the angle recommended by the manufacturer.

b. Connect gas and electrical power as required. All connections shall be made with flexible materials to allow for movement of the heater due to wind and other forces.
c. Provide combustion air and ventilation air, as recommended by the manufacturer, to dilute the products of combustion.
d. Place the heaters in operation and check for proper ignition and performance.

END OF SECTION 15616

SECTION 15626
****
ELECTRIC WALL HEATER

PART ONE - GENERAL

1.1 The GENERAL and SPECIAL CONDITIONS, Section 15100, are included as a part of this Section as though written in full in this document.

1.2 Scope of the Work shall include the furnishing and complete installation of the equipment covered by this Section, with all auxiliaries, ready for owner's use.

PART TWO - PRODUCTS

2.1 ELECTRIC WALL HEATERS

a. Electric wall heaters shall be commercial-grade, with an encased heating element, a quiet forced-air fan, controls, and an attractive housing. Capacity shall be as shown on the drawings. Units shall be UL-labeled.

b. Heating elements shall be enclosed type, cast in aluminum, or with sheathed-type elements. Voltage shall be as shown on the drawings and verified at the site.

c. Fan shall be quiet type to force air over the element and discharge the air downward, out the front of the heater.

d. Cabinet shall be designed for location in public areas and shall enclose all parts of the heater in a tamper-proof cover, be attractive in appearance, and be complete with wall frame, trim, etc., as required for a complete installation. Heater shall be semirecessed or surface-mounted, as shown on the drawings.

e. Unit shall include entrance terminals, circuit breaker breaking all legs, control transformer for larger units, and contactors, as required by code. Controls shall include a heavy-duty, snap-action- unit, mounted thermostat, located behind the cover and requiring some tool to reach for adjustment. Unit shall include a high-limit cutout for the element and a fan switch to delay fan operation until the element is warm and to prolong fan operation until the element is cooled.

2.2 ACCEPTABLE MANUFACTURERS

a. Acceptable manufacturers shall include the make and model number shown on the drawings or equivalent heaters by Nutone Markel (Model 3420/3450), Chromalox (Model AWH-4000), Electromode.

PART THREE - EXECUTION

3.1 INSTALLATION

a. Heater shall be installed in a heavy-gauge wall box for recessing in the wall or surface mounting on the wall. Provide trim for the installation required.

b. Install the heater in the box and make electrical connections to the terminals.

c. Place the heater in operation and check for proper operation. Set the thermostat for the temperature necessary for the location.

END OF SECTION 15626

SECTION 15627
****
ELECTRIC WALL HEATER
RADIANT TYPE

PART ONE - GENERAL

1.1 The GENERAL and SPECIAL CONDITIONS, Section 15100, are included as a part of this Section as though written in full in this document.

1.2 Scope of the Work shall included the furnishing and complete installation of the equipment covered by this Section, with all auxiliaries, ready for owner's use.

PART TWO - PRODUCTS

2.1 ELECTRIC WALL HEATERS

a. Electric wall heater shall be radiant convection type, with exposed wire-type resistor wound around a ceramic chimney. Capacity shall be as shown on the drawings. Heater shal be UL-listed.

b. Heating elements(s) shall be nickle-chromium wire, coiled and wound around ceramic chimney(s) operating at a high temperature, to radiate to the space.

c. Controls shall consist of a heavy-duty line voltage thermostat mounted on the front of the heater.

d. Cabinet shall be heavy-gauge steel with rust-resistant finish. Heaters located in wet areas shall have chrome-plated front cover. Cabinet shall restrict any foreign object from reaching the elements.

2.2 ACCEPTABLE MANUFACTURERS

a. Acceptable manufacturers shall include the make and model number shown on the drawings or equivalents accepted by the engineer.

PART THREE - EXECUTION

3.1 INSTALLATION

a. Mount heater enclosure on the wall as recommended by the manufacturer. Enclosure may be surface-mounted or semirecessed, as shown on the drawings.

b. Place heater in service and check for proper operation.

END OF SECTION 15627

SECTION 15630

****

ELECTRIC CONVECTOR

PART ONE - GENERAL

1.1 The GENERAL and SPECIAL CONDITIONS, Section 15100, are included as a part of this Section as though written in full in this document.

1.2 Scope of the Work shall include the furnishing and complete installation of the equipment covered by this Section, with all auxiliaries, ready for owner's use.

PART TWO - PRODUCTS

2.1 ELECTRIC CONVECTORE

a. Electric convector shall have a heavy-gauge furniture steel enclosure with an enclosed finned element to promote convective air-flow through the unit. Capacity shall be as shown on the drawings. Convector shall be UL-labeled.

b. Heating element shall be aluminum, finned, tubular type, rated for voltage shown on the drawings and confirmed at the site.

c. Enclosure shall be 16-gauge (1.5-mm) furniture steel, with stamped louvers and welded construction

d. Controls shall consist of full-length, thermal linear cutout with automatic reset and an integral thermostat with a removable knob (where unit must be tamper-proof).

PART THREE - EXECUTION

3.1 INSTALLATION

a. Convector shall be set on the floor and secured to the wall behind, for either semirecessed or surface mounting, as shown on the drawings.

b. Place unit in service and check for proper operation.

END OF SECTION 15630

SECTION 15632

****

ELECTRIC CEILING HEATER

COMMERCIAL TYPE

PART ONE - GENERAL

1.1 The GENERAL and SPECIAL CONDITIONS, Section 15100, are included as a part of this Section as though written in full in this document.

1.2 Scope of the Work shall include the furnishing and complete installation of the equipment covered by this Section, with all auxiliaries, ready for owner's use.

PART TWO - PRODUCTS

2.1 ELECTRIC CEILING HEATER

a. Electric ceiling heater shall be a commercial-grade heater for mounting at the ceiling, with enclosed finned element, quiet-type fan, and motor with heavy cabinet. Capacity shall be as shown on the drawings. Heater shall be UL-listed.

b. Heating element shall be totally enclosed, corrosion-resistant finned type with wide fin spacing, having a density of 60 watts per inch maximum.

c. Fan and motor shall be slow-speed, the same voltage as the heater. Motor shall be permanently lubricated.

d. Controls shall consist of a thermal high-limit cutout that is reset by turning the heater off for 5 min. A wall-mounted line voltage thermostat shall be provided for small-capacity heaters. A low-voltage wall thermostat with contactor shall be provided for heaters of higher capacity.

e. Cabinet shall be surface-mounted below the ceiling, with all exposed surfaces protected with baked enamel, or recess-mounted in a T-bar ceiling system with a finished return assembly and concentric supply grille exposed flush in the ceiling.

2.2 ACCEPTABLE MANUFACTURERS

a. Acceptable manufacturers shall include the make and model number shown on the drawings or equivalent heaters by Nutone Markel (Series 3470/3480).

PART THREE - EXECUTION

3.1 INSTALLATION

a. Surface-mounted heaters shall be securely fastened to the structure above the location, and flush-mounted heaters shall be mounted in the ceiling support T-bar system, with additional hangers to support the weight.

b. Provide electrical connections for power and controls as required by code and recommended by the manufacturer.

c. Place the heater in operation and check controls and heater for proper operation.

END OF SECTION 15632

SECTION 15633

****

ELECTRIC CEILING HEATER
MEDIUM CAPACITY - FAN-FORCED

PART ONE - GENERAL

1.1 The GENERAL and SPECIAL CONDITIONS, Section 15100, are included as a part of this Section as though written in full in this document.

1.2 Scope of the Work shall include the furnishing and complete installation of the equipment covered by this Section, with all auxiliaries, ready for owner's use.

PART TWO - PRODUCTS

2.1 ELECTRIC CEILING HEATER

a. Electric ceiling heater shall have an exposed radiant element and a slow-speed fan to circulate the warm air into the space. Heater capacity shall be as shown on the drawings. Heater shall be UL-listed.

b. Heating element shall be a radiant nickel-chromium alloy unit, supported on ceramic insulators.

c. Controls shall consist of a wall-mounted line voltage thermostat.

d. Fan and motor shall be slow-speed, quiet type to draw air in and direct it downward into the space.

2.2 ACCEPTABLE MANUFACTURERS

a. Acceptable manufacturers shall include the make and model number shown on the drawings and equivalent units by Nutone Markel (Series 540).

PART THREE - EXECUTION

3.1 INSTALLATION

a. Heater shall be surface-mounted at the ceiling as directed by the manufacturer.

b. Make electrical connections for power and control as required by the code and recommended by the manufacturer.

c. Place the heater in operation and check performance.

END OF SECTION 15633

SECTION 15634

****

ELECTRIC RADIANT HEATER
SMALL CAPACITY - HIGH WALL MOUNTING

PART ONE - GENERAL

1.1 The GENERAL and SPECIAL CONDITIONS, Section 15100, are included as a part of this Section as though written in full in this document.

1.2 Scope of the Work shall include the furnishing and complete installation of the equipment covered by this Section, with all auxiliaries, ready for owner's use.

PART TWO - PRODUCTS

2.1 RADIANT HEATERS

a. Heater shall be small-capacity radiant type, designed for mounting high on a wall. Capacity shall be as shown on the drawings. Heater shall be UL-listed.

b. Heating element shall be an exposed radiant nickel-chromium alloy element, coiled and supported on ceramic insulators. Voltage shall be as shown on drawings and as verified at the site.

c. Controls shall consist of built-in thermostat mounted on the bottom of the heater, accessible from the floor.

d. Cabinet shall be rust-protected with attractive finish.

e. Acceptable manufacturers shall be the make and model number shown on the drawings or equivalent by Nutone Markel (Series 510).

PART THREE - EXECUTION

3.1 INSTALLATION

a. Heater shall be mounted as directed by the manufacturer and as required by the code.

b. Place heater in operation and check for proper function.

END OF SECTION 15634

SECTION 15635
****
ELECTRIC CEILING HEAT PANELS
LAY-IN CEILING MOUNTED

PART ONE - GENERAL

1.1 The GENERAL and SPECIAL CONDITIONS, Section 15100, are included as a part of this Section as though written in full in this document.

1.2 Scope of the Work shall include the furnishing and complete installation of the equipment covered by this Section, with all auxiliaries, ready for owner's use.

PART TWO - PRODUCTS

2.1 CEILING HEAT PANELS

a. Electric ceiling heat panels shall be designed to produce low-density radiant heat. Panels shall be mounted in a standard lay-in T-bar system and shall match and line up with the standard ceiling tile. Number of panels and total capacity shall be as shown on the drawings. Panels shall be UL-listed for this application.

b. Panels shall be composed of a conductive coating laminated between two layers of dielectric polyester film, sealed and bonded to rigid fiberglass insulation board. The surface temperature of the heater shall not exceed 221 deg F (105 deg. C) with 90% of the input wattage radiated to the space. The face of the panel shall be provided with a coating to match the remaining ceiling tile.

c. Control of the panels shall be by a wall-mounted line voltage thermostat.

d. Acceptable manufacturers shall be the make and model number shown on the drawings or equivalent by Energy-Koat.

PART THREE - EXECUTION

3.1 INSTALLATION

a. The ceiling support system of T-bars shall be prepared and the panels installed in the position shown on the drawings.

b. The panels shall be connected by flexible connectors to the power leads.

c. Wall-mounted thermostat shall be located in an area as far from the panels as possible.

d. The panels shall be placed in service and checked for proper performance.

END OF SECTION 15635

SECTION 15648
****
GAS SERVICE

PART ONE - GENERAL

1.1 The GENERAL and SPECIAL CONDITIONS, Section 15100, are included as a part of this Section as though written in full in this document.

1.2 Scope of the Work shall include the furnishing and complete installation of the equipment covered by this Section, with all auxiliaries, ready for owner's use.

PART TWO - PRODUCTS

2.1 PIPE

a. Pipe above grade to be steel with screwed malleable iron fittings for 2 in. in size and below. Welded fittings above 2 in. in size.

b. Pipe underground to be plastic jacketed to provide electrolysys protection, with all fittings wrapped with matching tape to same thickness as jacket. Take care that jacket is not damaged.

c. Gas pipe below floor slab shall be run in cast iron pipe or steel pipe encased in no less than 4 in. of concrete as required by local gas code.

d. Provide medium- to low-pressure regulators having high-pressure and low-pressure safety trips where medium pressure exists. Vent to outside. Rockwell Model 143-6 or equivalent.

e. Gas cocks to be lubricated type for 2-in. sizes and larger.

PART THREE - EXECUTION

3.1 INSTALLATION

a. Ditches for underground pipe shall be straight and uniform in depth and no less than 24 in. deep. Backfill shall be clean dirt and sand (no rocks or trash) and compacted to prevent settlement.

b. Ditches near footings or under paved areas shall be tamped with a mechanical tamper to 90% compaction.

c. Any coating damaged while placing pipe in ditch shall be repaired to original thickness. Pipe shall be tested prior to coating fittings.

d. Gas pipe shall not enter any building below grade. Provide gas cock before entering any building, to provide vent for gas service.

e. This contractor shall arrange for gas service and bear all costs of meter and regulator, tap fees, service line extension, etc. If any deposits are involved, they shall be included and shall be made in the owner's name.

END OF SECTION 15648

CHAPTER 10

Pipe, Fittings, and Accessories

HOT WATER PIPE, FITTINGS, AND ACCESSORIES — SECTION 15706 — 9 Pages — 10-32

Pipe, fittings, and accessories for water systems, including valves, check valves, pipe supports, vents, drains, strainers, gauges, vacuum breakers, expansion compensation, sleeves, and escutcheons.

SECTION 15700

****

PIPE, FITTINGS, AND ACCESSORIES

PART ONE - GENERAL

1.1 The GENERAL and SPECIAL CONDITIONS, Section 15100, are included as a part of this Section as though written in full in this document.

1.2 Scope of the Work shall include the furnishing and complete installation of the equipment covered by this Section, with all auxiliaries, ready for owner's use.

PART TWO - PRODUCTS

2.1 PIPE

a. Materials for general service pipe systems, low-pressure, nominal corrosion, low-temperature service.

1. Steel pipe - black or galvanized - ASTM A-53 or A-120, Sch 40
2. Copper pipe (tube) - ASTM B-88, type M or L, soft or hard drawn
3. Plastic pipe - PVC Sch 40

b. Materials for higher pressure and/or temperature or corrosive service.

1. Steel pipe - black or galvanized - ASTM A-53 or A-120, Sch 80
2. Copper pipe (tube) - ASTM B-88, type K, soft or hard drawn
3. Plastic pipe - PVC Sch 80
4. Fiberglass-reinforced plastic (FRP) - ASTM 2310, ASTM 2396

2.2 FITTINGS: ELLS, TEES, CROSSES, BUSHINGS

a. Materials for fittings shall match the pipe system category for pressure, temperature, and corrosion.

b. Fittings for low pressure, and/or low temperature shall be:

1. Steel - galvanized or black
   - Size under 2 in. (50 mm)
     - Screwed, malleable iron fittings, Class 150
   - 2 in. (50 mm) and larger
     - Screwed, malleable iron fittings, Class 150
     - Welded, steel butt-welding fittings, Sch 40
     - Grooved couplings, ductile iron clamps, Sch 40 fittings
2. Copper - wrought or cast copper, solder type
3. PVC - PVC Sch 40, solvent-welded type

c. Fittings for more severe service shall match the pipe system.

4. Steel - screwed, ductile iron fittings, Class 300
   welded, steel butt-welding fittings, Sch 80
5. Copper - cast copper
6. PVC - PVC Sch 80, solvent welded type
7. FRP - FRP heat-cured bonding agent

2.3 UNIONS

a. Materials for unions shall match the pipe system category for pressure, temperature, and corrosion.

b. Unions for low pressure and/or low temperature shall be:

1. Steel - screwed, Class 150, malleable iron, O-ring or brass seat
   welded neck flanges with gaskets
2. Copper - cast copper to copper, metal seat
3. PVC - solvent welded, flange to flange, with gasket

c. Unions for more severe service shall match the pipe system for pressure, temperature, and corrosion.

1. Steel - screwed, Class 300, ductile iron, brass seat
   welded neck flanges with gaskets
2. Copper - cast brass flanges with gasket
3. FRP - FRP flanges with gasket

d. Dielectric unions shall separate all ferrous and nonferrous metals in every piping system. Unions shall match those above, except that metal-to-metal contact is to be avoided. Where flanges are used, the bolts shall be insulated from the body of the flange.

2.4 VALVES - GENERAL

a. Materials for small valves shall be all brass and for larger valves shall be cast iron, cast ductile iron, or cast steel, as required for the pressure and service of the valve and as listed below.

2.5 LOW-PRESSURE VALVES

a. Low-pressure valves for gravity and city water, low-pressure steam, and compressed air shall be as follows:

VALVES 2 in. (50 mm) AND SMALLER

1. Globe valves for throttling service shall be Class 150, straight through or angle type, Teflon disk, union bonnet, Stockham B-22 or B-232, or equivalent, threaded or solder ends.
2. Ball valves for throttling service shall be one-piece body, reduced port, Stockham S-127BR1, or equivalent.
3. Gate valves for shutoff service shall be Class 125, rising stem, screwed bonnet, split or solid wedge, Stockham B-114 or B-110, or equivalent, for screwed ends or equivalent solder ends.
4. Ball valves for shutoff service shall be two-piece body, full port, Stockham S-207BR1, or equivalent.
5. Fast-opening valves for drains, blow-off, etc., shall be Class 125, screwed bonnet, double-disk type with lever handle operator, Stockham B-116, or equivalent.

VALVES 2 in. (50 mm) THROUGH 8 in. (200 mm)

6. Gate valves for shutoff service shall be Class 125, iron body with brass trim, rising stem, bolted bonnet, solid wedge, OS&Y, Stockham G-625, or equivalent.
7. Butterfly valves for shutoff service or throttling, up to 150-psi (1035-kPa) working pressure shall be Stockham LD-711, or equivalent.

2.6 MEDIUM-PRESSURE VALVES

a. Valves for medium-pressure service for chilled and hot water, condenser water, and other such services shall be as follows:

VALVES 2 in. (50 mm) AND SMALLER

1. Globe valve for throttling service shall be Class 150, straight through or angle type, Teflon disk, union bonnet, Stockham B-37 or B-237, or equivalent, threaded or solder ends.
2. Gate valves for shutoff service shall be Class 150, rising stem, screwed bonnet, split or solid wedge, Stockham B-120 or B-124, or equivalent, for screwed ends, or equivalent solder ends.

VALVES LARGER THAN 2 in. (50 mm) THROUGH 10 in. (250 mm)

3. Gate valves for shutoff service shall be Class 150, ductile iron body with brass trim, rising stem, bolted bonnet, solid wedge, OS&Y, Stockham G-623, or equivalent.

2.7 VALVES FOR MEDIUM- AND/OR HIGH-PRESSURE STEAM

a. Valves for medium- and high-pressure steam up to 450 deg. F (232 deg. C) and pressure up to 250 psi (1750 kPa) shall be Class 250, iron body, brass trim, rising stem, bolted bonnet, OS&Y solid-wedge disk, Stockham F-666, or equivalent.

b. Valves for medium- and high-pressure steam up to 450 deg. F (232 deg. C) and pressure up to 250 psi (1750 kPa), located 7 ft (2.1 m) above floor or other inaccessible elevated locations, shall be provided with chain wheel with hammer blow operator. Valves shall be Class 150, ductile iron body, brass trim, rising stem, bolted bonnet, OS&Y solid-wedge disk, Stockham F-623, or equivalent.

2.8 CHECK VALVES

a. Materials for small valves shall be all brass and for larger valves shall be cast iron, cast ductile iron, or cast steel, as required for the pressure and service of the valve and as listed below.

b. Check valves for gravity systems, city water, low-pressure steam, and compresed air shall be as follows:

VALVES 2 in. (50 mm) AND SMALLER

1. For service less than 150 deg. F (65 deg. C): Class 125, bronze body with Buna-N disk seat, Stockham B-310B or B-320B, or equivalent.
2. For service up to 450 deg. F (232 deg. C): Class 125, bronze body with TFE disk, Stockham B-310T or B-320T, or equivalent.
3. For pressures up to 200 psi (1380 kPa): Class 150, bronze body with Teflon disk, Stockham B-322T, or equivalent.

VALVES LARGER THAN 2 in. (50 mm) THROUGH 12 in. (300 mm)

4. For service at or below 150 deg. F (65 deg. C): iron body, brass trim with Buna-N disk, Stockham G-932B.
5. For service up to 450 deg. F (232 deg. C): iron body, brass trim, bronze disk, Stockham D-931.
6. Check valves for service on air or steam with low pressure differential shall be provided with lever and spring for positive control in closing.

2.9 PIPE JOINT MATERIALS

a. Screwed pipe joints shall be made up using Teflon tape, Rectorseal No. 5 pipe dope, or other lubricants, as approved for the particular installation.

b. Soldered joints shall be made up with paste flux and 50/50 solder, 95/5 solder, or silver solder, as required under Part Three, Execution.

2.10 PIPE SUPPORT SYSTEM

a. Provide an adequate pipe suspension system in accordance with recognized engineering practices, using, where possible, standard, commercially accepted pipe hangers and accessories.

b. All pipe hangers and supports shall conform to the latest requirements of the ASA Code for Pressure Piping, B 31.1, and Manufacturers' Standardization Society documents MSS SP-58 and MSS SP-69.

c. The pipe hanger assembly must be capable of supporting the line in all operating conditions. Accurate weight balance calculations shall be made to determine the supporting force at each hanger in order to prevent excessive stress in either pipe or connected equipment.

d. Where references below refer to "type," these references shall be to Federal Specification WW-171. Where reference is to "Figure," it shall be to Fee & Mason designations used in their catalog. Equivalent products by Grinnell and others are acceptable.

1. Concrete Inserts
   Where piping is supported from a concrete structure, inserts shall be type 18, 19 shall be used, or structural shapes where provided where a continuous insert is required. Where support rod size exceeds 7/8 in. (22 mm) diameter or where the pipe load exceeds the recommended load for the insert, use two inserts with a trapeze-type connecting member below the concrete.
2. Beam Clamps
   Where piping is to be supported from structural steel, beam clamps, type 21, 28, 29, 30, or 31, shall be used. Beam clamp selection shall be on the basis of the required load to be supported. Where welded beam attachments are required, they shall be Figure 90, 131, 251, or 266. Holes drilled in structural steel for hanger support rods will not be permitted.

3. Riser Clamps
All vertical runs of piping shall be supported at each floor, and/or at specified intervals, by means of type 8 clamp for steel pipe, or Figure 368 clamp for copper tubing. For riser loadings in excess of the maximum recommended loads shown for the above items, clamps shall be designed in accordance with Figure 395 or 396.

4. Hanger Rods
Hanger rods shall be ASTM A-107 continuous-threaded rod. Eye rods shall be Figures 228 and 228 WL. Where hanger Rod sizes are catalog-listed for a specified hanger, these sizes shall govern. Where hanger rod sizes are not listed, the load on the hanger shall be the determining factor, and the maximum recommended hanger rod load as shown below shall govern.

5. Hanger Rod Loading
Maximum hanger rod load shall not exceed:

Rod diameter	Inches	3/8	1/2	5/8	3/4	7/8
	(mm)	9.5	12.7	15.8	19	22.2
Max load	Pounds	610	1130	1810	2710	3770
	(kg)	274	508	814	1219	1697

6. Hanger Spacing
The maximum allowable spacing for pipe hangers shall be in accordance with tabulation below. Where concentrated loads of valves, fittings, etc., occur, closer spacing will be necessary and shall be based on the weight to be supported and the maximum recommended loads for the hanger components.

STEEL PIPE

Nominal Pipe Size	Maximum Space Between Hangers
Up to 1 1/4 in. (30 mm)	7 feet (2 m)
1 1/2 in. (40 mm)	9 feet (3 m)
2 in. & 3 in. (50 & 75 mm)	10 feet (3 m)
4 in. & 5 in. (100 & 130 mm)	14 feet (4 m)
6 in. (150 mm)	17 feet (5 m)
8 in. (200 mm)	19 feet (6 m)
10 in. & 12 in. (250 & 300 mm)	22 feet (7 m)
14 in. (355 mm)	25 feet (7.5 m)
16 in. (410 mm)	27 feet (8 m)

COPPER PIPE

Nominal Tube Size	Maximum Space Between Hangers
Up to 1 in. (25 mm)	5 feet (1.5 m)
1 1/4 to 2 in. (30 to 50 mm)	7 feet (2 m)
2 1/2 in. (125 mm)	9 feet (3 m)
3 in. (150 mm)	10 feet (3 m)
3 1/2 & 4 in. (165 & 200 mm)	12 feet (3.6 m)

FRP and PVC pipe supports - Consult manufacturer's data for conditions and temperatures involved.

7. Hangers

(a) All hangers for piping 2 in. (50 mm) or larger shall be provided with means of vertical adjustment.

(b) On uninsulated steel pipe, hangers shall be type 1, 4, 6, or 11. On piping 2 in. (50 mm) and smaller, Figure 9, 10, or 25 will be permitted.

(c) On uninsulated copper tubing, hangers shall be Figure 307, 364, or 365, or type 10 or 11.

(d) On hot insulated steel pipe, hangers shall be Figure 261 or welded attachments Figure 90, 92, 94, or 96. Where thermal movement causes the hanger rod to deviate more than 5 deg. (0.09 r) from the vertical or where longitudinal expansion causes a movement of more than 1/2 in. (13 mm) in the piping supported from below, roller hangers, type 42, 44, 45, 47, or 48, shall be used in conjunction with a protection saddle, type 40, to suit the insulation thickness. On insulated steel pipe for chilled water or similar service, the hanger must be placed on the outside of the insulation with a type 41 shield.

(e) On insulated copper tubing, hangers shall be Figure 199, 201, 202, or 215 and shall be placed on the outside of the insulation with a type 41 shield. The type 41 shield shall be applied to distribute the hanger load over the insulation and to eliminate damage to the vapor barrier on the covering.

(f) Base supports shall be type 39.

8. Brackets and Racks

Where piping is run adjacent to walls or steel columns, welded steel brackets, types 32, 33, and 34, shall be used as base supports. Multiple pipe racks or trapeze hangers shall be fabricated from channel and accessories designed for this purpose.

9. Spring Hangers

Spring hangers shall be installed at hanger points where vertical thermal movement occurs. For light loads and noncritical movements in excess of 1/4 in. (6.35 mm), type 49, 50, or 51, variable spring supports shall be used.

10. Critical Systems

On critical systems, where movement is in excess of 1/2 in. (13 mm) constant supports, type 52, shall be used. For vibration and/or shock loadings, use Figure 470, 471, or 472, sway braces. Where it is necessary to reduce pipe vibration and sound transmission to building steel, Figure 403 or 404 vibration control hangers shall be used.

11. Anchors, Guides, and Sliding Supports

Anchors shall be installed as shown on the piping drawings. They may be Figure 140, 141, or 159. Guides shall be Figure 120, 121, 122, or 165. Sliding supports shall be Figure 143 or 145.

12. Auxiliary Steel

All auxiliary steel necessary for the installation of the pipe hangers and supports shall be designed in accordance with the AISC Steel Handbook, shall be furnished by the contractor, and shall receive one shop coat of primer paint.

13. Submittals

The contractor shall submit to the engineer, prior to installation, the following information and data for review and acceptance.

(a) Manufacturer's data sheets on all cataloged items to be used.

(b) Sketches covering all specially designed hanger assemblies and fabrications.

(c) Sketches showing locations, loads, calculated travel, type, and sizes of all spring hanger assemblies.

2.11 VENTS AND DRAINS

a. Air vents shall be of manual type where readily accessible; Dole No. 9 or acceptable equivalent.

b. Air vents above concealed ceiling systems shall be extension type; Dole No. 14-1 or acceptable equivalents.

c. Air vents in accessible areas, but not convenient for service, shall be automatic vent type with the vent pipe carried to the outside or to the nearest floor drain or acceptable receptacle; Dole No. 200 or acceptable equivalent.

d. Drain points, consisting of hose bibb at the low point of pipe systems less than 6 in. (150 mm) and 1-in. (25-mm) ball valve with hose-threaded coupling for larger systems, shall be installed for every low point of piping system for drainage.

2.12 STRAINERS

a. Strainers shall be installed as shown in details on drawings and before every control valve, metering valve, or orifice such as steam traps.

b. Strainer body shall be cast brass or cast iron to match the system pressure and temperature. The body shall provide for removal of the strainer element without interruption of the pipe. Each strainer shall have a blowdown valve installed.

c. Strainer element shall be 0.045-in. (1.15-mm) perforated stainless steel with effective screen area of no less than four times the pipe area.

2.13 THERMOMETERS

a. Thermometers shall be installed as shown in details on the drawings and on the inlet and outlet of every item of equipment where the fluid is either heated or cooled.

b. Thermometers shall be dial type, with glass face no less than 4 in. (100 mm) in diameter and with adjustable head for visibility. The body and stem shall be stainless steel.

c. The range of the thermometer shall be such that the normal operating point with the system in service shall be midrange on the dial.

d. Accuracy of the thermometer shall be such that the error will be no greater than 2% of the full scale value.

e. The thermometer shall be provided with a brass or stainless steel well, installed in a threaded coupling into the pipe being measured. Provide thermally conductive gel in the well for contact with the bulb.

2.14 PRESSURE GAUGES

a. Provide pressure gauges where shown in details on the drawings and at every item of equipment receiving or producing flow in or from the system.

b. Gauges shall be no less than 4 in. (100 mm) in diameter, with glass face and cast aluminum body.

c. Range shall be such that gauge shall operate in midrange during normal operation of the system. Gauges subject to vacuum conditions, such as pump suctions, shall be rated for vacuum service. Graduations between figure intervals shall not exceed 10% of the gauge range. Pointer shall be adjustable for calibration.

d. Accuracy shall be such that error is less than 1% over the midrange of the gauge and 2% out of the midrange area.
e. Every gauge shall be provided with a 1/4-in. (6.35-mm) needle-type valve and an impulse dampener. Gauges installed on steam lines shall be provided with a coil syphon.

2.15 VACUUM BREAKERS
a. Every steam-heated piece of equipment such as coils, converters, dryers, etc., shall be provided with a vacuum breaker designed to relieve the vacuum produced as steam is condensed.
b. Vacuum breaker shall be a spring-loaded, ball-type check valve matched to the service where used.
c. Water systems designed to automatically drain shall be provided with a vacuum breaker.

2.16 EXPANSION TANKS
a. Every closed system shall be provided with an expansion tank to prevent significant changes in system pressure caused by expansion and contraction of fluids due to temperature changes.
b. See Specifications for expansion tanks.

2.17 WATER MAKE-UP SYSTEMS
a. Every closed system shall be provided with a regulated water make-up system to automatically feed water to the system at a preset pressure.
b. Every water system not used for domestic water shall be provided with a reduced-pressure backflow preventer, certified for that service and acceptable to the local plumbing code administrator, to prevent backflow of contaminated water into the domestic water system.
c. See Specifications on water make-up system.

2.18 EXPANSION COMPENSATION
a. Pipe installation shall allow for expansion due to temperature differences. Provide expansion loops, or joints, as shown on the drawings, as follows.
b. Expansion joints shall be designed for the service, temperature, and pressure of the system and sized to accommodate the amount of movement required from maximum to minimum temperatures encountered by the system.
c. Joints for water systems shall be reinforced rubber corrugations secured by steel flanges. The number of corrugations shall be determined by the amount of movement required.
d. Joints for steam or higher-temperature fluids shall be composed of multiple layers of thin stainless steel, corrugated and mounted between flanges. The number of corrugations shall be selected for the movement involved.
e. Every expansion joint shall be provided with stay bolts to prevent expansion beyond the normal design limits. Rubber joints shall have stay bolts isolated in rubber where used to eliminate vibration.

2.19 PIPE SLEEVES
a. Provide pipe sleeve of galvanized steel at each wall or floor penetration. Sleeve shall be no lighter than 18 gauge and shall be built into the wall or floor during construction of the wall. Where pipes are insulated, the sleeve shall allow for insulation thickness.

b. Wall sleeves shall be even with both sides of the finished wall.
c. Floor sleeves shall project approximately 1/2 in. (13 mm) above the finished floor and be even with the underside of the floor. Floor sleeves shall be cast in place or permanently sealed into the floor structure to prevent any water on the floor above from following the pipe system.

2.20 ESCUTCHEONS

a. Provide escutcheon on each side of wall or floor penetrations to provide a finished appearance. For insulated pipes, the escutcheon shall surround the outside of the insulation.
b. Escutcheons for small pipes may be spring clip type. Escutcheons for larger pipes shall be held by setscrews.

PART THREE - EXECUTION

3.1 INSTALLATION OF PIPE SUPPORTS

a. Install concrete inserts, beam clamps, or other fixtures to support the pipe hangers acceptable to the engineer.
b. Provide hanger rods and loops, or clevises, to support the pipe at the height and grade required for proper drainage and air elimination.

3.2 INSTALLATION OF PIPE

a. Cut pipe accurately to measurements, and ream free of burrs and cutting splatter. Carefully align and grade pipe, and work accurately into place. Fittings shall be used for any change in direction. Make adequate provisions for expansion and contraction. Install anchors to prevent pipe movement, as shown on the drawings. Provide expansion loops, or joints, as shown on the drawings and where required to compensate for pipe expansion. Provide for expansion at every building expansion joint.
b. Protect open pipe ends to prevent trash being placed in the lines during installation. Clean all dirt and cutting debris from pipes before making the next joint.
c. Small pipe shall be screwed or soldered as required to produce a tight system with full joints and no leaks. Pipe joints showing seepage and drips shall be dismantled and remade in proper way, as required by proper installation.
d. Copper pipe shall be carefully reamed back to full inside diameter, and the mating surfaces shall be cleaned by brush or sandpaper. When clean, the paste flux shall be applied and the joint evenly heated and soldered. Any fittings discolored by heat shall be removed and replaced.
e. Solder used in making joints shall be 50/50 for small lines operating below 200 deg. F (93 deg. C) and 100 psi (670 kPa). Solder joints operating above these conditions shall be made with 95/5 solder unless silver solder is called for.
f. Solder joints in critical lines and all lines installed in inaccessible locations (under floor slabs, etc.) shall be soldered with a product having a melting point of 1100 deg. F (593 deg. C) silver solder or above, applied with the proper torch and flux.
g. All valves to be soldered into lines shall be dismantled to prevent the heat from destroying packing and seats.
h. Valves installed in screwed lines shall be properly supported and pipes carefully installed to prevent damage or distortion of the valve.

i. Grooved pipe shall be carefully prepared and all burrs removed inside, and outside of the pipe. The proper lubricant shall be applied and the gasket carefully placed prior to tightening the clamps to the correct torque.

j. Install drains at every low place and air vents at every high place. Pipe shall slope as shown on the drawings at 1 in. (25 mm) every 40 ft (12 m). If slope is not shown, slope in the direction of flow as required above. Install drain valves and air vents as specified.

k. Install pressure gauges and thermometers as shown in details and on drawings. Every pump and coil shall be provided with pressure gauge and thermometer. Every piece of equipment that produces chilled or hot water or steam or uses chilled or hot water or steam shall be provided with gauges and thermometers on entering and leaving sides.

3.3 CLEANING AND TREATING OF PIPE SYSTEMS

a. Every pipe system shall be cleaned to remove trash, mill scale, cutting oil, and welding and burning splatter from the lines before any control devices are installed. If such debris has collected in valves, the valves shall be disassembled and cleaned prior to closing for the first time.

b. After several hours of operation, each strainer shall be blown down. This shall be repeated as often as necessary to produce a clean discharge from the blowdown. Prior to turning system over to the owner, every strainer shall be removed and cleaned.

3.4 TESTING

a. Every pipe system shall be tested at 1.5 times its operating pressure, but no less than 125 psi (860 kPa) unless the engineer agrees to a lesser pressure.

b. Pipe and fittings shall be tested before any insulation or other covering is applied.

c. Testing may be performed in sections before vital equipment is connected if the test pressure is above the equipment rating.

d. Steam piping shall be checked again when hot, at the operating pressure and temperature.

e. Test medium shall be water under hydrostatic pressure with all air removed from the system. With engineer's consent, the test may be performed with compressed air to prevent danger from freezing. Hydrostatic pressure shall be held for no less than 2 hr with no drop in pressure. Air test shall be held for no less than 4 hr and the engineer may require longer test periods. Questionable joints shall be soaped to prove tightness.

f. The engineer, or representative, shall observe all tests. Notice to the engineer shall be given two full days before the testing is to be performed.

3.5 DESTRUCTIVE TESTING

a. The engineer reserves the right to select, at random, four fittings already completely installed for destructive testing. These joints shall be removed, by the contractor, with the connecting pipe, and the contractor shall replace these fittings at contractor's expense. The Engineer may destroy these joints by cutting them apart, separating soldered joints to check for full coverage, grinding the weld areas to observe voids, slag inclusions, or other defects. If major defects are noted in these joints, the contractor shall take corrective action to remedy the cause, and additional testing shall be performed to ensure that the system is adequate and complies with these Specifications and good workmanship.

END OF SECTION 15700

SECTION 15702

****

CHILLED/HOT WATER PIPE, FITTINGS, AND ACCESSORIES

PART ONE - GENERAL

1.1 The GENERAL and SPECIAL CONDITIONS, Section 15100, are included as a part of this Section as though written in full in this document.

1.2 Scope of the Work shall include the furnishing and complete installation of the equipment covered by this Section, with all auxiliaries, ready for owner's use.

PART TWO - PRODUCTS

2.1 PIPE

a. Materials for general service pipe systems, low-pressure, nominal corrosion, low-temperature service.

1. Steel pipe - black or galvanized - ASTM A-53 or A-120, Sch 40
2. Copper pipe (tube) - ASTM B-88, type M or L, soft or hard drawn
3. Plastic pipe - PVC Sch 40

2.2 FITTINGS: ELLS, TEES, CROSSES, BUSHINGS

a. Materials for fittings shall match the pipe system category for pressure, temperature, and corrosion.

b. Fittings for low pressure, and/or low temperature shall be:

1. Steel - galvanized or black
   - Size under 2 in. (50 mm)
     - Screwed, malleable iron fittings, Class 150
   - 2 in. (50 mm) and larger
     - Screwed, malleable iron fittings, Class 150
     - Welded, steel butt-welding fittings, Sch 40
     - Grooved couplings, ductile iron clamps, Sch 40 fittings
2. Copper - wrought or cast copper, solder type
3. PVC - PVC Sch 40, solvent-welded type

2.3 UNIONS

a. Materials for unions shall match the pipe system category for pressure, temperature, and corrosion.

b. Unions for low pressure and/or low temperature shall be:

1. Steel - screwed, Class 150, malleable iron, O-ring or brass seat
   - welded neck flanges with gaskets
2. Copper - cast copper to copper, metal seat
3. PVC - solvent-welded, flange to flange, with gasket

c. Dielectric unions shall separate all ferrous and nonferrous metals in every piping system. Unions shall match those above, except that metal-to-metal contact is to be avoided. Where flanges are used, the bolts shall be insulated from the body of the flange.

2.4 VALVES - GENERAL

a. Materials for small valves shall be all brass and for larger valves shall be cast iron, cast ductile iron, or cast steel, as required for the pressure and service of the valve and as listed below.

2.5 LOW-PRESSURE VALVES

a. Low-pressure valves for gravity and city water, low-pressure steam, and compressed air shall be as follows:

VALVES 2 in. (50 mm) AND SMALLER

1. Globe valves for throttling service shall be Class 150, straight through or angle type, Teflon disk, union bonnet, Stockham B-22 or B-232, or equivalent, threaded or solder ends.
2. Ball valves for throttling service shall be one-piece body, reduced port, Stockham S-127BR1, or equivalent.
3. Gate valves for shutoff service shall be Class 125, rising stem, screwed bonnet, split or solid wedge, Stockham B-114 or B-110, or equivalent, for screwed ends or solder ends.
4. Ball valves for shutoff service shall be two-piece body, full port, Stockham S-207BR1, or equivalent.

VALVES 2 in. (50 mm) THROUGH 8 in. (200 mm)

5. Gate valves for shutoff service shall be Class 125, cast iron body, bronze trim, rising stem, bolted bonnet, solid wedge, OS&Y, Stockham G-625, or equivalent. Valves located more than 7 ft (2.1 m) above floor or other inaccessible elevated locations, shall be provided with chain wheel with hammer-blow operator.
6. Butterfly valves for shutoff service or throttling, up to 150-psi (1035-kPa) working pressure shall be Stockham LD-711, or equivalent.

2.6 CHECK VALVES

a. Materials for small valves shall be all brass and for larger valves shall be cast iron, cast ductile iron, or cast steel, as required for the pressure and service of the valve and as listed below.

VALVES 2 in. (50 mm) AND SMALLER

1. For service less than 150 deg. F, (65 deg. C): Class 125, bronze body with Buna-N disk seat, Stockham B-310B or B-320B, or equivalent.
2. For service up to 250 deg. F (132 deg. C): Class 125, bronze body with TFE disk, Stockham B-310T or B-320T, or equivalent.
3. For pressures up to 200 psi (1380 kPa): Class 150, bronze body with Teflon disk, Stockham B-322T, or equivalent.

VALVES LARGER THAN 2 in. (50 mm) THROUGH 12 in. (300 mm)

4. For service at or below 150 deg. F (65 deg. C): iron body, brass trim with Buna-N disk, Stockham G-932B, or equivalent.
5. For service up to 250 deg. F (132 deg. C): iron body, brass trim, bronze disk, Stockham D-931, or equivalent.

2.7 PIPE JOINT MATERIALS

a. Screwed pipe joints shall be made up using Teflon tape, Rectorseal No. 5 pipe dope, or other lubricants, as approved for the particular installation.

b. Soldered joints shall be made up with paste flux and 95/5 solder, or silver solder, as required under Part Three, Execution.

2.8 PIPE SUPPORT SYSTEM

a. Provide an adequate pipe suspension system in accordance with recognized engineering practices, using, where possible, standard, commercially accepted pipe hangers and accessories.

b. All pipe hangers and supports shall conform to the latest requirements of the ASA Code for Pressure Piping, B 31.1, and Manufacturers' Standardization Society documents MSS SP-58 and MSS SP-69.

c. The pipe hanger assembly must be capable of supporting the line in all operating conditions. Accurate weight balance calculations shall be made to determine the supporting force at each hanger in order to prevent excessive stress in either pipe or connected equipment.

d. Where references below refer to "type," these references shall be to Federal Specification WW-171. Where reference is to "Figure," it shall be to Fee & Mason designations used in their catalog. Equivalent products by Grinnell and others are acceptable.

1. Concrete Inserts
   Where piping is supported from a concrete structure, inserts shall be type 18 or 19, or structural shapes where provided where a continuous insert is required. Where support rod size exceeds 7/8 in. (22 mm) diameter or where the pipe load exceeds the recommended load for the insert, use two inserts with a trapeze-type connecting member below the concrete.
2. Beam Clamps
   Where piping is to be supported from structural steel, beam clamps, type 21, 28, 29, 30, or 31, shall be used. Beam clamp selection shall be on the basis of the required load to be supported. Where welded beam attachments are required, they shall be Figure 90, 131, 251, or 266. Holes drilled in structural steel for hanger support rods will not be permitted.
3. Riser Clamps
   All vertical runs of piping shall be supported at each floor, and/or at specified intervals, by means of type 8 clamp for steel pipe, or Figure 368 clamp for copper tubing. For riser loadings in excess of the maximum recommended loads shown for the above items, clamps shall be designed in accordance with Figure 395 or 396.

4. Hanger Rods
Hanger rods shall be ASTM A-107 continuous-threaded rod. Eye rods shall be Figures 228 and 228 WL. Where hanger rod sizes are catalog-listed for a specified hanger, these sizes shall govern. Where hanger rod sizes are not listed, the load on the hanger shall be the determining factor, and the maximum recommended hanger rod load as shown below shall govern.

5. Hanger Rod Loading
Maximum hanger rod load shall not exceed:

Rod diameter	Inches	3/8	1/2	5/8	3/4	7/8
	(mm)	9.5	12.7	15.8	19	22.2
Max load	Pounds	610	1130	1810	2710	3770
	(kg)	274	508	814	1219	1697

6. Hanger Spacing
The maximum allowable spacing for pipe hangers shall be in accordance with tabulation below. Where concentrated loads of valves, fittings, etc., occur, closer spacing will be necessary and shall be based on the weight to be supported and the maximum recommended loads for the hanger components.

STEEL PIPE

Nominal Pipe Size	Maximum Space Between Hangers
Up to 1 1/4 in. (30 mm)	7 feet (2 m)
1 1/2 in. (40 mm)	9 feet (3 m)
2 in. & 3 in. (50 & 75 mm)	10 feet (3 m)
4 in. & 5 in. (100 & 130 mm)	14 feet (4 m)
6 in. (150 mm)	17 feet (5 m)
8 in. (200 mm)	19 feet (6 m)
10 in. & 12 in. (250 & 300 mm)	22 feet (7 m)
14 in. (355 mm)	25 feet (7.5 m)
16 in. (410 mm)	27 feet (8 m)

COPPER PIPE

Nominal Tube Size	Maximum Space Between Hangers
Up to 1 in. (25 mm)	5 feet (1.5 m)
1 1/4 to 2 in. (30 to 50 mm)	7 feet (2 m)
2 1/2 in. (125 mm)	9 feet (3 m)
3 in. (150 mm)	10 feet (3 m)
3 1/2 & 4 in. (165 & 200 mm)	12 feet (3.6 m)

FRP and PVC pipe supports - Consult manufacturer's data for conditions and temperatures involved.

7. Hangers
(a) All hangers for piping 2 in. (50 mm) or larger shall be provided with means of vertical adjustment.
(b) On uninsulated steel pipe, hangers shall be type 1, 4, 6, or 11. On piping 2 in. (50 mm) and smaller, Figure 9, 10, or 25 will be permitted.
(c) On uninsulated copper tubing, hangers shall be Figure 307, 364, or 365, or type 10 or 11.

(d) On hot insulated steel pipe, hangers shall be Figure 261 or welded attachments Figure 90, 92, 94, or 96. Where thermal movement causes the hanger rod to deviate more than 5 deg. (0.09 r) from the vertical or where longitudinal expansion causes a movement of more than 1/2 in. (13 mm) in the piping supported from below, roller hangers, type 42, 44, 45, 47, or 48, shall be used in conjunction with a protection saddle, type 40, to suit the insulation thickness. On insulated steel pipe for chilled water or similar service, the hanger must be placed on the outside of the insulation with a type 41 shield.

(e) On insulated copper tubing, hangers shall be Figure 199, 201, 202, or 215 and shall be placed on the outside of the insulation with a type 41 shield. The type 41 shield shall be applied to distribute the hanger load over the insulation and to eliminate damage to the vapor barrier on the covering.

(f) Base supports shall be type 39.

8. Brackets and Racks

Where piping is run adjacent to walls or steel columns, welded steel brackets, types 32, 33, and 34, shall be used as base supports. Multiple pipe racks or trapeze hangers shall be fabricated from channel and accessories designed for this purpose.

9. Spring Hangers

Spring hangers shall be installed at hanger points where vertical thermal movement occurs. For light loads and noncritical movements in excess of 1/4 in. (6.35 mm), type 49, 50, or 51, variable spring supports shall be used.

10. Critical Systems

On critical systems, where movement is in excess of 1/2 in. (13 mm) constant supports, type 52, shall be used. For vibration and/or shock loadings, use Figure 470, 471, or 472, sway braces. Where it is necessary to reduce pipe vibration and sound transmission to building steel, Figure 403 or 404 vibration control hangers shall be used.

11. Anchors, Guides, and Sliding Supports

Anchors shall be installed as shown on the piping drawings. They may be Figure 140, 141, or 159. Guides shall be Figure 120, 121, 122, or 165. Sliding supports shall be Figure 143 or 145.

12. Auxiliary Steel

All auxiliary steel necessary for the installation of the pipe hangers and supports shall be designed in accordance with the AISC Steel Handbook, shall be furnished by the contractor, and shall receive one shop coat of primer paint.

13. Submittals

The contractor shall submit to the engineer, prior to installation, the following information and data for review and acceptance.

(a) Manufacturer's data sheets on all cataloged items to be used.

(b) Sketches covering all specially designed hanger assemblies and fabrications.

(c) Sketches showing locations, loads, calculated travel, type, and sizes of all spring hanger assemblies.

2.9 VENTS AND DRAINS

a. Air vents shall be of manual type where readily accessible; Dole No. 9 or acceptable equivalent.

b. Air vents above concealed ceiling systems shall be extension type; Dole No. 14-1 or acceptable equivalents.

c. Air vents in accessible areas, but not convenient for service, shall be automatic vent type with the vent pipe carried to the outside or to the nearest floor drain or acceptable receptacle; Dole No. 200 or acceptable equivalent.

d. Drain points, consisting of hose bibb at the low point of pipe systems less than 6 in. (150 mm) and 1-in. (25-mm) ball valve with hose-threaded coupling for larger systems, shall be installed for every low point of piping system for drainage.

2.10 STRAINERS

a. Strainers shall be installed as shown in details on drawings and before every control valve, metering valve, or orifice such as steam traps.

b. Strainer body shall be cast brass or cast iron to match the system pressure and temperature. The body shall provide for removal of the strainer element without interruption of the pipe. Each strainer shall have a blowdown valve installed.

c. Strainer element shall be 0.045-in. (1.15-mm) perforated stainless steel with effective screen area of no less than four times the pipe area.

2.11 THERMOMETERS

a. Thermometers shall be installed as shown 1n details on the drawings and on the inlet and outlet of every item of equipment where the fluid is either heated or cooled.

b. Thermometers shall be dial type, with glass face no less than 4 in. (100 mm) in diameter and with adjustable head for visibility. The body and stem shall be stainless steel.

c. The range of the thermometer shall be such that the normal operating point with the system in service shall be midrange on the dial.

d. Accuracy of the thermometer shall be such that the error will be no greater than 2% of the full scale value.

e. The thermometer shall be provided with a brass or stainless steel well, installed in a threaded coupling into the pipe being measured. Provide thermally conductive gel in the well for contact with the bulb.

2.12 PRESSURE GAUGES

a. Provide pressure gauges where shown in details on the drawings and at every item of equipment receiving or producing flow in or from the system.

b. Gauges shall be no less than 4 in. (100 mm) in diameter, with glass face and cast aluminum body.

c. Range shall be such that gauge shall operate in midrange during normal operation of the system. Gauges subject to vacuum conditions, such as pump suctions, shall be rated for vacuum service. Graduations between figure intervals shall not exceed 10% of the gauge range. Pointer shall be adjustable for calibration.

d. Accuracy shall be such that error is less than 1% over the midrange of the gauge and 2% out of the midrange area.

e. Every gauge shall be provided with a 1/4-in. (6.35-mm) needle-type valve and an impulse dampener. Gauges installed on steam lines shall be provided with a coil syphon.

2.13 VACUUM BREAKERS

a. Water systems designed to automatically drain shall be provided with a vacuum breaker.

2.14 EXPANSION TANKS

a. Every closed system shall be provided with an expansion tank to prevent significant changes in system pressure caused by expansion and contraction of fluids due to temperature changes.

b. See Specifications for expansion tanks.

2.15 WATER MAKE-UP SYSTEMS

a. Every closed system shall be provided with a regulated water make-up system to automatically feed water to the system at a preset pressure.

b. Every water system not used for domestic water shall be provided with a reduced-pressure backflow preventer, certified for that service and acceptable to the local plumbing code administrator, to prevent backflow of contaminated water into the domestic water system.

c. See Specifications on water make-up system.

2.16 EXPANSION COMPENSATION

a. Pipe installation shall allow for expansion due to temperature differences. Provide expansion loops, or joints, as shown on the drawings, as follows.

b. Expansion joints shall be designed for the service, temperature, and pressure of the system and sized to accommodate the amount of movement required from maximum to minimum temperatures encountered by the system.

c. Joints for water systems shall be reinforced rubber corrugations secured by steel flanges. The number of corrugations shall be determined by the amount of movement required.

d. Every expansion joint shall be provided with stay bolts to prevent expansion beyond the normal design limits. Rubber joints shall have stay bolts isolated in rubber where used to eliminate vibration.

2.17 PIPE SLEEVES

a. Provide pipe sleeve of galvanized steel at each wall or floor penetration. Sleeve shall be no lighter than 18 gauge and shall be built into the wall or floor during construction of the wall. Where pipes are insulated, the sleeve shall allow for insulation thickness.

b. Wall sleeves shall be even with both sides of the finished wall.

c. Floor sleeves shall project approximately 1/2 in. (13 mm) above the finished floor and be even with the underside of the floor. Floor sleeves shall be cast in place or permanently sealed into the floor structure to prevent any water on the floor above from following the pipe system.

2.18 ESCUTCHEONS

a. Provide escutcheon on each side of wall or floor penetrations to provide a finished appearance. For insulated pipes, the escutcheon shall surround the outside of the insulation.

b. Escutcheons for small pipes may be spring clip type. Escutcheons for larger pipes shall be held by setscrews.

PART THREE - EXECUTION

3.1 INSTALLATION OF PIPE SUPPORTS

a. Install concrete inserts, beam clamps, or other fixtures to support the pipe hangers acceptable to the engineer.
b. Provide hanger rods and loops, or clevises, to support the pipe at the height and grade required for proper drainage and air elimination.

3.2 INSTALLATION OF PIPE

a. Cut pipe accurately to measurements, and ream free of burrs and cutting splatter. Carefully align and grade pipe, and work accurately into place. Fittings shall be used for any change in direction. Make adequate provisions for expansion and contraction. Install anchors to prevent pipe movement, as shown on the drawings. Provide expansion loops, or joints, as shown on the drawings and where required to compensate for pipe expansion. Provide for expansion at every building expansion joint.
b. Protect open pipe ends to prevent trash being placed in the lines during installation. Clean all dirt and cutting debris from pipes before making the next joint.
c. Small pipe shall be screwed or soldered as required to produce a tight system with full joints and no leaks. Pipe joints showing seepage and drips shall be dismantled and remade in proper way, as required by proper installation.
d. Copper pipe shall be carefully reamed back to full inside diameter, and the mating surfaces shall be cleaned by brush or sandpaper. When clean, the paste flux shall be applied and the joint evenly heated and soldered. Any fittings discolored by heat shall be removed and replaced.
e. Solder used in making joints shall be 50/50 for small lines operating below 200 deg. F (93 deg. C) and 100 psi (670 kPa). Solder joints operating above these conditions shall be made with 95/5 solder unless silver solder is called for.
f. Solder joints in critical lines, lines subject to temperatures above 200 deg. F (93 deg. C), and all lines installed in inaccessible locations (under floor slabs, etc.) shall be soldered with a product having a melting point of 1100 deg. F (593 deg. C) silver solder or above, applied with the proper torch and flux.
g. All valves to be soldered into lines shall be dismantled to prevent the heat from destroying packing and seats.
h. Valves installed in screwed lines shall be properly supported and pipes carefully installed to prevent damage or distortion of the valve.
i. Grooved pipe shall be carefully prepared and all burrs removed, inside and outside of the pipe. The proper lubricant shall be applied and the gasket carefully placed prior to tightening the clamps to the correct torque.
j. Install drains at every low place and air vents at every high place. Pipe shall slope as shown on the drawings at 1 in. (25 mm) every 40 ft (12 m). If slope is not shown, slope in the direction of flow as required above. Install drain valves and air vents as specified.
k. Install pressure gauges and thermometers as shown in details and on drawings. Every pump and coil shall be provided with pressure gauge and thermometer. Every piece of equipment that produces chilled or hot water or steam or uses chilled or hot water or steam shall be provided with gauges and thermometers on entering and leaving sides.

3.3 CLEANING AND TREATING OF PIPE SYSTEMS

a. Every pipe system shall be cleaned to remove trash, mill scale, cutting oil, and welding and burning splatter from the lines before any control devices are installed. If such debris has collected in valves, the valves shall be disassembled and cleaned prior to closing for the first time.

b. After several hours of operation, each strainer shall be blown down. This shall be repeated as often as necessary to produce a clean discharge from the blowdown. Prior to turning system over to the owner, every strainer shall be removed and cleaned.

3.4 TESTING

a. Every pipe system shall be tested at 1.5 times its operating pressure, but no less than 125 psi (860 kPa) unless the engineer agrees to a lesser pressure.

b. Pipe and fittings shall be tested before any insulation or other covering is applied.

c. Testing may be performed in sections before vital equipment is connected if the test pressure is above the equipment rating.

d. Test medium shall be water under hydrostatic pressure with all air removed from the system. With engineer's consent, the test may be performed with compressed air to prevent danger from freezing. Hydrostatic pressure shall be held for no less than 2 hr with no drop in pressure. Air test shall be held for no less than 4 hr and the engineer may require longer test periods. Questionable joints shall be soaped to prove tightness.

e. The engineer, or representative, shall observe all tests. Notice to the engineer shall be given two full days before the testing is to be performed.

3.5 DESTRUCTIVE TESTING

a. The engineer reserves the right to select, at random, four fittings already completely installed for destructive testing. These joints shall be removed, by the contractor, with the connecting pipe, and the contractor shall replace these fittings at contractor's expense. The Engineer may destroy these joints by cutting them apart, separating soldered joints to check for full coverage, grinding the weld areas to observe voids, slag inclusions, or other defects. If major defects are noted in these joints, the contractor shall take corrective action to remedy the cause, and additional testing shall be performed to ensure that the system is adequate and complies with these Specifications and good workmanship.

END OF SECTION 15702

SECTION 15704

****

CHILLED WATER PIPE, FITTINGS, AND ACCESSORIES

PART ONE - GENERAL

1.1 The GENERAL and SPECIAL CONDITIONS, Section 15100, are included as a part of this Section as though written in full in this document.

1.2 Scope of the Work shall include the furnishing and complete installation of the equipment covered by this Section, with all auxiliaries, ready for owner's use.

PART TWO - PRODUCTS

2.1 PIPE

a. Materials for general service chilled water pipe systems, low-pressure, nominal corrosion, low-temperature service not to exceed 150 deg. F (65 deg. C) under any conditions.

1. Steel pipe - black or galvanized - ASTM A-53 or A-120, Sch 40
2. Copper pipe (tube) - ASTM B-88, type M or L, soft or hard drawn
3. Plastic pipe - PVC Sch 40

2.2 FITTINGS: ELLS, TEES, CROSSES, BUSHINGS

a. Materials for fittings shall match the pipe system category for pressure, and temperature.

b. Fittings for low pressure, and/or low temperature shall be:

1. Steel - galvanized or black
   - Size under 2 in. (50 mm)
     - Screwed, malleable iron fittings, Class 150
   - 2 in. (50 mm) and larger
     - Screwed, malleable iron fittings, Class 150
     - Welded, steel butt-welding fittings, Sch 40
     - Grooved couplings, ductile iron clamps, Sch 40 fittings
2. Copper - wrought or cast copper, solder type
3. PVC - PVC Sch 40, solvent-welded type

2.3 UNIONS

a. Materials for unions shall match the pipe system category for pressure, temperature, and corrosion.

b. Unions for low pressure and/or low temperature shall be:

1. Steel - screwed, Class 150, malleable iron, O-ring or brass seat
   welded neck flanges with gaskets
2. Copper - cast copper to copper, metal seat
3. PVC - solvent-welded, flange to flange, with gasket

c. Dielectric unions shall separate all ferrous and nonferrous metals in every piping system. Unions shall match those above, except that metal-to-metal contact is to be avoided. Where flanges are used, the bolts shall be insulated from the body of the flange.

2.4 VALVES - GENERAL

a. Materials for small valves shall be all brass and for larger valves shall be cast iron, PVC, or as required for the pressure and service of the valve and as listed below.

2.5 LOW-PRESSURE VALVES

a. Low-pressure valves for chilled water shall be as follows:

VALVES 2 in. (50 mm) AND SMALLER

1. Globe valves for throttling service shall be Class 150, straight through or angle type, Teflon disk, union bonnet, Stockham B-22 or B-232, or equivalent, threaded or solder ends.
2. Ball valves for throttling service shall be one-piece body, reduced port, Stockham S-127BR1, or equivalent.
3. Gate valves for shutoff service shall be Class 125, rising stem, screwed bonnet, split or solid wedge, Stockham B-114 or B-110, or equivalent, for screwed ends or solder ends.
4. Ball valves for shutoff service shall be two-piece body, full port, Stockham S-207BR1, or equivalent.

VALVES 2 in. (50 mm) THROUGH 8 in. (200 mm)

5. Gate valves for shutoff service shall be Class 125, cast iron body, bronze trim, rising stem, bolted bonnet, solid wedge, OS&Y, Stockham G-625, or equivalent. Valves located more than 7 ft (2.1 m) above floor or other inaccessible elevated locations, shall be provided with chain wheel with hammer-blow operator.
6. Butterfly valves for shutoff service or throttling, up to 150-psi (1035-kPa) working pressure shall be Stockham LD-711, or equivalent.

2.6 CHECK VALVES

a. Materials for small valves shall be all brass and cast iron with brass trim for larger valves.

VALVES 2 in. (50 mm) AND SMALLER

1. For service less than 150 deg. F (65 deg. C): Class 125, bronze body with Buna-N disk seat, Stockham B-310B or B-320B, or equivalents.
2. For pressures up to 200 psi (1380 kPa): Class 150, bronze body with Teflon disk, Stockham B-322T, or equivalent.

VALVES LARGER THAN 2 in. (50 mm) THROUGH 12 in. (300 mm)

3. For service at or below 150 deg. F (65 deg. C): iron body, brass trim with Buna-N disk, Stockham G-932B, or equivalent.

2.7 PIPE JOINT MATERIALS

a. Screwed pipe joints shall be made up using Teflon tape, Rectorseal No. 5 pipe dope, or other lubricants, as approved for the particular installation.

b. Soldered joints shall be made up with paste flux and 95/5 solder, or silver solder, as required under Part Three, Execution.

2.8 PIPE SUPPORT SYSTEM

a. Provide an adequate pipe suspension system in accordance with recognized engineering practices, using, where possible, standard, commercially accepted pipe hangers and accessories.

b. All pipe hangers and supports shall conform to the latest requirements of the ASA Code for Pressure Piping, B 31.1, and Manufacturers' Standardization Society documents MSS SP-58 and MSS SP-69.

c. The pipe hanger assembly must be capable of supporting the line in all operating conditions. Accurate weight balance calculations shall be made to determine the supporting force at each hanger in order to prevent excessive stress in either pipe or connected equipment.

d. Where references below refer to "type," these references shall be to Federal Specification WW-171. Where reference is to "Figure," it shall be to Fee & Mason designations used in their catalog. Equivalent products by Grinnell and others are acceptable.

1. Concrete Inserts
   Where piping is supported from a concrete structure, inserts shall be type 18 or 19, or structural shapes where provided where a continuous insert is required. Where support rod size exceeds 7/8 in. (22 mm) diameter or where the pipe load exceeds the recommended load for the insert, use two inserts with a trapeze-type connecting member below the concrete.
2. Beam Clamps
   Where piping is to be supported from structural steel, beam clamps, type 21, 28, 29, 30, or 31, shall be used. Beam clamp selection shall be on the basis of the required load to be supported. Where welded beam attachments are required, they shall be Figure 90, 131, 251, or 266. Holes drilled in structural steel for hanger support rods will not be permitted.
3. Riser Clamps
   All vertical runs of piping shall be supported at each floor, and/or at specified intervals, by means of type 8 clamp for steel pipe, or Figure 368 clamp for copper tubing. For riser loadings in excess of the maximum recommended loads shown for the above items, clamps shall be designed in accordance with Figure 395 or 396.
4. Hanger Rods
   Hanger rods shall be ASTM A-107 continuous-threaded rod. Eye rods shall be Figures 228 and 228 WL. Where hanger rod sizes are catalog-listed for a specified hanger, these sizes shall govern. Where hanger rod sizes are not listed, the load on the hanger shall be the determining factor, and the maximum recommended hanger rod load as shown below shall govern.

5. Hanger Rod Loading
Maximum hanger rod load shall not exceed:

Rod diameter	Inches	3/8	1/2	5/8	3/4	7/8
	(mm) 9.5	12.7	15.8	19	22.2	
Max load	Pounds	610	1130	1810	2710	3770
	(kg) 274	508	814	1219	1697	

6. Hanger Spacing
The maximum allowable spacing for pipe hangers shall be in accordance with tabulation below. Where concentrated loads of valves, fittings, etc., occur, closer spacing will be necessary and shall be based on the weight to be supported and the maximum recommended loads for the hanger components.

STEEL PIPE

Nominal Pipe Size	Maximum Space Between Hangers
Up to 1 1/4 in. (30 mm)	7 feet (2 m)
1 1/2 in. (40 mm)	9 feet (3 m)
2 in. & 3 in. (50 & 75 mm)	10 feet (3 m)
4 in. & 5 in. (100 & 130 mm)	14 feet (4 m)
6 in. (150 mm)	17 feet (5 m)
8 in. (200 mm)	19 feet (6 m)
10 in. & 12 in. (250 & 300 mm)	22 feet (7 m)
14 in. (355 mm)	25 feet (7.5 m)
16 in. (410 mm)	27 feet (8 m)

COPPER PIPE

Nominal Tube Size	Maximum Space Between Hangers
Up to 1 in. (25 mm)	5 feet (1.5 m)
1 1/4 to 2 in. (30 to 50 mm)	7 feet (2 m)
2 1/2 in. (125 mm)	9 feet (3 m)
3 in. (150 mm)	10 feet (3 m)
3 1/2 & 4 in. (165 & 200 mm)	12 feet (3.6 m)

FRP and PVC pipe supports - Consult manufacturer's data for conditions and temperatures involved.

7. Hangers
(a) All hangers for piping 2 in. (50 mm) or larger shall be provided with means of vertical adjustment.
(b) On steel pipe for chillled water service, the hanger shall be placed on the outside of the insulation with a type 41 shield.
(c) On insulated copper tubing, hangers shall be Figure 199, 201, 202, or 215 and shall be placed on the outside of the insulation with a type 41 shield. The type 41 shield shall be applied to distribute the hanger load over the insulation and to eliminate damage to the vapor barrier on the covering.
(d) Base supports shall be type 39.

8. Brackets and Racks
Where piping is run adjacent to walls or steel columns, welded steel brackets, types 32, 33, and 34, shall be used as base supports. Multiple pipe racks or trapeze hangers shall be fabricated from channel and accessories designed for this purpose.

9. Spring Hangers
Spring hangers shall be installed at hanger points where vertical thermal movement occurs. For light loads and noncritical movements in excess of 1/4 in. (6.35 mm), type 49, 50, or 51, variable spring supports shall be used.

10. Critical Systems
On critical systems, where movement is in excess of 1/2 in. (13 mm) constant supports, type 52, shall be used. For vibration and/or shock loadings, use Figure 470, 471, or 472 sway braces. Where it is necessary to reduce pipe vibration and sound transmission to building steel, Figure 403 or 404 vibration control hangers shall be used.

11. Anchors, Guides, and Sliding Supports
Anchors shall be installed as shown on the piping drawings. They may be Figure 140, 141, or 159. Guides shall be Figure 120, 121, 122, or 165. Sliding supports shall be Figure 143 or 145.

12. Auxiliary Steel
All auxiliary steel necessary for the installation of the pipe hangers and supports shall be designed in accordance with the AISC Steel Handbook, shall be furnished by the contractor, and shall receive one shop coat of primer paint.

13. Submittals
The contractor shall submit to the engineer, prior to installation, the following information and data for review and acceptance.
(a) Manufacturer's data sheets on all cataloged items to be used.
(b) Sketches covering all specially designed hanger assemblies and fabrications.
(c) Sketches showing locations, loads, calculated travel, type, and sizes of all spring hanger assemblies.

2.9 VENTS AND DRAINS

a. Air vents shall be of manual type where readily accessible; Dole No. 9 or acceptable equivalent.
b. Air vents above concealed ceiling systems shall be extension type; Dole No. 14-1 or acceptable equivalents.
c. Air vents in accessible areas, but not convenient for service, shall be automatic vent type with the vent pipe carried to the outside or to the nearest floor drain or acceptable receptacle; Dole No. 200 or acceptable equivalent.
d. Drain points, consisting of hose bibb at the low point of pipe systems less than 6 in. (150 mm) and 1-in. (25-mm) ball valve with hose-threaded coupling for larger systems, shall be installed for every low point of piping system for drainage.

2.10 STRAINERS

a. Strainers shall be installed as shown in details on drawings and before every control valve, metering valve, or orifice such as steam traps.
b. Strainer body shall be cast brass or cast iron to match the system pressure and temperature. The body shall provide for removal of the strainer element without interruption of the pipe. Each strainer shall have a blowdown valve installed.
c. Strainer element shall be 0.045-in. (1.15-mm) perforated stainless steel with effective screen area of no less than four times the pipe area.

2.11 THERMOMETERS

a. Thermometers shall be installed as shown in details on the drawings and on the inlet and outlet of every item of equipment where the fluid is either heated or cooled.

b. Thermometers shall be dial type, with glass face no less than 4 in. (100 mm) in diameter and with adjustable head for visibility. The body and stem shall be stainless steel.

c. The range of the thermometer shall be such that the normal operating point with the system in service shall be midrange on the dial.

d. Accuracy of the thermometer shall be such that the error will be no greater than 2% of the full scale value.

e. The thermometer shall be provided with a brass or stainless steel well, installed in a threaded coupling into the pipe being measured. Provide thermally conductive gel in the well for contact with the bulb.

2.12 PRESSURE GAUGES

a. Provide pressure gauges where shown in details on the drawings and at every item of equipment receiving or producing flow in or from the system.

b. Gauges shall be no less than 4 in. (100 mm) in diameter, with glass face and cast aluminum body.

c. Range shall be such that gauge shall operate in midrange during normal operation of the system. Gauges subject to vacuum conditions, such as pump suctions, shall be rated for vacuum service. Graduations between figure intervals shall not exceed 10% of the gauge range. Pointer shall be adjustable for calibration.

d. Accuracy shall be such that error is less than 1% over the midrange of the gauge and 2% out of the midrange area.

e. Every gauge shall be provided with a 1/4-in. (6.35-mm) needle-type valve and an impulse dampener. Gauges installed on steam lines shall be provided with a coil syphon.

2.13 VACUUM BREAKERS

a. Water systems designed to automatically drain shall be provided with a vacuum breaker.

2.14 EXPANSION TANKS

a. Every closed system shall be provided with an expansion tank to prevent significant changes in system pressure caused by expansion and contraction of fluids due to temperature changes.

b. See Specifications for expansion tanks.

2.15 WATER MAKE-UP SYSTEMS

a. Every closed system shall be provided with a regulated water make-up system to automatically feed water to the system at a preset pressure.

b. Every water system not used for domestic water shall be provided with a reduced-pressure backflow preventer, certified for that service and acceptable to the local plumbing code administrator, to prevent backflow of contaminated water into the domestic water system.

c. See Specifications on water make-up system.

2.16 EXPANSION COMPENSATION

a. Pipe installation shall allow for expansion due to temperature differences. Provide expansion loops, or joints, as shown on the drawings, as follows.

b. Expansion joints shall be designed for the service, temperature, and pressure of the system and sized to accommodate the amount of movement required from maximum to minimum temperatures encountered by the system.

c. Joints for chilled water systems shall be reinforced rubber corrugations secured by steel flanges. The number of corrugations shall be determined by the amount of movement required.

d. Every expansion joint shall be provided with stay bolts to prevent expansion beyond the normal design limits. Rubber joints shall have stay bolts isolated in rubber where used to eliminate vibration.

2.17 PIPE SLEEVES

a. Provide pipe sleeve of galvanized steel at each wall or floor penetration. Sleeve shall be no lighter than 18 gauge and shall be built into the wall or floor during construction of the wall. Where pipes are insulated, the sleeve shall allow for insulation thickness.

b. Wall sleeves shall be even with both sides of the finished wall.

c. Floor sleeves shall project approximately 1/2 in. (13 mm) above the finished floor and be even with the underside of the floor. Floor sleeves shall be cast in place or permanently sealed into the floor structure to prevent any water on the floor above from following the pipe system.

2.18 ESCUTCHEONS

a. Provide escutcheon on each side of wall or floor penetrations to provide a finished appearance. For insulated pipes, the escutcheon shall surround the outside of the insulation.

b. Escutcheons for small pipes may be spring clip type. Escutcheons for larger pipes shall be held by setscrews.

PART THREE - EXECUTION

3.1 INSTALLATION OF PIPE SUPPORTS

a. Install concrete inserts, beam clamps, or other fixtures to support the pipe hangers acceptable to the engineer.

b. Provide hanger rods and loops, or clevises, to support the pipe at the height and grade required for proper drainage and air elimination.

3.2 INSTALLATION OF PIPE

a. Cut pipe accurately to measurements, and ream free of burrs and cutting splatter. Carefully align and grade pipe, and work accurately into place. Fittings shall be used for any change in direction. Make adequate provisions for expansion and contraction. Install anchors to prevent pipe movement, as shown on the drawings. Provide expansion loops, or joints, as shown on the drawings and where required to compensate for pipe expansion. Provide for expansion at every building expansion joint.

b. Protect open pipe ends to prevent trash being placed in the lines during installation. Clean all dirt and cutting debris from pipes before making the next joint.

c. Small pipe shall be screwed or soldered as required to produce a tight system with full joints and no leaks. Pipe joints showing seepage and drips shall be dismantled and remade in proper way, as required by proper installation.
d. Copper pipe shall be carefully reamed back to full inside diameter, and the mating surfaces shall be cleaned by brush or sandpaper. When clean, the paste flux shall be applied and the joint evenly heated and soldered. Any fittings discolored by heat shall be removed and replaced.
e. Solder used in making joints shall be 50/50 for small lines operating below 150 deg. F (65 deg. C) and 100 psi (670 kPa). Solder joints operating above these conditions and larger lines shall be made with 95/5 solder unless silver solder is called for.
f. Solder joints in critical lines, lines subject to temperatures above 150 deg. F (65 deg. C), and all lines installed in inaccessible locations (under floor slabs, etc.) shall be soldered with a product having a melting point of 1100 deg. F (593 deg. C) silver solder or above, applied with the proper torch and flux.
g. All valves to be soldered into lines shall be dismantled to prevent the heat from destroying packing and seats.
h. Valves installed in screwed lines shall be properly supported and pipes carefully installed to prevent damage or distortion of the valve.
i. Grooved pipe shall be carefully prepared and all burrs removed, inside and outside of the pipe. The proper lubricant shall be applied and the gasket carefully placed prior to tightening the clamps to the correct torque.
j. Install drains at every low place and air vents at every high place. Pipe shall slope as shown on the drawings at 1 in. (25 mm) every 40 ft (12 m). If slope is not shown, slope in the direction of flow as required above. Install drain valves and air vents as specified.
k. Install pressure gauges and thermometers as shown in details and on drawings. Every pump and coil shall be provided with pressure gauge and thermometer. Every piece of equipment that produces chilled or hot water or steam or uses chilled or hot water or steam shall be provided with gauges and thermometers on entering and leaving sides.

3.3 CLEANING AND TREATING OF PIPE SYSTEMS

a. Every pipe system shall be cleaned to remove trash, mill scale, cutting oil, and welding and burning splatter from the lines before any control devices are installed. If such debris has collected in valves, the valves shall be disassembled and cleaned prior to closing for the first time.
b. After several hours of operation, each strainer shall be blown down. This shall be repeated as often as necessary to produce a clean discharge from the blowdown. Prior to turning system over to the owner, every strainer shall be removed and cleaned.

3.4 TESTING

a. Every pipe system shall be tested at 1.5 times its operating pressure, but no less than 125 psi (860 kPa) unless the engineer agrees to a lesser pressure.
b. Pipe and fittings shall be tested before any insulation or other covering is applied.
c. Testing may be performed in sections before vital equipment is connected if the test pressure is above the equipment rating.

d. Test medium shall be water under hydrostatic pressure with all air removed from the system. With engineer's consent, the test may be performed with compressed air to prevent danger from freezing. Hydrostatic pressure shall be held for no less than 2 hr with no drop in pressure. Air test shall be held for no less than 4 hr and the engineer may require longer test periods. Questionable joints shall be soaped to prove tightness.

e. The engineer, or representative, shall observe all tests. Notice to the engineer shall be given two full days before the testing is to be performed.

3.5 DESTRUCTIVE TESTING

a. The engineer reserves the right to select, at random, four fittings already completely installed for destructive testing. These joints shall be removed, by the contractor, with the connecting pipe, and the contractor shall replace these fittings at contractor's expense. The Engineer may destroy these joints by cutting them apart, separating soldered joints to check for full coverage, grinding the weld areas to observe voids, slag inclusions, or other defects. If major defects are noted in these joints, the contractor shall take corrective action to remedy the cause, and additional testing shall be performed to ensure that the system is adequate and complies with these Specifications and good workmanship.

END OF SECTION 15704

SECTION 15706

****

HOT WATER PIPE, FITTINGS, AND ACCESSORIES

PART ONE - GENERAL

1.1 The GENERAL and SPECIAL CONDITIONS, Section 15100, are included as a part of this Section as though written in full in this document.

1.2 Scope of the Work shall include the furnishing and complete installation of the equipment covered by this Section, with all auxiliaries, ready for owner's use.

PART TWO - PRODUCTS

2.1 PIPE

a. Materials for general service pipe systems, low-pressure, nominal corrosion, low-temperature service.

1. Steel pipe - black or galvanized - ASTM A-53 or A-120, Sch 40
2. Copper pipe (tube) - ASTM B-88, type M or L, soft or hard drawn

2.2 FITTINGS: ELLS, TEES, CROSSES, BUSHINGS

a. Materials for fittings shall match the pipe system category for pressure, temperature, and corrosion.

b. Fittings for low pressure, and/or low temperature shall be:

1. Steel - galvanized or black
   Size under 2 in. (50 mm)
   Screwed, malleable iron fittings, Class 150
   2 in. (50 mm) and larger
   Screwed, malleable iron fittings, Class 150
   Welded, steel butt-welding fittings, Sch 40
2. Copper - wrought or cast copper, solder type

2.3 UNIONS

a. Materials for unions shall match the pipe system category for pressure, temperature, and corrosion.

b. Unions for low pressure and/or low temperature shall be:

1. Steel - screwed, Class 150, malleable iron, O-ring or brass seat
   welded neck flanges with gaskets
2. Copper - cast copper to copper, metal seat

c. Dielectric unions shall separate all ferrous and nonferrous metals in every piping system. Unions shall match those above, except that metal-to-metal contact is to be avoided. Where flanges are used, the bolts shall be insulated from the body of the flange.

2.4 VALVES - GENERAL

a. Materials for small valves shall be all brass and for larger valves shall be cast iron, cast ductile iron, or cast steel, as required for the pressure and service of the valve and as listed below.

2.5 LOW-PRESSURE VALVES

a. Low-pressure valves for hot water systems shall be as follows:

VALVES 2 in. (50 mm) AND SMALLER

1. Globe valves for throttling service shall be Class 150, straight through or angle type, Teflon disk, union bonnet, Stockham B-22 or B-232, or equivalent, threaded or solder ends.
2. Ball valves for throttling service shall be one-piece body, reduced port, Stockham S-127BR1, or equivalent.
3. Gate valves for shutoff service shall be Class 125, rising stem, screwed bonnet, split or solid wedge, Stockham B-114 or B-110, or equivalent, for screwed ends or solder ends.
4. Ball valves for shutoff service shall be two-piece body, full port, Stockham S-207BR1, or equivalent.

VALVES 2 in. (50 mm) THROUGH 8 in. (200 mm)

5. Gate valves for shutoff service shall be Class 125, cast iron body, bronze trim, rising stem, bolted bonnet, solid wedge, OS&Y, Stockham G-625, or equivalent. Valves located more than 7 ft (2.1 m) above floor or other inaccessible elevated locations, shall be provided with chain wheel with hammer-blow operator.
6. Butterfly valves for shutoff service or throttling, up to 150-psi (1035-kPa) working pressure shall be Stockham LD-711, or equivalent.

2.6 CHECK VALVES

a. Materials for small valves shall be all brass and for larger valves shall be cast iron, cast ductile iron, or cast steel, as required for the pressure and service of the valve and as listed below.

VALVES 2 in. (50 mm) AND SMALLER

1. For service up to 250 deg. F (232 deg. C): Class 125, bronze body with TFE disk, Stockham B-310T or B-320T, or equivalent.
2. For pressures up to 200 psi (1380 kPa): Class 150, bronze body with Teflon disk, Stockham B-322T, or equivalent.

VALVES LARGER THAN 2 in. (50 mm) THROUGH 12 in. (300 mm)

3. For service up to 250 deg. F (132 deg. C): iron body, brass trim, bronze disk, Stockham D-931, or equivalent.

2.7 PIPE JOINT MATERIALS

a. Screwed pipe joints shall be made up using Teflon tape, Rectorseal No. 5 pipe dope, or other lubricants, as approved for the particular installation.

b. Soldered joints shall be made up with paste flux and 95/5 solder, or silver solder, as required under Part Three, Execution.

2.8 PIPE SUPPORT SYSTEM

a. Provide an adequate pipe suspension system in accordance with recognized engineering practices, using, where possible, standard, commercially accepted pipe hangers and accessories.

b. All pipe hangers and supports shall conform to the latest requirements of the ASA Code for Pressure Piping, B 31.1, and Manufacturers' Standardization Society documents MSS SP-58 and MSS SP-69.

c. The pipe hanger assembly must be capable of supporting the line in all operating conditions. Accurate weight balance calculations shall be made to determine the supporting force at each hanger in order to prevent excessive stress in either pipe or connected equipment.

d. Where references below refer to "type," these references shall be to Federal Specification WW-171. Where reference is to "Figure," it shall be to Fee & Mason designations used in their catalog. Equivalent products by Grinnell and others are acceptable.

1. Concrete Inserts
   Where piping is supported from a concrete structure, inserts shall be type 18 or 19, or structural shapes where provided where a continuous insert is required. Where support rod size exceeds 7/8 in. (22 mm) diameter or where the pipe load exceeds the recommended load for the insert, use two inserts with a trapeze-type connecting member below the concrete.
2. Beam Clamps
   Where piping is to be supported from structural steel, beam clamps, type 21, 28, 29, 30, or 31, shall be used. Beam clamp selection shall be on the basis of the required load to be supported. Where welded beam attachments are required, they shall be Figure 90, 131, 251, or 266. Holes drilled in structural steel for hanger support rods will not be permitted.
3. Riser Clamps
   All vertical runs of piping shall be supported at each floor, and/or at specified intervals, by means of type 8 clamp for steel pipe, or Figure 368 clamp for copper tubing. For riser loadings in excess of the maximum recommended loads shown for the above items, clamps shall be designed in accordance with Figure 395 or 396.
4. Hanger Rods
   Hanger rods shall be ASTM A-107 continuous-threaded rod. Eye rods shall be Figures 228 and 228 WL. Where hanger rod sizes are catalog-listed for a specified hanger, these sizes shall govern. Where hanger rod sizes are not listed, the load on the hanger shall be the determining factor, and the maximum recommended hanger rod load as shown below shall govern.

5. Hanger Rod Loading
Maximum hanger rod load shall not exceed:

Rod diameter	Inches	3/8	1/2	5/8	3/4	7/8
	(mm)	9.5	12.7	15.8	19	22.2
Max load	Pounds	610	1130	1810	2710	3770
	(kg)	274	508	814	1219	1697

6. Hanger Spacing
The maximum allowable spacing for pipe hangers shall be in accordance with tabulation below. Where concentrated loads of valves, fittings, etc., occur, closer spacing will be necessary and shall be based on the weight to be supported and the maximum recommended loads for the hanger components.

STEEL PIPE

Nominal Pipe Size	Maximum Space Between Hangers
Up to 1 1/4 in. (30 mm)	7 feet (2 m)
1 1/2 in. (40 mm)	9 feet (3 m)
2 in. & 3 in. (50 & 75 mm)	10 feet (3 m)
4 in. & 5 in. (100 & 130 mm)	14 feet (4 m)
6 in. (150 mm)	17 feet (5 m)
8 in. (200 mm)	19 feet (6 m)
10 in. & 12 in. (250 & 300 mm)	22 feet (7 m)
14 in. (355 mm)	25 feet (7.5 m)
16 in. (410 mm)	27 feet (8 m)

COPPER PIPE

Nominal Tube Size	Maximum Space Between Hangers
Up to 1 in. (25 mm)	5 feet (1.5 m)
1 1/4 to 2 in. (30 to 50 mm)	7 feet (2 m)
2 1/2 in. (125 mm)	9 feet (3 m)
3 in. (150 mm)	10 feet (3 m)
3 1/2 & 4 in. (165 & 200 mm)	12 feet (3.6 m)

7. Hangers
(a) All hangers for piping 2 in. (50 mm) or larger shall be provided with means of vertical adjustment.
(b) On uninsulated steel pipe, hangers shall be type 1, 4, 6, or 11. On piping 2 in. (50 mm) and smaller, Figure 9, 10, or 25 will be permitted.
(c) On uninsulated copper tubing, hangers shall be Figures 307, 364, or 365, or type 10 or 11.
(d) On hot insulated steel pipe, hangers shall be Figure 261 or welded attachments Figure 90, 92, 94, or 96. Where thermal movement causes the hanger rod to deviate more than 5 deg. (0.09 r) from the vertical or where longitudinal expansion causes a movement of more than 1/2 in. (13 mm) in the piping supported from below, roller hangers, type 42, 44, 45, 47, or 48, shall be used in conjunction with a protection saddle, type 40, to suit the insulation thickness.

(e) On insulated copper tubing, hangers shall be Figure 199, 201, 202, or 215 and shall be placed on the outside of the insulation with a type 41 shield. The type 41 shield shall be applied to distribute the hanger load over the insulation and to eliminate damage to the vapor barrier on the covering.
(f) Base supports shall be type 39.

8. Brackets and Racks
Where piping is run adjacent to walls or steel columns, welded steel brackets, types 32, 33, and 34, shall be used as base supports. Multiple pipe racks or trapeze hangers shall be fabricated from channel and accessories designed for this purpose.

9. Spring Hangers
Spring hangers shall be installed at hanger points where vertical thermal movement occurs. For light loads and noncritical movements in excess of 1/4 in. (6.35 mm), type 49, 50, or 51, variable spring supports shall be used.

10. Critical Systems
On critical systems, where movement is in excess of 1/2 in. (13 mm) constant supports, type 52, shall be used. For vibration and/or shock loadings, use Figure 470, 471, or 472, sway braces. Where it is necessary to reduce pipe vibration and sound transmission to building steel, Figure 403 or 404 vibration control hangers shall be used.

11. Anchors, Guides, and Sliding Supports
Anchors shall be installed as shown on the piping drawings. They may be Figure 140, 141, or 159. Guides shall be Figure 120, 121, 122, or 161. Sliding supports shall be Figure 143 or 145.

12. Auxiliary Steel
All auxiliary steel necessary for the installation of the pipe hangers and supports shall be designed in accordance with the AISC Steel Handbook, shall be furnished by the contractor, and shall receive one shop coat of primer paint.

13. Submittals
The contractor shall submit to the engineer, prior to installation, the following information and data for review and acceptance.
(a) Manufacturer's data sheets on all cataloged items to be used.
(b) Sketches covering all specially designed hanger assemblies and fabrications.
(c) Sketches showing locations, loads, calculated travel, type, and sizes of all spring hanger assemblies.

2.9 VENTS AND DRAINS

a. Air vents shall be of manual type where readily accessible; Dole No. 9 or acceptable equivalent.

b. Air vents above concealed ceiling systems shall be extension type; Dole No. 14-1 or acceptable equivalent.

c. Air vents in accessible areas, but not convenient for service, shall be automatic vent type with the vent pipe carried to the outside or to the nearest floor drain or acceptable receptacle; Dole No. 200 or acceptable equivalent.

d. Drain points, consisting of hose bibb at the low point of pipe systems less than 6 in. (150 mm) and 1-in. (25-mm) ball valve with hose-threaded coupling for larger systems, shall be installed for every low point of piping system for drainage.

2.10 STRAINERS

a. Strainers shall be installed as shown in details on drawings and before every control valve, metering valve, or orifice such as steam traps.

b. Strainer body shall be cast brass or cast iron to match the system pressure and temperature. The body shall provide for removal of the strainer element without interruption of the pipe. Each strainer shall have a blowdown valve installed.

c. Strainer element shall be 0.045-in. (1.15-mm) perforated stainless steel with effective screen area of no less than four times the pipe area.

2.11 THERMOMETERS

a. Thermometers shall be installed as shown in details on the drawings and on the inlet and outlet of every item of equipment where the fluid is either heated or cooled.

b. Thermometers shall be dial type, with glass face no less than 4 in. (100 mm) in diameter and with adjustable head for visibility. The body and stem shall be stainless steel.

c. The range of the thermometer shall be such that the normal operating point with the system in service shall be midrange on the dial.

d. Accuracy of the thermometer shall be such that the error will be no greater than 2% of the full scale value.

e. The thermometer shall be provided with a brass or stainless steel well, installed in a threaded coupling into the pipe being measured. Provide thermally conductive gel in the well for contact with the bulb.

2.12 PRESSURE GAUGES

a. Provide pressure gauges where shown in details on the drawings and at every item of equipment receiving or producing flow in or from the system.

b. Gauges shall be no less than 4 in. (100 mm) in diameter, with glass face and cast aluminum body.

c. Range shall be such that gauge shall operate in midrange during normal operation of the system. Gauges subject to vacuum conditions, such as pump suctions, shall be rated for vacuum service. Graduations between figure intervals shall not exceed 10% of the gauge range. Pointer shall be adjustable for calibration.

d. Accuracy shall be such that error is less than 1% over the midrange of the gauge and 2% out of the midrange area.

e. Every gauge shall be provided with a 1/4-in. (6.35-mm) needle-type valve and an impulse dampener.

2.13 VACUUM BREAKERS

a. Water systems designed to automatically drain shall be provided with a vacuum breaker.

2.14 EXPANSION TANKS

a. Every closed system shall be provided with an expansion tank to prevent significant changes in system pressure caused by expansion and contraction of fluids due to temperature changes.

b. See Specifications for expansion tanks.

2.15 WATER MAKE-UP SYSTEMS

a. Every closed system shall be provided with a regulated water make-up system to automatically feed water to the system at a preset pressure.

b. Every water system not used for domestic water shall be provided with a reduced-pressure backflow preventer, certified for that service and acceptable to the local plumbing code administrator, to prevent backflow of contaminated water into the domestic water system.

c. See Specifications on water make-up system.

2.16 EXPANSION COMPENSATION

a. Pipe installation shall allow for expansion due to temperature differences. Provide expansion loops, or joints, as shown on the drawings, as follows.

b. Expansion joints shall be designed for the service, temperature, and pressure of the system and sized to accommodate the amount of movement required from maximum to minimum temperatures encountered by the system.

c. Joints for water systems shall be reinforced rubber corrugations secured by steel flanges. The number of corrugations shall be determined by the amount of movement required.

d. Every expansion joint shall be provided with stay bolts to prevent expansion beyond the normal design limits. Rubber joints shall have stay bolts isolated in rubber where used to eliminate vibration.

2.17 PIPE SLEEVES

a. Provide pipe sleeve of galvanized steel at each wall or floor penetration. Sleeve shall be no lighter than 18 gauge and shall be built into the wall or floor during construction of the wall. Where pipes are insulated, the sleeve shall allow for insulation thickness.

b. Wall sleeves shall be even with both sides of the finished wall.

c. Floor sleeves shall project approximately 1/2 in. (13 mm) above the finished floor and be even with the underside of the floor. Floor sleeves shall be cast in place or permanently sealed into the floor structure to prevent any water on the floor above from following the pipe system.

2.18 ESCUTCHEONS

a. Provide escutcheon on each side of wall or floor penetrations to provide a finished appearance. For insulated pipes, the escutcheon shall surround the outside of the insulation.

b. Escutcheons for small pipes may be spring clip type. Escutcheons for larger pipes shall be held by setscrews.

PART THREE - EXECUTION

3.1 INSTALLATION OF PIPE SUPPORTS

a. Install concrete inserts, beam clamps, or other fixtures to support the pipe hangers acceptable to the engineer.
b. Provide hanger rods and loops, or clevises, to support the pipe at the height and grade required for proper drainage and air elimination.

3.2 INSTALLATION OF PIPE

a. Cut pipe accurately to measurements, and ream free of burrs and cutting splatter. Carefully align and grade pipe, and work accurately into place. Fittings shall be used for any change in direction. Make adequate provisions for expansion and contraction. Install anchors to prevent pipe movement, as shown on the drawings. Provide expansion loops, or joints, as shown on the drawings and where required to compensate for pipe expansion. Provide for expansion at every building expansion joint.
b. Protect open pipe ends to prevent trash being placed in the lines during installation. Clean all dirt and cutting debris from pipes before making the next joint.
c. Small pipe shall be screwed or soldered as required to produce a tight system with full joints and no leaks. Pipe joints showing seepage and drips shall be dismantled and remade in proper way, as required by proper installation.
d. Copper pipe shall be carefully reamed back to full inside diameter, and the mating surfaces shall be cleaned by brush or sandpaper. When clean, the paste flux shall be applied and the joint evenly heated and soldered. Any fittings discolored by heat shall be removed and replaced.
e. Solder used in making joints shall be 50/50 for small lines operating below 200 deg. F (93 deg. C) and 100 psi (670 kPa). Solder joints operating above these conditions shall be made with 95/5 solder unless silver solder is called for.
f. Solder joints in critical lines, lines subject to temperatures above 200 deg. F (93 deg. C), and all lines installed in inaccessible locations (under floor slabs, etc.) shall be soldered with a product having a melting point of 1100 deg. F (593 deg. C) silver solder or above, applied with the proper torch and flux.
g. All valves to be soldered into lines shall be dismantled to prevent the heat from destroying packing and seats.
h. Valves installed in screwed lines shall be properly supported and pipes carefully installed to prevent damage or distortion of the valve.
i. Install drains at every low place and air vents at every high place. Pipe shall slope as shown on the drawings at 1 in. (25 mm) every 40 ft (12 m). If slope is not shown, slope in the direction of flow as required above. Install drain valves and air vents as specified.
j. Install pressure gauges and thermometers as shown in details and on drawings. Every pump and coil shall be provided with pressure gauge and thermometer. Every piece of equipment that produces hot water or steam or uses hot water or steam shall be provided with gauges and thermometers on entering and leaving sides.

3.3 CLEANING AND TREATING OF PIPE SYSTEMS

a. Every pipe system shall be cleaned to remove trash, mill scale, cutting oil, and welding and burning splatter from the lines before any control devices are installed. If such debris has collected in valves, the valves shall be disassembled and cleaned prior to closing for the first time.

b. After several hours of operation, each strainer shall be blown down. This shall be repeated as often as necessary to produce a clean discharge from the blowdown. Prior to turning system over to the owner, every strainer shall be removed and cleaned.

3.4 TESTING

a. Every pipe system shall be tested at 1.5 times its operating pressure, but no less than 125 psi (860 kPa) unless the engineer agrees to a lesser pressure.

b. Pipe and fittings shall be tested before any insulation or other covering is applied.

c. Testing may be performed in sections before vital equipment is connected if the test pressure is above the equipment rating.

d. Test medium shall be water under hydrostatic pressure with all air removed from the system. With engineer's consent, the test may be performed with compressed air to prevent danger from freezing. Hydrostatic pressure shall be held for no less than 2 hr with no drop in pressure. Air test shall be held for no less than 4 hr and the engineer may require longer test periods. Questionable joints shall be soaped to prove tightness.

e. The engineer, or representative, shall observe all tests. Notice to the engineer shall be given two full days before the testing is to be performed.

3.5 DESTRUCTIVE TESTING

a. The engineer reserves the right to select, at random, four fittings already completely installed for destructive testing. These joints shall be removed, by the contractor, with the connecting pipe, and the contractor shall replace these fittings at contractor's expense. The Engineer may destroy these joints by cutting them apart, separating soldered joints to check for full coverage, grinding the weld areas to observe voids, slag inclusions, or other defects. If major defects are noted in these joints, the contractor shall take corrective action to remedy the cause, and additional testing shall be performed to ensure that the system is adequate and complies with these Specifications and good workmanship.

END OF SECTION 15706

CHAPTER 11

Pumps and Accessories

SECTION 15750

****

BOOSTER PUMP

PART ONE - GENERAL

1.1 The GENERAL and SPECIAL CONDITIONS, Section 15100, are included as a part of this Section as though written in full in this document.

1.2 Scope of the Work shall include the furnishing and complete installation of the equipment covered by this Section, with all auxiliaries, ready for owner's use.

PART TWO - PRODUCTS

2.1 BOOSTER PUMP

a. Pump shall be designed for mounting in the pipe system and shall have capacity shown on the drawings at a speed and power input no greater than that shown.

2.2 PUMP CASING

a. Pump casing shall be cast iron, but all internal wetted surfaces, including the impeller, shall be stainless steel. Pump shall be lubricated by the liquid being pumped. Seal between motor and impeller shall not be required.

b. Companion flanges shall be furnished with the pump.

2.3 MOTOR

a. Motor shall have adjustable speed and power by means of external plug connectors. The motor bearings shall be tungsten carbide.

2.4 AUXILIARY EQUIPMENT

a. Provide manual or magnetic overload protection. If automatic control of the pump is required, provide magnetic starter.

2.5 ACCEPTABLE MANUFACTURERS

a. Pump shall be the model number shown on the drawings or equivalent by Grundfos (Series 6000).

PART THREE - EXECUTION

3.1 INSTALLATION

a. Install the pump as recommended by the manufacturer, using pipe flanges to support the pump. Install shutoff valve on both sides of the pump for isolation. Provide pressure gauges and thermometer as specified and shown on details.

b. Connect the electrical service to the pump terminal block as shown by manufacturer and required by codes. If automatic control of the booster pump is required, wire the internal automatically resetting overload thermal contacts in the motor in series with the external overload contacts to stop the pump in case of high motor temperature or excessive current draw.

c. Fill the system and vent it of all air. Purge the pump of air as recommended by manufacturer; then check for proper rotation.
d. Place the pump in service and check power draw, voltage, and proper system operation.
e. Report the actual current draw, flow, and other information required by Balancing and Testing, Section 15111.

END OF SECTION 15750

SECTION 15762

****

CIRCULATOR PUMP

PART ONE - GENERAL

1.1 The GENERAL and SPECIAL CONDITIONS, Section 15100, are included as a part of this Section as though written in full in this document.

1.2 Scope of the Work shall include the furnishing and complete installation of the equipment covered by this Section, with all auxiliaries, ready for owner's use.

PART TWO - PRODUCTS

2.1 PUMP

a. Pump shall be designed for mounting in the pipe system and shall have capacity shown on the drawings at a speed and power input no greater than that shown.

2.2 PUMP CASING

a. Pump casing shall be cast iron, bronze-fitted. Pump shall be oil-lubricated. Shaft shall be polished steel with integral thrust collar with mechanical seal of carbon on ceramic.

b. Companion flanges shall be furnished with the pump.

2.3 MOTOR

a. Motor shall be open drip-proof type with sleeve bearings, and shall be mounted in rubber for quiet operation. Motor shall have built-in thermal overload protection or shall be provided with a starter with thermal overload protection. Motor shall be connected to the pump through a flexible coupling.

2.4 AUXILIARY EQUIPMENT

a. Provide starter or contactor and external overload for single-phase motor if automatic control of the pump is required.

2.5 ACCEPTABLE MANUFACTURERS

a. Pump shall be the model number shown on the drawings or equivalent by Bell & Gossett (Series 100, 75, PR, HV, LD, HD, or PD).

PART THREE - EXECUTION

3.1 INSTALLATION

a. Install the pump as recommended by the manufacturer, using pipe flanges to support the pump. Install shutoff valve on both sides of the pump for isolation. Provide pressure gauges and thermometer as specified and shown on details.

b. Connect the electrical service to the pump terminal block as shown by manufacturer and required by codes. If automatic control of the circulator is required, provide motor starter or contactor.

c. Fill the system and vent it of all air. Purge the pump of air as recommended by manufacturer; then check for proper rotation.

d. Place the pump in service and check power draw, voltage, and proper system operation.

e. Report the actual current draw, pump flow, and other information required by Balancing and Testing, Section 15111.

END OF SECTION 15762

SECTION 15764
****
IN-LINE PUMP

PART ONE - GENERAL

1.1 The GENERAL and SPECIAL CONDITIONS, Section 15100, are included as a part of this Section as though written in full in this document.

1.2 Scope of the Work shall include the furnishing and complete installation of the equipment covered by this Section, with all auxiliaries, ready for owner's use.

PART TWO - PRODUCTS

2.1 PUMP

a. Pump shall be single-stage vertical split case type, designed for mounting in the pipe system, and shall have capacity shown on the drawings at a speed and power input no greater than that shown.

2.2 PUMP CASING

a. Pump casing shall be cast iron, bronze-fitted. Pump shaft shall be oil-lubricated. Shaft shall be polished steel with integral thrust collar with mechanical seal of carbon on ceramic.

b. Companion flanges shall be furnished with the pump.

2.3 MOTOR

a. Motor and pump shall be connected with a coupler assembly to reduce vibration and allow minor misalignment. Motor shall be open, drip-proof type with oil-lubricated sleeve bearings. Motor shall be mounted in rubber for quiet operation. Motor shall be provided with a magnetic starter with thermal overload protection.

2.4 AUXILIARY EQUIPMENT

a. Pump body shall be all bronze if used for domestic water service and cast iron, bronze-fitted for other applications.

2.5 ACCEPTABLE MANUFACTURERS

a. Pump shall be the model number shown on the drawings or equivalent by Bell & Gossett (Series 60).

PART THREE - EXECUTION

3.1 INSTALLATION

a. Install the pump as recommended by the manufacturer, using pipe hangers on either side to support the pump. Provide additional support under motor if required for stability.

b. Install shutoff valve on both sides of the pump for isolation. Provide pressure gauges and thermometer as specified and shown on details.

c. Connect the electrical service to the pump terminal block as shown by manufacturer and required by codes. If automatic control of the circulator is required, provide motor starter or contactor.

d. Fill the system and vent it free of all air. Purge the pump of air as recommended by manufacturer; then check for proper rotation.
e. Place the pump in service and check power draw, voltage, and proper system operation.
f. Report the actual current draw and pump flow and other information required by Balancing and Testing, Section 15111.

END OF SECTION 15764

SECTION 15766

****

DIRECT-CONNECTED PUMP

PART ONE - GENERAL

1.1 The GENERAL and SPECIAL CONDITIONS, Section 15100, are included as a part of this Section as though written in full in this document.

1.2 Scope of the Work shall include the furnishing and complete installation of the equipment covered by this Section, with all auxiliaries, ready for owner's use.

PART TWO - PRODUCTS

2.1 PUMP

a. Pump shall be single-stage vertical split case type, designed for mounting the impeller on the motor shaft, and shall have capacity shown on the drawings at speed and power input no greater than that shown.

2.2 PUMP CASING

a. Pump casing shall be cast iron, bronze-fitted. Pump shaft shall be oil-lubricated. Shaft shall be polished steel with integral thrust collar with mechanical seal of carbon on ceramic.

b. Companion flanges shall be furnished with the pump.

2.3 MOTOR

a. Motor and pump shall be connected with a coupler assembly to reduce vibration and allow minor misalignment. Motor shall be open drip-proof type, with sealed ball bearings. Motor shall be provided with a magnetic starter with thermal overload protection.

2.4 AUXILIARY EQUIPMENT

a. Pump body shall be all bronze if used for domestic water service.

2.5 ACCEPTABLE MANUFACTURERS

a. Pump shall be the model number shown on the drawings or equivalent by Weinman (Series AC or AE).

PART THREE - EXECUTION

3.1 INSTALLATION

a. Install the pump as recommended by the manufacturer. Install shutoff valves on both sides of the pump for isolation. Provide pressure gauges and thermometer as specified and shown on details.

b. Connect the electrical service to the pump terminal block as shown by manufacturer and required by codes. If automatic control of the circulator is required, provide motor starter or contactor.

c. Fill and vent the system of all air. Purge the pump of air as recommended by manufacturer; then check for proper rotation.
d. Place the pump in service and check power draw, voltage, and proper system operation.
e. Report the actual current draw and pump flow and other information required by Balancing and Testing, Section 15111.

END OF SECTION 15766

SECTION 15768

****

CLOSE-COUPLED PUMP

PART ONE - GENERAL

1.1 The GENERAL and SPECIAL CONDITIONS, Section 15100, are included as a part of this Section as though written in full in this document.

1.2 Scope of the Work shall include the furnishing and complete installation of the equipment covered by this Section, with all auxiliaries, ready for owner's use.

PART TWO - PRODUCTS

2.1 PUMP

a. Pump shall be single-stage vertical split case type, designed for mounting on the motor bracket, and shall have capacity shown on the drawings at a speed and power input no greater than that shown.

2.2 PUMP CASING

a. Pump casing shall be cast iron, bronze-fitted. Pump shaft shall be oil-lubricated. Shaft shall be polished steel with integral thrust collar with mechanical seal of carbon on ceramic.

b. Companion flanges shall be furnished with the pump.

2.3 MOTOR

a. Motor and pump shall be connected with a coupler assembly to reduce vibration and allow minor misalignment. Motor shall be open drip-proof type with oil-lubricated sleeve bearings and shall be mounted in rubber for quiet operation. Motor shall be provided with a magnetic starter with thermal overload protection.

2.4 AUXILIARY EQUIPMENT

a. Pump body shall be all bronze if used for domestic water service.

b. Provide external motor overload protection by use of manual or magnetic starters.

c. Motor shall be provided with ball bearings for 3500-rpm operation.

2.5 ACCEPTABLE MANUFACTURERS

a. Pump shall be the model number shown on the drawings or equivalent by Bell & Gossett (Series 1535).

PART THREE - EXECUTION

3.1 INSTALLATION

a. Install the pump as recommended by the manufacturer. Install shutoff valves on both sides of the pump for isolation. Provide pressure gauges and thermometer as specified and shown on details.

b. Connect the electrical service to the pump terminal block as shown by manufacturer and required by codes. If automatic control of the circulator is required, provide motor starter or contactor.

c. Fill and vent the system of all air. Purge the pump of air as recommended by manufacturer; then check for proper rotation.

d. Place the pump in service and check power draw, voltage, and proper system operation.

e. Report the actual current draw and pump flow and other information required by Balancing and Testing, Section 15111.

END OF SECTION 15768

SECTION 15770
****
BASE-MOUNTED PUMP

PART ONE - GENERAL

1.1 The GENERAL and SPECIAL CONDITIONS, Section 15100, are included as a part of this Section as though written in full in this document.

1.2 Scope of the Work shall include the furnishing and complete installation of the equipment covered by this Section, with all auxiliaries, ready for owner's use.

PART TWO - PRODUCTS

2.1 PUMP
a. Pump shall be single-stage vertical split case type, designed for mounting on fabricated steel base, and shall have capacity and speed shown on the drawings and power input no greater than that shown.

2.2 PUMP CASING AND SEAL
a. Pump casing shall be cast iron, bronze-fitted. Pump bearings shall be oil-lubricated. Shaft shall be polished steel with integral thrust collar with mechanical seal of carbon on ceramic or tungsten carbide. Pump shaft shall be provided with a replaceable sleeve to cover all wetted area of the shaft under the seal. Pump impeller shall be enclosed type, balanced and keyed to the shaft.

2.3 MOTOR
a. Motor and pump shall be connected with a coupler assembly to reduce vibration and allow minor misalignment. Motor shall be open drip-proof type, with oil-lubricated sleeve bearings, and shall be mounted in rubber for quiet operation. Motor shall be provided with a magnetic starter with thermal overload protection.

2.4 AUXILIARY EQUIPMENT
a. Pump body shall be all bronze if used for domestic water service and cast iron, bronze-fitted for other applications.
b. Seal shall be stuffing box-type with impregnated asbestos packing, plus lantern ring for flushing.
c. Provide magnetic starter for motor overload protection.

2.5 ACCEPTABLE MANUFACTURERS
a. Pump shall be the model number shown on the drawings or equivalent by Bell & Gossett (Series 1510) or Weinman.

PART THREE - EXECUTION

3.1 INSTALLATION

a. Install the pump as recommended by the manufacturer and as shown in details on the drawings.
b. Connect the electrical service to the pump terminal block as shown by manufacturer and required by codes. If automatic control of the circulator is required, provide motor starter or contactor.
c. Fill the system and vent it of all air. Purge the pump of air as recommended by manufacturer; then check for proper rotation.
d. Place the pump in service and check power draw, voltage, and proper system operation.
e. Report the actual current draw and pump flow, and other information required by Balancing and Testing, Section 15111.

END OF SECTION 15770

SECTION 15772
****
SINGLE-STAGE DOUBLE-SUCTION PUMP

PART ONE - GENERAL

1.1 The GENERAL and SPECIAL CONDITIONS, Section 15100, are included as a part of this Section as though written in full in this document.

1.2 Scope of the Work shall include the furnishing and complete installation of the equipment covered by this Section, with all auxiliaries, ready for owner's use.

PART TWO - PRODUCTS

2.1 PUMP

a. Pump shall be single-stage, double-suction radially split case type, designed for mounting pump and motor on fabricated steel base, and shall have capacity and speed shown on the drawings and power input no greater than that shown.

2.2 PUMP CASING AND SEAL

a. Pump casing shall be cast iron, bronze-fitted. Pump bearings shall be regreaseable ball bearings. Shaft shall be 18-8 stainless steel with mechanical seal of carbon on ceramic or tungsten carbide. Pump impellers shall be double-inlet, enclosed type, balanced and keyed to the shaft.

2.3 COUPLING

a. Motor and pump shall be connected with a coupler assembly to reduce vibration and allow minor misalignment.

2.4 MOTOR

a. Motor shall be open drip-proof type with regreaseable ball bearings.

2.5 AUXILIARY EQUIPMENT

a. Seals shall be stuffing-box type with impregnated asbestos packing, plus lantern ring for flushing. Provide 18-8 stainless steel shaft sleeve with stuffing boxes.

b. Provide magnetic starter for motor overload protection.

2.6 ACCEPTABLE MANUFACTURERS

a. Pump shall be the model number shown on the drawings or equivalent by Bell & Gossett (Series VSC) or acceptable alternates.

PART THREE - EXECUTION

3.1 INSTALLATION

a. Install the pump as recommended by the manufacturer and as shown on the drawings, with concrete base and vibration pads under the base. Install shutoff valve on both sides of the pump for isolation. Provide pressure gauges and thermometer as specified and shown on details.

b. Connect the electrical service to the pump terminal block as shown by manufacturer and required by codes.

c. Fill and vent the system of all air. Purge the pump of air as recommended by manufacturer; then check for proper rotation.

d. Place the pump in service and check power draw, voltage, and proper system operation.

e. Report the actual current draw and pump flow and other information required by Balancing and Testing, Section 15111.

END OF SECTION 15772

SECTION 15774
****
HORIZONTAL SPLIT CASE DOUBLE-SUCTION PUMP

PART ONE - GENERAL

1.1 The GENERAL and SPECIAL CONDITIONS, Section 15100, are included as a part of this Section as though written in full in this document.

1.2 Scope of the Work shall include the furnishing and complete installation of the equipment covered by this Section, with all auxiliaries, ready for owner's use.

PART TWO - PRODUCTS

2.1 a. Pump shall be single-stage, double-suction horizontal split case type, designed for mounting pump and motor on fabricated steel base, and shall have capacity and speed shown on the drawings and power input no greater than that shown.

2.2 PUMP CASING AND SEAL
a. Pump casing shall be cast iron, bronze-fitted. Pump bearings shall be regreaseable ball bearings. Shaft shall be 18-8 stainless steel with mechanical seal of carbon on ceramic or tungsten carbide. Pump impeller shall be double-inlet enclosed type, balanced and keyed to the shaft.

2.3 COUPLING
a. Motor and pump shall be connected with a coupler assembly to reduce vibration and allow minor misalignment.

2.4 MOTOR
a. Motor shall be open drip-proof type with regreaseable ball bearings.

2.5 AUXILIARY EQUIPMENT
a. Seals shall be stuffing-box type with impregnated asbestos packing, plus lantern ring for flushing. Provide 18-8 stainless steel shaft sleeve with stuffing boxes.

2.6 ACCEPTABLE MANUFACTURERS
a. Pump shall be the model number shown on the drawings or equivalents by Weinman (Type L), Armstrong, Aurora, or acceptable alternates.

PART THREE - EXECUTION

3.1 INSTALLATION
a. Install the pump as recommended by the manufacturer and as shown on the drawings, with concrete base and vibration pads under the base. Install shutoff valves on both sides of the pump for isolation. Provide pressure gauges and thermometer as specified and shown on details.

b. Connect the electrical service to the pump terminal block as shown by manufacturer and required by codes.
c. Fill and vent the system of all air. Purge the pump of air as recommended by manufacturer; then check for proper rotation.
d. Place the pump in service and check power draw, voltage, and proper system operation.
e. Report the actual current draw and pump flow and other information required by Balancing and Testing, Section 15111.

END OF SECTION 15774

SECTION 15776

****

HORIZONTAL SPLIT CASE MULTISTAGE PUMP

PART ONE - GENERAL

1.1 The GENERAL and SPECIAL CONDITIONS, Section 15100, are included as a part of this Section as though written in full in this document.

1.2 Scope of the Work shall include the furnishing and complete installation of the equipment covered by this Section, with all auxiliaries, ready for owner's use.

PART TWO - PRODUCTS

2.1 PUMP

a. Pump shall be multistage, single-suction, horizontal split case type, designed for mounting pump and motor on fabricated steel base, and shall have capacity and speed shown on the drawings and power input no greater than that shown.

2.2 PUMP CASING AND SEAL

a. Pump casing shall be cast iron, bronze-fitted. Pump bearings shall be regreaseable ball bearings. Shaft shall be 18-8 stainless steel with mechanical seal of carbon on ceramic or tungsten carbide. Pump impeller shall be double-inlet, enclosed type, balanced and keyed to the shaft.

2.3 COUPLING

a. Motor and pump shall be connected with a coupler assembly to reduce vibration and allow minor misalignment.

2.4 MOTOR

a. Motor shall be open drip-proof type with regreaseable ball bearings.

2.5 AUXILIARY EQUIPMENT

a. Seals shall be stuffing-box type with impregnated asbestos packing, plus lantern ring for flushing. Provide 18-8 stainless steel shaft sleeve with stuffing boxes.

2.6 ACCEPTABLE MANUFACTURERS

a. Pump shall be the model number shown on the drawings or equivalents by Weinman (Type JD), Armstrong, Aurora.

PART THREE - EXECUTION

3.1 INSTALLATION

a. Install the pump as recommended by the manufacturer and as shown on the drawings, with concrete base and vibration pads under the base. Install shutoff valves on both sides of the pump for isolation. Provide pressure gauges and thermometer as specified and shown on details.

b. Connect the electrical service to the pump terminal block as shown by manufacturer and required by codes.

c. Fill and vent the system of all air. Purge the pump of air as recommended by manufacturer; then check for proper rotation.

d. Place the pump in service and check power draw, voltage, and proper system operation.

e. Report the actual current draw and pump flow, and other information required by Balancing and Testing, Section 15111.

END OF SECTION 15776

SECTION 15780

****

ACCESSORIES FOR PUMPED SYSTEMS

PART ONE - GENERAL

1.1 The GENERAL and SPECIAL CONDITIONS, Section 15100, are included as a part of this Section as though written in full in this document.

1.2 Scope of the Work shall include the furnishing and complete installation of the equipment covered by this Section, with all auxiliaries, ready for owner's use.

PART TWO - PRODUCTS

2.1 a. Accessories for pumped systems shall be furnished as required below:

2.1.1 BACKFLOW PREVENTER FOR MAKE-UP WATER SYSTEM

a. Provisions shall be made for filling each water system from a source of clean domestic water at the pressure required to force water into the static system and to provide make-up water to the system.

b. If a nonpotable water system is not provided by others, provide a reduced-pressure-zone backflow preventer at the point of connection to the domestic water system. This backflow preventer shall be acceptable to the plumbing official in the area where the installation is to be made.

c. The size and capacity of the backflow preventer shall match the system in which installed, but the capacity shall be not less than 1% of the flow of the system pump at normal operating pressures.

d. The backflow preventer shall be as shown on the drawings or acceptable equivalent by Watts Regulator Co. (900 Series).

2.1.2 INSTALLATION

a. The backflow preventer shall be installed in an area with an acceptable drain to accept the small quantity of water normally released from such a device. If location of installation is not suitable for an exposed drip, provide a galvanized drain pan located under the vent opening but not interfering with this "open sight" drain, as required by the plumbing code. Pipe the drain from the pan to an acceptable drain point with type M copper or DWV plastic pipe, no less than 1 in. (25 mm) in size. Slope drain in direction of flow.

b. Secure acceptance by the Plumbing Official.

2.2 PRESSURE-REGULATING SYSTEM

2.2.1 PRESSURE REGULATOR

a. The make-up water to every closed system shall be provided through a water pressure regulator to maintain a minimum/maximum pressure in the water system sufficient to force water to the top of the system, with a residual pressure at the uppermost fitting of 10% of the pressure required.

b. The closed system shall be provided with an ASME-stamped relief valve set at approximately 20% above the fill pressure.
c. A combination reducing valve with built-in relief valve may be used to meet the above requirements.
d. The capacity of the regulator shall be no less than 1% of the pump volume at system operating pressure.
e. The capacity of the relief valve shall be determined by the type of system, fired or nonfired, and the applicable code.
f. The pressure regulator shall be as shown on the drawings, or approved equivalent by Bell & Gossett or Watts Regulator.

2.2.2 INSTALLATION
a. Install the make-up assembly generally as shown on the drawings in a location acceptable for this type of equipment.
b. Provide a strainer on the inlet side and a shutoff valve and union on either side of the strainer/regulator/relief valve combination for service of the system.
c. Provide a bypass line with a throttling valve to allow manual operation of the system with the make-up assembly removed and to allow quick filling of the system.
d. Fill the system, check operation of the pressure regulator, and adjust to the correct pressure if necessary. Check the relief valve by manually opening the valve and seeing that it reseats properly.

2.3 OPERATING PRESSURE CONTROL

2.3.1 EXPANSION TANK
a. Provide a bladder type expansion tank which will allow the extra volume of water created by increase in temperature to flow into a bladder inside a steel tank under an adjustable and regulated air pressure.
b. The capacity of the bladder and the tank shall be matched to restrict pressure increase in the system to 10% over the static fill pressure.
c. Tank shall be ASME-stamped and provided with a steel ring base for mounting on the floor and shall include a suitable handhole for replacement of the bladder.
d. Tank shall be equivalent to the tank shown on the drawings or acceptable equal by Amtrol or Bell & Gossett.

2.3.2 INSTALLATION
a. Install the expansion tank generally as shown on the drawings and as recommended by the manufacturer.
b. Pipe the tank to the water system near the suction inlet of the system pump with pipe sized to match the system.
c. Fill the system and adjust the air in the tank to the proper pressure to match the static system pressure.

2.3.3 COMPRESSION TANK (ALTERNATE TO EXPANSION TANK)

a. Provide an ASME-stamped steel compression tank, sized to receive the excess water created by the expansion of the system with an increase in system pressure of no more than 10% of the system static pressure.
b. The compression tank shall be provided with red-line sight glass, with gage glass cocks at top and bottom.
c. Provide an air separator and a tank fitting, as recommended by the manufacturer for the system size involved.
d. Compression tank, air separator, and tank fitting shall be as shown on the drawings or equivalents by Bell & Gossett.

2.3.4 INSTALLATION

a. Install the compression tank at or above the ceiling of the mechanical room being served. Supports shall be capable of supporting the tank when full of water.
b. Install the air separator in the system near and just ahead of the pump suction. Connect the compression tank to the air separator with a line sloped up to the tank. Connect the water make-up into this line so that any air brought in with the water will rise to the tank.
c. Fill the system with water and observe the water level in the tank to ensure that the top of the gauge glass is not leaking air.

2.3.5 AIR ELIMINATION

2.3.5.1 AIR VENTS (FOR USE WITH DIAPHRAGM EXPANSION TANKS)

a. Air eliminator shall provide a means of separating the air from the fluid and of venting the air from the system automatically.
b. Main air vent shall be as shown on the drawings or equivalent by Bell & Gossett, or accepted equivalents.

2.3.5.2 INSTALLATION

a. Provide main air vent in system, after heat source but before the pump suction, to eliminate air circulating in the system.
b. Install the air vent in an accessible location with a drain, or provide a vent tube from the air eliminator to a satisfactory location where the vent can be installed.

2.3.6 AIR SEPARATORS (FOR USE WITH COMPRESSION TANKS)

2.3.6.1 SEPARATORS

a. Air separators shall be designed to remove the air from the fluid in the system and divert the air to the compression tank.
b. Air separator may be of centrifugal type incorporating a strainer in the same system component, or may be in the form of a dip tube installed in the proper fittings in the heating source such as a boiler.

2.3.6.2 INSTALLATION

a. Install the separation device as generally shown on the drawings and as recommended by the manufacturer.
b. Do not install any valves between the separator and the compression tank.

2.4 PUMP VALVES AND FITTINGS

2.4.1 SUCTION DIFFUSER

a. Provide a combination strainer and straightening device on the suction of each pump to reduce cavitation due to vortex at the pump.

b. Strainer shall be removable and shall be provided with a ball-type blowdown valve.

c. Provide start-up strainer to protect pump from debris.

d. Strainer shall be 16-mesh bronze.

e. Maximum pressure drop through the suction diffuser and fine-mesh strainer shall not exceed 1% of the system final pressure.

f. Suction diffuser shall be as shown on the drawings or equivalent to Bell & Gossett Suction Diffuser (Type X, Y, or Z).

2.4.2 INSTALLATION

a. Provide support leg for the suction diffuser to prevent stress on the pump housing.

b. Locate the pump and suction diffuser so that strainer can be removed for cleaning. Install the strainer blow-off ball valve and pipe to an adequate open site drain.

c. If pipe is insulated, install removable blanket-type insulation on the suction diffuser for easy removal and replacement.

d. Prior to acceptance by the owner, remove the start-up strainer, and clean the fine mesh strainer.

2.5 DISCHARGE VALVES

2.5.1 VALVE

a. Install a combination valve in every pump discharge performing the functions of a shutoff valve, a spring-loaded check valve, and a balancing valve with a memory stop. Valve shall be angle or straight-through pattern, with pressure loss through the valve of no more than 2% of the total pump head.

b. Valve shall be as shown on the drawings or equivalent by Bell & Gossett (Triple Duty), or accepted alternates.

2.5.2 INSTALLATION

a. Install the valve as recommended by the manufacturer. Provide access to remove the internal parts for service.

b. Check the operation of the valve and the pressure drop across the valve, and regulate the flow as required for the system.

c. Read the pressure drop carefully and determine the system flow from the manufacturer's pressure drop charts.

d. Report the pressure drop and flow as required in Balancing and Testing, Section 15111.

2.6 VIBRATION ISOLATION

2.6.1 FLEXIBLE CONNECTORS

a. Provide a rubber-type flexible connector in every pump suction and discharge, as close to the pump as possible to absorb vibration and prevent noise transmission.
b. Flex connector shall be reinforced Butyl rubber, designed and cataloged for this service and temperature, equivalent to Keyflex (Ke-Max) or other acceptable alternates.
c. Provide stay bolts cushioned in rubber to limit travel of the joint.

2.6.2 INSTALLATION

2.6.2.1 MOUNTING

a. Mount the flex connectors in the suction and discharge piping as recommended by the manufacturer. Prevent any tension, torsion, or compression on the connectors from the pipe system.
b. Tighten the connecting bolts with a torsion wrench as recommended by the manufacturer.
c. Check the installation for leaks and remake any leaking joints.

2.7 FLOOR-MOUNTED PUMP BASE

2.7.1 CONCRETE BASE

a. Form and pour a concrete base 6 in. (150 mm) thick for each floor-mounted pump. Base shall be larger than the pump dimensions by no less than 2 in. (50 mm) in each direction. Provide a recess in the top surface to act as a sump to collect any condensate or fluid drip from the pump.

2.7.2 INSTALLATION

a. Mount the concrete base on four corner pads of vibration pad material, sized for the weight of the pump and base.
b. Extend the drain to the nearest acceptable open site drain.
c. Secure the pump to the base with anchor bolts as recommended by the manufacturer. Pour cement grout between the pump frame and the base to provide a uniform, level mounting surface for the pump frame.
d. Check the pump for alignment and place in service.

END OF SECTION 15780

CHAPTER 12

Boilers

SECTION 15800

****

HOT WATER BOILER
FINNED WATER TUBE TYPE

PART ONE - GENERAL

1.1 The GENERAL and SPECIAL CONDITIONS, Section 15100, are included as a part of this Section as though written in full in this document.

1.2 Scope of the Work shall include the furnishing and complete installation of the equipment covered by this Section, with all auxiliaries, ready for owner's use.

PART TWO - PRODUCTS

2.1 HOT WATER BOILER

a. Boiler shall be natural (LP) gas-fired water tube type, designed for hot water service, with capacity as shown on the drawings.

b. Boiler shall be AGA-certified. Burners shall be natural draft, stainless steel, and mounted in a removable drawer for service.

c. All wet surfaces shall be brass or bronze. Heat exchanger tubes shall be cleanable, with extended external surface with bronze headers. The boiler shall bear the National Board ASME stamp for 160 psig (1100 kPa).

d. Combustion chamber shall be insulated with a lightweight cast refractory backed up with fiberglass insulation.

e. Boiler jacket shall be heavy-gauge galvanized (or equal) steel with factory-applied baked enamel.

f. Boiler controls shall be provided with a 110- to 24-volt transformer. The main gas valve shall be 24 volts.

g. Boiler controls shall provide for 80% modulation in firing rate and shall allow a leaving water temperature as low as 105 deg. F (40 deg. C) without condensation. Provide electronic ignition of the pilot. Gas safety controls shall prove pilot flame prior to opening of the main gas valve. The pilot shall be lit only when the boiler is calling for heat.

h. Burner capacity reduction shall be controlled by an electric (self-contained) sensing element reset by the outdoor air temperature, so that at 0 deg.F (-18 deg. C) outside the water temperature shall be 200 deg. F (105 deg. C) and at 70 deg. F (21 deg. C) outside the water temperature shall be 110 deg. F (43 deg. C). Provide a mechanical preset high-limit control set at 245 deg. F (118 deg. C) and an electric high-limit automatic reset control set at 242 deg. F (116 deg. C).

i. Provide a low-water cutoff to shut the burner down if the header is not completely full of water.

j. Provide a flow switch, or pressure differential switch, to prove water flow before the controls can be energized. The boiler controls shall be interlocked through the controls of the pump that is providing water flow for the boiler.

k. Provide draft hood as required for AGA approval. Provide ASME pressure relief valve, combination temperature and pressure gauge, drain valve, and other accessories recommended by the manufacturer and required for a complete installation.

2.2 AUXILIARY EQUIPMENT

a. Provide high-pressure gas regulator, or additional gas regulator, where the available gas pressure is above the manufacturer's standards.

b. Provide controls to continue pump operation until the heat exchanger is cooled and refractory heat is absorbed. Provide pump interlock wiring with the controls.

c. Provide antifreeze in the system, as noted on the drawings, for freeze protection down to -20 deg. F (-30 deg. C). Capacity of the boiler shall be adjusted to meet the output required on the drawings with the derating required by the antifreeze.

d. Provide a one-shot (batch) feeder for the addition of water treatment and antifreeze to the system.

2.3 ACCEPTABLE MANUFACTURERS

a. Boiler shall be the make and model number shown on the drawings or equivalent by Raypak (Model T), Lochinvar (Model CHW), Teledyne Laars (Series H).

PART THREE - EXECUTION

3.1 MECHANICAL SPACE AND INSTALLATION

a. The boiler shall be installed in a space reserved for the boiler and provided with combustion air and ventilation required by the codes and shown on the drawings.

b. The boiler flue pipe shall be UL-listed and shall be the full size of the draft hood outlet. Provide a UL-listed vent cap, and install cap and pipe as required by the code.

c. The floor of the mechanical space shall be noncombustible, or a noncombustible pad shall be installed under the boiler, as required by the boiler manufacturer's AGA rating.

d. Provide city water for make-up to the boiler. Include a water make-up assembly which includes a reduced-pressure type of backflow preventer acceptable to the local authorities. The mechanical room must be provided with an adequate drain and the floor must be waterproof. (Or the boiler shall be installed in a 20-gauge metal pan with soldered joints. Connect the pan to an adequate drain with 1-in. (25-mm) type M copper).

e. Connect the gas with a gas cock installed above ground and outside of the mechanical space, in addition to the gas cock furnished with each boiler gas train. Install gas regulator as required to reduce gas pressure to that required by the boiler. (Do not connect gas if the mechanical room is not ventilated according to code!)

f. Connect hot water lines and pump to the boiler with adequate drains to flush and drain the system. Install the relief valve and carry full size to a point acceptable for discharge of hot water.

g. Make the proper electrical connections for the boiler and equipment. Provide a disconnect means for the boiler controls, to turn off power to the main gas valve from an easily accessible point inside the mechanical room door.

h. Fill the boiler with water, and pressure-test the boiler and system up to the rating of the relief valve. Flush the system to remove all trash and dirt, and refill the system, including inhibitor and antifreeze, as specified.

i. Bleed the gas line in a safe manner and energize the boiler controls.

3.2 TEST

a. Observe the ignition of the pilot and the main burner to make sure they are smooth and complete.

b. Check out the safety controls and verify that they are functioning properly. Report the actual and design pressure drop through the boiler.

c. Set the operating controls for the proper temperature.

d. Operate the boiler for no less than 2 hr, or return at the beginning of the heating season for completion of this test. Vary the outdoor sensor temperature from 32 deg. F (0 deg. C) to 70 deg. F (21 deg. C) and observe the reset of the output.

e. Provide a written report, through channels and on company letterhead, that ignition is proper, safety controls have been checked, and operating controls are set and functioning properly. Report the actual and design pressure drop through the boiler. The system is not complete until this report has been received.

3.3 OWNER'S INSTRUCTIONS

a. Provide the owner's operator with three copies of written instructions for operation and maintenance of the system. A letter shall be sent, through channels, listing the materials given to the operator, the operator's name, and the date of this transmittal.

b. The operator shall be instructed carefully in the methods, and necessity, for blowing down the boiler to remove the accumulated mud on a regular basis.

END OF SECTION 15800

SECTION 15801

****

HOT WATER BOILER
FINNED WATER TUBE TYPE
OUTDOOR INSTALLATION

PART ONE - GENERAL

1.1 The GENERAL and SPECIAL CONDITIONS, Section 15100, are included as a part of this Section as though written in full in this document.

1.2 Scope of the Work shall include the furnishing and complete installation of the equipment covered by this Section, with all auxiliaries, ready for owner's use.

PART TWO - PRODUCTS

2.1 a. Boiler shall be natural (LP) gas-fired water tube type, designed for hot water service, with capacity as shown on the drawings.

b. Boiler shall be AGA-certified for outdoor installation without additional wind or weather protection. Burners shall be natural draft, stainless steel, and mounted in a removable drawer for service.

c. All wet surfaces shall be brass or bronze. Heat exchanger tubes shall be cleanable, with extended external surface with bronze headers. The boiler shall bear the National Board ASME stamp for 160 psig (1100 kPa).

d. Combustion chamber shall be insulated with a lightweight cast refractory and fiberglass insulation.

e. Boiler jacket shall be heavy-gauge galvanized (or equal) steel with factory-applied baked enamel.

f. Boiler controls shall be provided with a 110- to 24-volt transformer. The main gas valve shall be 24 volts. All controls shall be inside the boiler casing or housed in weatherproof electrical boxes with restricted access.

g. Boiler controls shall provide for 80% modulation in firing rate and shall allow a leaving water temperature as low as 105 deg. F (40 deg. C) without condensation. Provide electronic ignition of the pilot. Gas safety controls shall prove pilot flame prior to opening of the main gas valve. The pilot shall be lit only when the boiler is calling for heat.

h. Burner capacity reduction shall be controlled by an electric (self-contained) sensing element reset by the outdoor air temperature, so that at 0 deg. F (-18 deg. C) outside, the water temperature shall be 200 deg. F (105 deg. C); and at 70 deg. F (21 deg. C) outside, the water temperature shall be 110 deg. F (43 deg. C). (Ratio shall be adjustable.) Provide a preset mechanical high-limit control set at 245 deg. F (118 deg. C) and an electric high-limit automatic reset control set at 242 deg. F (116 deg. C).

i. Provide a flow switch, or pressure differential switch, to prove water flow before the controls can be energized. The boiler controls shall be interlocked through the controls of the pump that is providing water flow for the boiler.

j. Provide-storm proof draft hood, as required for AGA approval for outside installation. Provide ASME pressure relief valve, combination temperature and pressure gauge, drain valve, and other accessories recommended by the manufacturer and required for a complete installation.

2.2 AUXILIARY EQUIPMENT

a. Provide high-pressure gas regulator, or additional gas regulator, where the available gas pressure is above the manufacturer's standards.

b. Provide controls to continue pump operation until the heat exchanger is cooled and refractory heat is absorbed. Provide pump interlock wiring with the controls.

c. Provide antifreeze in the system, as noted on the drawings for freeze protection down to -20 deg. F (-30 deg. C). Capacity of the boiler shall be adjusted to meet the output required on the drawings with the derating required by the antifreeze.

d. Provide a one-shot (batch) feeder for the addition of water treatment and antifreeze to the system.

2.3 ACCEPTABLE MANUFACTURERS

a. Boiler shall be the make and model number shown on the drawings or equivalent by Raypak (Model TP), Lochinvar (Model CHW), Teledyne Laars (Series HJ).

PART THREE - EXECUTION

3.1 INSTALLATION

a. The boiler shall be installed on a concrete pad of sufficient size to allow a 24-in. (60-cm) pad extension on all sides. Pad shall be no less than 6 in. (15 cm) above adjoining grade on any side.

b. Provide city water for make-up to the boiler. Locate in a heated space and connect the make-up water to the return side of the pump. Include a water make-up assembly which includes a reduced-pressure-type backflow preventer acceptable to the local authorities.

c. Provide gas regulator to reduce the gas pressure to that required by the boiler.

d. Connect hot water lines and pump to the boiler with adequate drains to flush and drain the system. Install the relief valve and carry full size to a point acceptable for discharge of hot water.

e. Make the proper electrical connections for the boiler and equipment in weatherproof boxes. Provide a disconnect means for the boiler controls to turn off power to the main gas valve from an easily accessible point near the boiler.

f. Fill the boiler with water, and pressure-test the boiler and system up to the rating of the relief valve. Clean the system with trisodium phosphate or equal cleaner, flush the system to remove all trash and dirt, and refill the system, including inhibitor and antifreeze as specified.

g. Bleed the gas line in a safe manner and energize the boiler controls.

3.2 TEST

a. Observe the ignition of the pilot and the main burner to make sure they are smooth and complete.

b. Check out the safety controls and verify that they are functioning properly. Report the actual and design pressure drop through the boiler.

c. Set the operating controls for the proper temperature.

d. Operate the boiler for no less than 2 hr or return at the beginning of the heating season for completion of this test. Vary the outdoor sensor temperature from 32 deg. F (0 deg. C) to 70 deg. F (21 deg. C) and observe the reset of the output.

e. Provide a written report, through channels and on company letterhead, that ignition is proper, safety controls have been checked, and operating controls are set and functioning properly. Report the actual and design pressure drop through the boiler. The system is not complete until this report has been received.

3.3 OWNER'S INSTRUCTIONS

a. Provide the owner's operator with three copies of written instructions for operation and maintenance of the system. A letter shall be sent, through channels, listing the materials given to the operator, the operator's name, and the date of this transmittal.

b. The operator shall be instructed carefully in the methods, and necessity, for blowing down the boiler to remove the accumulated mud on a regular basis.

END OF SECTION 15801

SECTION 15802

****

PACKAGED HOT WATER BOILER
FINNED WATER TUBE TYPE

PART ONE - GENERAL

1.1 The GENERAL and SPECIAL CONDITIONS, Section 15100, are included as a part of this Section as though written in full in this document.

1.2 Scope of the Work shall include the furnishing and complete installation of the equipment covered by this Section, with all auxiliaries, ready for owner's use.

PART TWO - PRODUCTS

2.1 a. Packaged boiler shall be natural (LP) gas-fired water tube type, designed for hot water service, including circulating pump and expansion tank, with capacity as shown on the drawings.

b. Boiler shall be AGA-certified. Burners shall be natural draft, stainless steel, and mounted in a removable drawer for service.

c. All wet surfaces shall be brass or bronze. Heat exchanger tubes shall be cleanable, with extended external surface with bronze headers. The boiler shall bear the National Board ASME stamp for 160 psig (1100 kPa).

d. Pump shall be centrifugal type with cast iron body and bronze-fitted for hot water service. See drawings for type and capacity.

e. Expansion tank shall be bladder type with the water contained inside the bladder, sized to accept the expanded water without pressure increase of more than 10% of the fill pressure.

f. Combustion chamber shall be insulated with a lightweight cast refractory backed up with fiberglass insulation.

g. Boiler jacket shall be heavy-gauge galvanized (or equal) steel with factory-applied baked enamel.

h. Boiler controls shall be provided with a 110- to 24-volt transformer. The main gas valve shall be 24 volts.

i. Boiler controls shall provide for 80% modulation in firing rate and shall allow a leaving water temperature as low as 105 deg. F (40 deg. C) without condensation. Provide electronic ignition of the pilot. Gas safety controls shall prove pilot flame prior to opening of the main gas valve. The pilot shall be lit only when the boiler is calling for heat.

j. Burner capacity reduction shall be controlled by an electric (self-contained) sensing element reset by the outdoor air temperature, so that at 0 deg. F (-18 deg. C) outside, the water temperature shall be 200 deg. F (105 deg. C), and at 70 deg. F (21 deg. C) outside, the water temperature shall be 110 deg. F (43 deg. C). Provide a mechanical preset high-limit control set at 245 deg. F (118 deg. C) and an electric high-limit automatic reset control set at 242 deg. F (116 deg. C).

k. Package shall include a flow switch, or pressure differential switch, to prove water flow before the controls can be energized. The boiler controls shall be interlocked through the controls of the pump that is providing water flow for the boiler.

1. Provide draft hood as required for AGA approval. Provide ASME pressure relief valve, combination temperature and pressure gauge, drain valve, and other accessories recommended by the manufacturer and required for a complete installation.

2.2 AUXILIARY EQUIPMENT

a. Provide high-pressure gas regulator, or additional gas regulator, where the available gas pressure is above the manufacturer's standards.

b. Provide controls to continue pump operation until the heat exchanger is cooled and refractory heat is absorbed. Provide pump interlock wiring with the controls.

c. Provide antifreeze in the system, as noted on the drawings, for freeze protection down to -20 deg. F (-30 deg. C). Capacity of the boiler shall be adjusted to meet the output required on the drawings with the derating required by the antifreeze.

d. Provide a one-shot (batch) feeder for the addition of water treatment and antifreeze to the system.

e. Provide pressure gauges for inlet and outlet of boiler and the pump, reading in small increments for measurement of pressure drop.

2.3 ACCEPTABLE MANUFACTURERS

a. Boiler shall be the make and model number shown on the drawings or equivalent by Raypak (Model TP), Lochinvar (Model CHW), Teledyne Laars (Series H).

PART THREE - EXECUTION

3.1 MECHANICAL SPACE AND INSTALLATION

a. The boiler shall be installed in a space reserved for the boiler and provided with combustion air and ventilation required by the codes and shown on the drawings.

b. The boiler flue pipe shall be UL-listed and shall be the full size of the draft hood outlet. Provide a UL-listed vent cap, and install cap and pipe as required by the code.

c. The floor of the mechanical space shall be noncombustible, or a noncombustible pad shall be installed under the boiler, as required by the boiler manufacturer's AGA rating.

d. Provide city water for make-up to the boiler. Include a water make-up assembly which includes a reduced-pressure type of backflow preventer acceptable to the local authorities. The mechanical room must be provided with an adequate drain and the floor must be waterproof. (Or install the boiler in a 20-gauge metal pan with soldered joints, and connect the pan to the drain with 1-in. (25-mm) type M copper.)

e. Connect the gas with a gas cock installed above ground and outside of the mechanical space, in addition to the gas cock furnished with each boiler gas train. Install gas regulator as required to reduce gas pressure to that required by the boiler. (Do not connect gas if the mechanical room is not ventilated according to code!)

f. Connect hot water lines and pump to the boiler with adequate drains to flush and drain the system. Install pressure gauges on inlet and outlet of boiler and pump. Install the relief valve and carry full size to a point acceptable for discharge of hot water.

g. Make the proper electrical connections for the boiler and equipment. Provide a disconnect means for the boiler controls, to turn off power to the main gas valve from an easily accessible point inside the mechanical room door.
h. Fill the boiler with water, and pressure-test the boiler and system up to the rating of the relief valve. Clean the system with trisodium phosphate or equal cleaner, flush the system to remove all trash and dirt, and refill the system, including inhibitor and antifreeze as specified.
i. Bleed the gas line in a safe manner and energize the boiler controls.

3.2 TEST

a. Observe the ignition of the pilot and the main burner to make sure they are smooth and complete.
b. Check out the safety controls and verify that they are functioning properly. Report the actual and design pressure drop through the boiler.
c. Set the operating controls for the proper temperature.
d. Operate the boiler for no less than 2 hr, or return at the beginning of the heating season for completion of this test. Vary the outdoor sensor temperature from 32 deg. F (0 deg. C) to 70 deg. F (21 deg. C) and observe the reset of the output. Read and record the pressure drop through the boiler and across the pump, and compare them to the design pressure drops.
e. Observe the pressure increase as the system heats up, and verify that the increase in pressure does not exceed 10%.
f. Provide a written report, through channels and on company letterhead, that ignition is proper, safety controls have been checked, and operating controls are set and functioning properly. Report the actual pressure drop and the design pressure drop and the pump head reading. The system is not complete until this report has been received.

3.3 OWNER'S INSTRUCTIONS

a. Provide the owner's operator with three copies of written instructions for operation and maintenance of the system. A letter should be sent, through channels, listing the materials given to the operator, the operator's name, and the date of this transmittal.
b. The operator should be instructed carefully in the methods, and necessity, for blowing down the boiler to remove the accumulated mud on a regular basis.

END OF SECTION 15802

SECTION 15803

****

WEATHERPROOF PACKAGED HOT WATER BOILER
FINNED WATER TUBE TYPE

PART ONE - GENERAL

1.1 The GENERAL and SPECIAL CONDITIONS, Section 15100, are included as a part of this Section as though written in full in this document.

1.2 Scope of the Work shall include the furnishing and complete installation of the equipment covered by this Section, with all auxiliaries, ready for owner's use.

PART TWO - PRODUCTS

2.1 a. Packaged boiler shall be natural (LP) gas-fired water tube type, designed for hot water service, including circulating pump and expansion tank, with capacity as shown on the drawings.

b. Boiler shall be AGA-certified for outdoor installation without additional wind or weather protection. Burners shall be natural draft, stainless steel, and mounted in a removable drawer for service.

c. All wet surfaces shall be brass or bronze. Heat exchanger tubes shall be cleanable, with extended external surface with bronze headers. The boiler shall bear the National Board ASME stamp for 160 psig (1100 kPa).

d. Pump shall be centrifugal type with cast iron body and bronze-fitted for hot water service. See drawings for type and capacity.

e. Expansion tank shall be bladder type with the water contained inside the bladder, which shall be sized to accept the expanded water without pressure increase of more than 10% of the fill pressure.

f. Combustion chamber shall be insulated with a lightweight cast refractory backed up with fiberglass insulation.

g. Boiler jacket shall be heavy-gauge galvanized (or equal) steel with factory-applied baked enamel.

h. Boiler controls shall be provided with a 110- to 24-volt transformer. The main gas valve shall be 24 volts.

i. Boiler controls shall provide for 80% modulation in firing rate and shall allow a leaving water temperature as low as 105 deg. F (40 deg. C) without condensation. Provide electronic ignition of the pilot. Gas safety controls shall prove pilot flame prior to opening of the main gas valve. The pilot shall be lit only when the boiler is calling for heat.

j. Burner capacity reduction shall be controlled by an electric (self-contained) sensing element reset by the outdoor air temperature, so that at 0 deg. F (-18 deg. C) outside the water temperature shall be 200 deg. F (105 deg. C) and at 70 deg. F (21 deg. C) outside the water temperature shall be 110 deg. F (43 deg. C). Provide a mechanical preset high-limit control set at 245 deg. F (118 deg. C) and an electric high-limit automatic reset control set at 242 deg. F (116 deg. C).

k. Package shall include a flow switch, or pressure differential switch, to prove water flow before the controls can be energized. The boiler controls shall be interlocked through the controls of the pump that is providing water flow for the boiler.

1. Provide storm proof draft hood as required for AGA approval for outside installation. Provide ASME pressure relief valve, combination temperature and pressure gauge, drain valve, and other accessories recommended by the manufacturer and required for a complete installation.

2.2 AUXILIARY EQUIPMENT

a. Provide high-pressure gas regulator, or additional gas regulator, where the available gas pressure is above the manufacturer's standards.
b. Provide controls to continue pump operation until the heat exchanger is cooled and refractory heat is absorbed. Provide pump interlock wiring with the controls.
c. Provide antifreeze in the system, as noted on the drawings, for freeze protection down to -20 deg. F (-30 deg. C). Capacity of the boiler shall be adjusted to meet the output required on the drawings with the derating required by the antifreeze.
d. Provide a one-shot (batch) feeder for the addition of water treatment and antifreeze to the system.
e. Provide pressure gauges for inlet and outlet of boiler and the pump, reading in small increments for measurement of pressure drop.

2.3 ACCEPTABLE MANUFACTURERS

a. Boiler shall be the make and model number shown on the drawings or equivalent by Raypak (Model TP), Lochinvar (Model CHW), Teledyne Laars (Series H).

PART THREE - EXECUTION

3.1 MECHANICAL SPACE AND INSTALLATION

a. The boiler shall be installed on a concrete pad of sufficient size to allow a 24-in. (60-cm) pad extension on all sides. Pad shall be no less than 6 in. (13 cm) above adjoining grade on all sides.
b. Provide city water for make-up to the boiler. Locate the make-up assembly in a heated space and connect to the return side of the pump. The water make-up assembly shall include a reduced-pressure-type backflow preventer acceptable to the local authorities.
c. Install gas regulator as required to reduce gas pressure to that required.
d. Connect hot water lines to the pump and to the boiler, with adequate drains to flush and drain the system. Install the relief valve and carry full size to a point acceptable for discharge of hot water.
e. Make the proper electrical connections for the boiler and equipment. Provide a disconnect means for the boiler controls to turn off power to the main gas valve from an easily accessible point near the boiler.
f. Fill the boiler with water, and pressure-test the boiler and system up to the rating of the relief valve. Clean the system with trisodium phosphate or equal cleaner, flush the system to remove all trash and dirt, and refill the system, including inhibitor and antifreeze as specified.
g. Bleed the gas line in a safe manner and energize the boiler controls.

3.2 TEST

a. Observe the ignition of the pilot and the main burner to make sure they are smooth and complete.

b. Check out the safety controls and verify that they are functioning properly.

c. Set the operating controls for the proper temperature.

d. Operate the boiler for no less than 2 hr, or return at the beginning of the heating season for completion of this test. Vary the outdoor sensor temperature from 32 deg. F (0 deg. C) to 70 deg. F (21 deg. C) and observe the reset of the output.

e. Observe the pressure increase as the system heats up, and verify that the increase in pressure does not exceed 10%.

f. Provide a written report, through channels and on company letterhead, that ignition is proper, safety controls have been checked, and operating controls are set and functioning properly. Report the actual and design boiler pressure drop and the pump head. The system is not complete until this report has been received.

3.3 OWNER'S INSTRUCTIONS

a. Provide the owner's operator with three copies of written instructions for operation and maintenance of the system. A letter should be sent, through channels, listing the materials given to the operator, the operator's name, and the date of this transmittal.

b. The operator should be instructed carefully in the methods, and necessity, for blowing down the boiler to remove the accumulated mud on a regular basis.

END OF SECTION 15803

SECTION 15804

****

MULTIPLE HOT WATER BOILERS
FINNED WATER TUBE TYPE WITH SEQUENCER

PART ONE - GENERAL

1.1 The GENERAL and SPECIAL CONDITIONS, Section 15100, are included as a part of this Section as though written in full in this document.

1.2 Scope of the Work shall include the furnishing and complete installation of the equipment covered by this Section, with all auxiliaries, ready for owner's use.

PART TWO - PRODUCTS

2.1 BOILERS

a. Boilers shall be multiple and staged, natural (LP) gas-fired water tube design for hot water service, with total capacity as shown on the drawings. Boilers shall be AGA-certified.

( ) Burners shall be natural draft gas-fired, stainless steel, and mounted in a removable drawer for service.

b. Burners shall be power burners designed to burn gas or light oil, as selected by controls.

c. All wet surfaces shall be brass or bronze. Heat exchanger tubes shall be cleanable, with extended external surface with bronze headers. The boilers shall bear the National Board ASME stamp for 160 psig (1100 kPa).

d. Combustion chamber shall be insulated with a lightweight cast refractory backed up with fiberglass insulation.

e. Boiler jackets shall be heavy-gauge galvanized (or equal) steel with factory-applied baked enamel.

f. Boiler controls shall be provided with a sequencer to fire the boilers only as needed to meet the load. Boilers not in use shall be at room temperature until the controls require their capacity. The last boiler to come on shall modulate to maintain the leaving water temperature as required. As the load falls off, the last boiler shall modulate down to 20% of load and then go off.

g. Individual boiler controls shall include a 110- to 24-volt transformer. The main gas valve shall be 24 volts.

h. Boiler controls shall provide for 80% modulation in firing rate and shall allow a leaving water temperature as low as 105 deg. F (40 deg. C) without condensation. Provide electronic ignition of the pilot. Gas safety controls shall prove pilot flame prior to opening of the main gas valve. The pilot shall be lit only when the boiler is calling for heat.

i. Overall capacity reduction shall be controlled by an electronic controller with reset by the outdoor air temperature, so that at 0 deg. F (-18 deg. C) outside the supply water temperature shall be 200 deg. F (105 deg. C) and at 70 deg. F (21 deg. C) outside the water temperature shall be 110 deg. F (43 deg. C). Provide a mechanical preset high-limit control on each boiler set at 245 deg. F (118 deg. C) and an electric high-limit automatic reset control set at 242 deg. F (116 deg. C).

j. Provide a flow switch, or pressure differential switch, in each boiler to prove water flow before that boiler can be energized. The boiler controls shall be interlocked through the controls of the pump that is providing water flow for the boiler.
k. For each boiler, provide draft hood as required for AGA approval. Provide ASME pressure relief valve, combination temperature and pressure gauge, drain valve, and other accessories recommended by the manufacturer and required for a complete installation.

2.2 AUXILIARY EQUIPMENT
a. Provide high-pressure gas regulator, or additional gas regulator, where the available gas pressure is above the manufacturer's standards.
b. Provide controls to continue pump operation until the heat exchanger is cooled and refractory heat is absorbed. Provide pump interlock wiring with the controls.
c. Provide a one-shot (batch) feeder for the addition of water treatment to the system.

2.3 ACCEPTABLE MANUFACTURERS
a. Boilers shall be the make and model number shown on the drawings or equivalent by Raypak (Model T), Lochinvar (Model CHW), Teledyne Laars (Series H).

PART THREE - EXECUTION

3.1 MECHANICAL SPACE AND INSTALLATION
a. The boiler shall be installed in a space reserved for the boiler and provided with combustion air and ventilation required by the codes and shown on the drawings.
b. The boiler flue pipe shall be UL-listed and shall be the full size of the draft outlet. Provide a UL-listed vent cap, and install cap and pipe as required by the code.
c. The floor of the mechanical space shall be noncombustible, or a noncombustible pad shall be installed under the boiler, as required by the boiler manufacturer's AGA rating.
d. Provide city water for make-up to the boiler. Include a water make-up assembly which includes a reduced-pressure type of backflow preventer acceptable to the local authorities. The mechanical room must be provided with an adequate drain and the floor must be waterproof. (Or install the boiler in a 20-gauge metal pan with soldered joints, and connect the pan to the drain with 1-in. (25-mm) type M copper.)
e. Connect the gas to each boiler. In addition, provide a gas cock for the mechanical room, installed above ground and outside of the mechanical space. Install gas regulator as required to reduce gas pressure from supply pressure to that required by the boilers. (Do not connect gas if the mechanical room is not ventilated according to code!)
f. Connect hot water lines and pumps to each of the boilers with adequate drains to flush and drain the system. Install pressure gauges on inlet and outlet of boilers and pumps. Install the relief valves and carry full size to a point acceptable for discharge of hot water.

g. Make the proper electrical connections for the boilers and equipment. Provide a disconnect means for the boiler controls, to turn off power to the main gas valve of all boilers from an easily accessible point inside the mechanical room door.
h. Fill the boiler with water, and pressure-test the boiler and system up to the rating of the relief valve. Clean the system with trisodium phosphate or equal cleaner, flush the system to remove all trash and dirt, and refill the system, including inhibitor and antifreeze as specified.
i. Bleed the gas line in a safe manner and energize the boiler controls.

3.2 TEST

a. Observe the ignition of the pilot and the main burner of each boiler to make sure they are smooth and complete.
b. Check out the safety controls and verify that they are functioning properly. Report the actual and design pressure drop through each of the boilers.
c. Set the operating controls for the proper temperatures.
d. Operate each of the boilers for no less than 2 hr, or return at the beginning of the heating season for completion of this test. Vary the outdoor sensor temperature from 32 deg. F (0 deg. C) to 70 deg. F (21 deg. C) and observe the reset of the output of the boilers as a group.
e. Provide a written report, through channels and on company letterhead, for each boiler that ignition is proper, safety controls have been checked, and operating controls are set and functioning properly. Report the actual and design pressure drop through the boiler. The system is not complete until this report has been received.

3.3 OWNER'S INSTRUCTIONS

a. Provide the owner's operator with three copies of written instructions for operation and maintenance of the system. A letter shall be sent, through channels, listing the materials given to the operator, the operator's name, and the date of this transmittal.
b. The operator shall be instructed carefully in the methods, and necessity, for blowing down the boilers to remove the accumulated mud on a regular basis.

END OF SECTION 15804

SECTION 15806

****

HOT WATER BOILER
INCLINED-STEEL WATER TUBE TYPE
HIGH PRESSURE - POWER BURNER

PART ONE - GENERAL

1.1 The GENERAL and SPECIAL CONDITIONS, Section 15100, are included as a part of this Section as though written in full in this document.

1.2 Scope of the Work shall include the furnishing and complete installation of the equipment covered by this Section, with all auxiliaries, ready for owner's use.

PART TWO - PRODUCTS

2.1 HOT WATER BOILER

a. Boiler shall be inclined-steel water tube type, with power burner for gas/light oil, designed for hot water service and with output capacity shown on the drawings.

b. Burner shall be pressure-atomizing forced-draft type, designed for No. 2 (light) oil and natural gas. Oil pump shall be integral with the burner or as a separate unit mounted on the boiler. Burner shall be full modulating type, on either fuel, with flame safeguard controls to meet or exceed UL requirements. Controls shall be 110-volt single-phase and shall include an operating thermostat and a manual reset high-limit control. Provide the following trim:

1. Main gas shutoff cock.
2. Pilot gas cock, regulator, and solenoid valve.
3. Gas pressure regulator.
4. Motorized main gas valve.
5. Auxiliary motorized main gas valve.
6. Motorized modulating valve (gas and oil).
7. Auxiliary oil valve, piped in series.
8. Low-water cutoff with manual reset.

c. Water tubes and headers shall be high-grade steel. Tubes shall be 2 in. (50 mm) in diameter, 13 gauge, 0.095 in. (2.413 mm) rolled into the headers. Boiler shall be ASME-stamped for 125 psi (1060 kPa). Boiler head plates shall be removable for tube inspection, replacement, and cleaning.

d. Firebox shall be insulated with high-temperature castable refractory or heavy-duty fire brick, backed up with 3-in. (75-mm) high-density rockwool insulation. Refractory shall be field-replaceable.

e. Boiler jacket shall be no less than 16-gauge (1.5-mm) steel with baked-enamel finish. Jacket shall be removable for access to fire side of tubes. Jacket shall be insulated to reduce heat loss.

f. Boiler trim shall include the following:

1. ASME full-capacity relief valve.
2. Pressure and temperature gauge.
3. Built-in air eliminator fitting and dip tube.

g. Provide a draft hood as required for AGA approval or a barometric damper to serve the same purpose.

2.2 AUXILIARY EQUIPMENT

a. Provide high-pressure gas regulator, or additional regulator, where the gas pressure is above the manufacturer's standard rating.

b. Provide boiler water chemical treatment and a one-shot feeder, including chemicals for complete treatment of the system. Provide initial 50-pound (22-kg) treatment.

c. Provide an induced-draft fan, furnished as a part of the boiler by the manufacturer, as shown on the drawings.

2.3 ACCEPTABLE MANUFACTURERS

a. Boiler shall be the make and model number shown on the drawings or equivalent by Ajax (Model SGX), Rite, ThermoPak.

PART THREE - EXECUTION

3.1 MECHANICAL SPACE AND INSTALLATION

a. The boiler shall be installed in a space reserved for the boiler and provided with combustion air and ventilation required by the code and as shown on the drawings.

b. The boiler flue pipe shall be UL-listed and shall be the full size of the draft hood outlet or the barometric damper. If an induced-draft fan is provided, the flue shall be designed by an engineer for the specific application. Provide a UL-listed flue cap and install both pipe and cap as required by the code.

c. The boiler room shall be provided with a drain adequate for the blowdown, draining, and washdown of the boiler.

d. Provide city water for make-up to the boiler and the boiler feedwater system. Include a water make-up assembly which includes a backflow preventer of the reduced-pressure type, acceptable to the local authorities. Provide necessary hose connections for washing down the boiler.

e. Provide boiler pump(s) as shown and specified. Connect to the boiler as recommended by the manufacturer and as shown on the drawings.

f. Connect gas to the gas train with a gas cock installed above ground and outside the mechanical space, in addition to the gas cock in the boiler gas train. Install gas regulator as required to reduce the gas pressure to that required by the equipment. (Do not connect gas if the mechanical room is not ventilated as required by code.)

g. Install header as shown on the drawings. Install the relief valve and carry to an acceptable discharge point. See detail on drawings.

h. Make the proper electrical connections for the boiler and equipment. Provide a disconnect means for the boiler controls, to turn off power to the main burner gas valve from an easily accessible point inside the mechanical room door.

i. Fill the boiler with water, and pressure-test the boiler and system up to the rating of the relief valve. Clean the boiler and system with trisodium phosphate or equivalent cleaner, and flush the system to remove all trash and dirt; then chemically clean the system as specified and refill the system with clean water and the proper boiler water treatment.

j. Bleed the gas line in a safe manner and energize the boiler controls.

3.2 TEST

a. Observe the ignition of the pilot and the main burner to make sure they are smooth and complete.

b. Check out the safety controls and verify that they are functioning properly.

c. Set the operating controls for the proper temperatures.

d. Operate the boiler for no less than 2 hr.

e. Provide a written report, through channels and on company letterhead, stating that ignition is proper, safety controls have been checked, and operating controls are set and functioning properly. The system is not complete until this report has been received.

3.3 OWNER'S INSTRUCTIONS

a. Instruct the operators in proper care and operation of the boiler and controls.

b. Provide the owner's operator with three copies of written instructions for operation and maintenance of the system. A letter shall be sent, through channels, listing the materials given to the operator, the operator's name, and the date of this transmittal.

c. The operator shall be instructed carefully in the methods, and necessity, for blowing down the boiler to remove the accumulated mud on a regular basis.

END OF SECTION 15806

SECTION 15807

****

LOW-PRESSURE STEAM BOILER
INCLINED-STEEL WATER TUBE TYPE
NATURAL DRAFT

PART ONE - GENERAL

1.1 The GENERAL and SPECIAL CONDITIONS, Section 15100, are included as a part of this Section as though written in full in this document.

1.2 Scope of the Work shall include the furnishing and complete installation of the equipment covered by this Section, with all auxiliaries, ready for owner's use.

PART TWO - PRODUCTS

2.1 STEAM BOILER

a. Boiler shall be inclined-steel water tube type, with natural (LP) gas atmospheric natural draft burner, designed for hot water service and with output capacity shown on the drawings.

b. Boiler unit shall be AGA-certified and UL-labeled. Burners shall be raised-port cast iron. Gas train shall include main gas cock, pilot gas cock, main gas regulator, electronic flame safety supervision with intermittent pilot, automatic ignition, and 100% shutoff. Controls shall be 110-volt single-phase and shall include an operating pressure control in the boiler and a manual reset high-limit pressure control.

c. Tubes and headers shall be high-grade steel. Tubes shall be 2 in. in diameter (50 mm) 13 gauge, 0.095 in. (2.413 mm) rolled into the headers. Boiler shall be ASME-stamped for 15-psi (100-kPa) working pressure. Boiler head plates shall be removable for tube inspection and cleaning.

d. Firebox shall be insulated with high-temperature castable refractory or fire brick, backed up with 3-in. (75-mm) high-density rockwool insulation. Refractory shall be field-replaceable.

e. Boiler jacket shall be no less than 16-gauge (1.5-mm) steel with baked-enamel finish. Jacket shall be removable for access to fire side of tubes. Jacket shall be insulated to reduce heat loss.

f. Boiler trim shall include the following:

1. ASME full-capacity relief valve.
2. Pressure gauge.
3. Water column with try cocks and red-line water gauge glass.
4. Low-water cutoff with boiler feed pump control.
5. Second low-water cutoff with alarm contact, alarm horn, alarm red light, alarm panel including horn silencer, etc., all as required for a complete system.

g. Controls shall operate as follows:

( ) High-Low-Off operation by two-stage aquastat. (Ignition shall take place on low fire, then proceed to high fire.)

( ) High-Low-Off operation with modulation from low to high fire as required to maintain pressure. (Start on low fire.)

h. Provide a draft hood as required for AGA approval or a barometric damper to serve the same purpose.

2.2 AUXILIARY EQUIPMENT

a. Provide high-pressure gas regulator, or additional regulator, where the gas pressure is above the manufacturer's standard rating.

b. Provide a boiler water chemical treatment system, including treatment tank, pump, piping, etc., for a complete system. Provide initial 50 pounds (22 kg) of treatment.

c. Provide an induced-draft fan, furnished as a part of the boiler by the manufacturer, as shown on the drawings.

2.3 ACCEPTABLE MANUFACTURERS

a. Boiler shall be the make and model number shown on the drawings or equivalent by Ajax (Model SGX), Rite, ThermoPak.

PART THREE - EXECUTION

3.1 MECHANICAL SPACE AND INSTALLATION

a. The boiler shall be installed in a space reserved for the boiler and provided with combustion air and ventilation required by the code and as shown on the drawings.

b. The boiler flue pipe shall be UL-listed and shall be the full size of the draft hood outlet or the barometric damper. If an induced-draft fan is provided, the flue shall be designed by an engineer for the specific application. Provide a UL-listed flue cap and install both pipe and cap as required by the code.

c. The boiler room shall be provided with a drain adequate for the blowdown, draining, and washdown of the boiler.

d. Provide city water for make-up to the boiler and the boiler feedwater system. Include a water make-up assembly which includes a backflow preventer of the reduced-pressure type, acceptable to the local authorities. Provide necessary hose connections for washing down the boiler.

e. Install the low-water cutoffs and feed units in the following (descending) order:

1. First stage - bring on feedwater pump.
2. Second stage - low-water cutoff with automatic reset.
3. Third stage - low-water cutoff with manual reset and alarm circuit and system.
4. Fourth stage - automatic manual water feeder.

First stage shall be set at 1/2 in. (13 mm) below normal water level.

Second stage shall be set at 1 1/2 in. (30 mm) below normal water level.

Third stage shall be set at 2 1/2 in. (63 mm) below normal water level.

Fourth stage shall be set at 3 in. (75 mm) below normal water level.

f. Provide boiler feedwater storage and pump(s) as shown and specified. Connect to the boiler as recommended by the manufacturer and as shown on the drawings.

g. Connect gas to the gas train with a gas cock installed above ground and outside the mechanical space, in addition to the gas cock in the boiler gas train. Install gas regulator as required to reduce the gas pressure to that required by the equipment. (Do not connect gas if the mechanical room is not ventilated as required by code.)

h. Install steam header as shown on the drawings. Install the relief valve and carry full size through the roof. See detail on drawings.

i. Make the proper electrical connections for the boiler and equipment. Provide a disconnect means for the boiler controls, to turn off power to the main burner gas valve from an easily accessible point inside the mechanical room door.

j. Fill the boiler with water, and pressure-test the boiler and system up to the rating of the relief valve. Clean the boiler and system with trisodium phosphate or equivalent cleaner, and flush the system to remove all trash and dirt; then chemically clean the system as specified and refill the system with clean water and the proper boiler water treatment.

k. Bleed the gas line in a safe manner and energize the boiler controls.

3.2 TEST

a. Observe the ignition of the pilot and the main burner to make sure they are smooth and complete.

b. Check out the safety controls and verify that they are functioning properly.

c. Set the operating controls for the proper temperatures.

d. Operate the boiler for no less than 2 hr.

e. Provide a written report, through channels and on company letterhead, stating that ignition is proper, safety controls have been checked, and operating controls are set and functioning properly. The system is not complete until this report has been received.

3.3 OWNER'S INSTRUCTIONS

a. Instruct the operators in proper care and operation of the boiler and controls.

b. Provide the owner's operator with three copies of written instructions for operation and maintenance of the system. A letter shall be sent, through channels, listing the materials given to the operator, the operator's name, and the date of this transmittal.

c. The operator shall be instructed carefully in the methods, and necessity, for blowing down the boiler to remove the accumulated mud on a regular basis.

END OF SECTION 15807

SECTION 15808
****
LOW-PRESSURE STEAM BOILER
INCLINED-STEEL WATER TUBE TYPE
POWER BURNER

PART ONE - GENERAL

1.1 The GENERAL and SPECIAL CONDITIONS, Section 15100, are included as a part of this Section as though written in full in this document.

1.2 Scope of the Work shall include the furnishing and complete installation of the equipment covered by this Section, with all auxiliaries, ready for owner's use.

PART TWO - PRODUCTS

2.1 STEAM BOILER

a. Boiler shall be inclined-steel water tube type, with natural (LP) gas atmospheric natural draft burner, designed for hot water service and with output capacity shown on the drawings.

b. Boiler unit shall be AGA-certified and UL-labeled. Burners shall be raised-port cast iron. Gas train shall include main gas cock, pilot gas cock, main gas regulator, electronic flame safety supervision with intermittent pilot, automatic ignition, and 100% shutoff. Controls shall be 110-volt single-phase and shall include an operating pressure control in the boiler and a manual reset high-limit pressure control.

c. Tubes and headers shall be high-grade steel. Tubes shall be 2 in. (50 mm) in diameter, 13 gauge, 0.095 in. (2.413 mm) rolled into the headers. Boiler shall be ASME-stamped for 15-psi (100-kPa) working pressure. Boiler head plates shall be removable for tube inspection and cleaning.

d. Firebox shall be insulated with high-temperature castable refractory or fire brick, backed up with 3-in. (75-mm) high-density, rockwool insulation. Refractory shall be field-replaceable.

e. Boiler jacket shall be no less than 16-gauge (1.5-mm) steel with baked-enamel finish. Jacket shall be removable for access to fire side of tubes. Jacket shall be insulated to reduce heat loss.

f. Boiler trim shall include the following:

1. ASME full-capacity relief valve.
2. Pressure gauge, 6 in. (150 mm) minimum.
3. Water column with try cocks and red-line water gauge glass.
4. Low-water cutoff with boiler feed pump control.
5. Second low-water cutoff with alarm contact, alarm horn, alarm red light, alarm panel including horn silencer, etc., all as required for a complete system.

g. Controls shall operate as follows:

( ) High-Low-Off operation by two-stage aquastat. (Ignition shall take place on low fire, then proceed to high fire.)

( ) High-Low-Off operation with modulation from low to high fire as required to maintain pressure. (Start on low fire.)

h. Provide a draft hood as required for AGA approval or a barometric damper to serve the same purpose.

2.2 AUXILIARY EQUIPMENT

a. Provide high-pressure gas regulator, or additional regulator, where the gas pressure is above the manufacturer's standard rating.
b. Provide a boiler water chemical treatment system, including treatment tank, pump, piping, etc., for a complete system. Provide initial 50 pounds (22 kg) of treatment.
c. Provide an induced-draft fan, furnished as a part of the boiler by the manufacturer, as shown on the drawings.

2.3 ACCEPTABLE MANUFACTURERS

a. Boiler shall be the make and model number shown on the drawings or equivalent by Ajax (Model SGX), Rite, ThermoPak.

PART THREE - EXECUTION

3.1 MECHANICAL SPACE AND INSTALLATION

a. The boiler shall be installed in a space reserved for the boiler and provided with combustion air and ventilation required by the code and as shown on the drawings.
b. The boiler flue pipe shall be UL-listed and shall be the full size of the draft hood outlet or the barometric damper. If an induced-draft fan is provided, the flue shall be designed by an engineer for the specific application. Provide a UL-listed flue cap and install both pipe and cap as required by the code.
c. The boiler room shall be provided with a drain adequate for the blowdown, draining, and washdown of the boiler.
d. Provide city water for make-up to the boiler and the boiler feedwater system. Include a water make-up assembly which includes a backflow preventer of the reduced-pressure type, acceptable to the local authorities. Provide necessary hose connections for washing down the boiler.
e. Install the low-water cutoffs and feed units in the following (descending) order:

1. First stage - bring on feedwater pump.
2. Second stage - low-water cutoff with automatic reset.
3. Third stage - low-water cutoff with manual reset and alarm circuit and system.

First stage shall be set at 1/2 in. (13 mm) below normal water level.

Second stage shall be set at 1 1/2 in. (30 mm) below normal water level.

Third stage shall be set at 2 1/2 in. (63 mm) below normal water level.

f. Provide boiler feedwater pump(s) as shown and specified. Connect to the boiler as recommended by the manufacturer and as shown on the drawings.
g. Connect gas to the gas train with a gas cock installed above ground and outside the mechanical space, in addition to the gas cock in the boiler gas train. Install gas regulator as required to reduce the gas pressure to that required by the equipment. (Do not connect gas if the mechanical room is not ventilated as required by code.)
h. Install steam header as shown on the drawings. Install the relief valve and carry full size through the roof. See detail on drawings.

i. Make the proper electrical connections for the boiler and equipment. Provide a disconnect means for the boiler controls, to turn off power to the main burner gas valve from an easily accessible point inside the mechanical room door.
j. Fill the boiler with water, and pressure-test the boiler and system up to the rating of the relief valve. Clean the boiler and system with trisodium phosphate or equivalent cleaner, and flush the system to remove all trash and dirt; then chemically clean the system as specified and refill the system with clean water and the proper boiler water treatment.
k. Bleed the gas line in a safe manner and energize the boiler controls.

3.2 TEST

a. Observe the ignition of the pilot and the main burner to make sure they are smooth and complete.
b. Check out the safety controls and verify that they are functioning properly.
c. Set the operating controls for the proper temperature.
d. Operate the boiler for no less than 2 hr.
e. Provide a written report, through channels and on company letterhead, stating that ignition is proper, safety controls have been checked, and operating controls are set and functioning properly. The system is not complete until this report has been received.

3.3 OWNER'S INSTRUCTIONS

a. Instruct the operators in proper care and operation of the boiler and controls.
b. Provide the owner's operator with three copies of written instructions for operation and maintenance of the system. A letter shall be sent, through channels, listing the materials given to the operator, the operator's name, and the date of this transmittal.
c. The operator shall be instructed carefully in the methods, and necessity, for blowing down the boiler to remove the accumulated mud on a regular basis.

END OF SECTION 15808

SECTION 15810

****

HOT WATER BOILER-BURNER UNIT

CAST IRON SECTIONAL TYPE - POWER BURNER

PART ONE - GENERAL

1.1 The GENERAL and SPECIAL CONDITIONS, Section 15100, are included as a part of this Section as though written in full in this document.

1.2 Scope of the Work shall include the furnishing and complete installation of the equipment covered by this Section, with all auxiliaries, ready for owner's use.

PART TWO - PRODUCTS

2.1 SECTIONAL BOILER

a. Boiler shall be low-pressure, sectional cast iron type, fired with power-type gas (gas/light oil) burner, designed for hot water service and with output capacity as shown on the drawings.

b. Boiler unit shall be AGA-certified, UL-labeled, and constructed and tested in accordance with ASME Pressure Vessel Code, Section IV, for maximum working pressure of 40 (80) psig (275 (550) kPa). Controls shall be 110-volt single-phase and shall include an operating thermostat in the boiler and a manual reset high-limit control.

c. Sections shall be fine-grain cast iron, of one-piece design, with perimeter joints permanently pressure-sealed with high-temperature ceramic fiber rope. Cleanout covers, smoke hood, slide damper, and burner mounting plate shall be cast iron. Burner mounting plate shall be insulated with high-temperature ceramic-faced insulation block.

d. Provide pressure-tight flame observation ports with cast iron covers and frames in front and rear of boiler. Provide cast iron cleanout covers to allow full access to the heating surfaces for cleaning.

e. Boiler jacket shall be not less than 20-gauge (.965-mm) steel with baked-enamel finish. Jacket shall be removable for access for cleaning. Jacket shall be insulated to reduce heat loss.

f. Burner shall be pressure-atomizing forced-draft type, designed for No. 2 (light) oil and natural gas. Oil pump shall be integral with the burner or as a separate unit mounted on the boiler. Burner shall be full modulating type on either fuel, with flame safeguard controls to meet or exceed UL requirements. Controls shall be 110-volt single-phase and shall include an operating thermostat and a manual reset high-limit control. Provide the following trim:

1. Main gas shutoff cock.
2. Pilot gas cock, regulator, and solenoid valve.
3. Motorized main gas valve.
4. Auxiliary motorized main gas valve.
5. Motorized modulating valve (gas and oil).
6. Auxiliary oil valve, piped in series.
7. Combination low-water cutoff and safety feeder.

g. Boiler trim shall include the following:
   1. ASME full-capacity relief valve.
   2. Pressure and temperature gauge.
   3. Built-in air eliminator fitting and dip tube.
   4. Low-water cutoff.

h. Controls shall operate as follows:
   ( ) High-Low-Off operation by two-stage aquastat. (Ignition shall take place on low fire, then proceed to high fire.)
   ( ) High-Low-Off operation with modulation from low to high fire as required to meet capacity required. (Start on low fire.)

2.2 AUXILIARY EQUIPMENT

a. Provide high-pressure gas regulator, or additional regulator, where the gas pressure is above the manufacturer's standard rating.
b. Provide a one-shot (batch) feeder for the addition of water treatment to the system.

2.3 ACCEPTABLE MANUFACTURERS

a. Boiler shall be the make and model number shown on the drawings or equivalent by H.B. Smith (Models 2500L, 28, or 3500.)

PART THREE - EXECUTION

3.1 MECHANICAL SPACE AND INSTALLATION

a. The boiler shall be installed in a space reserved for the boiler and provided with combustion air and ventilation required by the code and as shown on the drawings.
b. The boiler vent pipe shall be UL-listed for gas/oil use and shall be the full size of the boiler outlet. If an induced-draft fan is provided, the flue shall be designed by an engineer for the specific application. Provide a UL-listed flue cap and install both pipe and cap as required by the code.
c. The boiler room shall be provided with a drain adequate for the draining and washdown of the boiler.
d. Provide city water for make-up to the boiler. Include a water make-up assembly which includes a backflow preventer of the reduced-pressure type, acceptable to the local authorities.
e. Connect gas to the gas train with a gas cock installed above ground and outside the mechanical space, in addition to the gas cock in the boiler gas train. Install gas regulator as required to reduce the gas pressure to that required by the equipment. (Do not connect gas if the mechanical room is not ventilated as required by code!)
f. Connect hot water lines and pump to the boiler with adequate drains to flush and drain the system. Install the relief valve and carry full size to a point acceptable for discharge of the hot water.
g. Make the proper electrical connections for the boiler and equipment. Provide a disconnect means for the boiler controls, to turn off power to the main gas valve from an easily accessible point inside the mechanical room door.

h. Fill the boiler with water, and pressure-test the boiler and system up to the rating of the relief valve. Clean the system with trisodium phosphate or equal cleaner, flush the system to remove all trash and dirt, and refill the system, including inhibitor as specified.
i. Bleed the gas line in a safe manner and energize the boiler controls.

3.2 TEST

a. Observe the ignition of the pilot and the main burner to make sure they are smooth and complete.
b. Check out the safety controls and verify that they are functioning properly.
c. Set the operating controls for the proper temperature.
d. Operate the boiler for no less than 2 hr., or return at the beginning of the heating season for completion of this test.
e. Provide a written report, through channels and on company letterhead, stating that ignition is proper, safety controls have been checked, and operating controls are set and functioning properly. The system is not complete until this report has been received.

3.3 OWNER'S INSTRUCTIONS

a. Provide the owner's operator with three copies of written instructions for operation and maintenance of the system. A letter shall be sent, through channels, listing the materials given to the operator, the operator's name, and the date of this transmittal.
b. The operator shall be instructed carefully in the methods, and necessity, for blowing down the boiler to remove the accumulated mud on a regular basis.

END OF SECTION 15810

SECTION 15811

****

LOW-PRESSURE STEAM BOILER-BURNER UNIT
CAST IRON SECTIONAL TYPE - POWER BURNER

PART ONE - GENERAL

1.1 The GENERAL and SPECIAL CONDITIONS, Section 15100, are included as a part of this Section as though written in full in this document.

1.2 Scope of the Work shall include the furnishing and complete installation of the equipment covered by this Section, with all auxiliaries, ready for owner's use.

PART TWO - PRODUCTS

2.1 SECTIONAL BOILER

a. Boiler shall be low-pressure, sectional cast iron type, fired with power-type gas (gas/light oil) burner, designed for hot water service and with output capacity as shown on the drawings.

b. Boiler unit shall be AGA-certified, UL-labeled, and constructed and tested in accordance with ASME Pressure Vessel Code, Section IV, for maximum working pressure of 40 (80) psig (275 (550) kPa). Controls shall be 110-volt single-phase and shall include an operating thermostat in the boiler and a manual reset high-limit control.

c. Sections shall be fine-grain cast iron of one-piece design, with perimeter joints permanently pressure-sealed with high-temperature ceramic fiber rope. Cleanout covers, smoke hood, slide damper, and burner mounting plate shall be cast iron. Burner mounting plate shall be insulated with high-temperature ceramic-faced insulation block.

d. Provide pressure-tight flame observation ports with cast iron covers and frames in front and rear of boiler. Provide cast iron cleanout covers to allow full access to the heating surfaces for cleaning.

e. Boiler jacket shall be no less than 20-gauge (0.965-mm) steel with baked-enamel finish. Jacket shall be removable for access for cleaning. Jacket shall be insulated to reduce heat loss.

f. Burner shall be pressure-atomizing forced-draft type, designed for No. 2 (light) oil and natural gas. Oil pump shall be integral with the burner or as a separate unit mounted on the boiler. Burner shall be full modulating type on either fuel, with flame safeguard controls to meet or exceed UL requirements. Controls shall be 110-volt single-phase and shall include an operating thermostat and a manual reset high-limit control. Provide the following trim:

1. Main gas shutoff cock.
2. Pilot gas cock, regulator, and solenoid valve.
3. Motorized main gas valve.
4. Auxiliary motorized main gas valve.
5. Motorized modulating valve (gas and oil).
6. Auxiliary oil valve, piped in series.
7. Combination low-water cutoff and safety feeder.

g. Boiler trim shall include the following:
   1. ASME full-capacity relief valve.
   2. Pressure temperature gauge.
   3. Built-in air eliminator fitting and dip tube.
   4. Low-water cutoff.

h. Controls shall operate as follows:
   ( ) High-Low-Off operation by two-stage aquastat. (Ignition shall take place on low fire, then proceed to high fire.)
   ( ) High-Low-Off operation with modulation from low to high fire as required to meet capacity required. (Start on low fire.)

2.2 AUXILIARY EQUIPMENT

a. Provide high-pressure gas regulator, or additional regulator, where the gas pressure is above the manufacturer's standard rating.
b. Provide a one-shot (batch) feeder for the addition of water treatment to the system.

2.3 ACCEPTABLE MANUFACTURERS

a. Boiler shall be the make and model number shown on the drawings or equivalent by H.B. Smith (Model 2500L, 28 or 3500).

PART THREE - EXECUTION

3.1 MECHANICAL SPACE AND INSTALLATION

a. The boiler shall be installed in a space reserved for the boiler and provided with combustion air and ventilation required by the code and as shown on the drawings.
b. The boiler vent pipe shall be UL-listed for gas/oil use and shall be the full size of the boiler outlet. If an induced-draft fan is provided, the flue shall be designed by an engineer for the specific application. Provide a UL-listed flue cap and install both pipe and cap as required by the code.
c. The boiler room shall be provided with a drain adequate for the draining and washdown of the boiler.
d. Provide city water for make-up to the boiler. Include a water make-up assembly which includes a backflow preventer of the reduced-pressure type, acceptable to the local authorities.
e. Connect gas to the gas train with a gas cock installed above ground and outside the mechanical space, in addition to the gas cock in the boiler gas train. Install gas regulator as required to reduce the gas pressure to that required by the equipment. (Do not connect gas if the mechanical room is not ventilated as required by code!)
f. Connect hot water lines and pump to the boiler with adequate drains to flush and drain the system. Install the relief valve and carry full size to a point acceptable for discharge of the hot water.
g. Make the proper electrical connections for the boiler and equipment. Provide a disconnect means for the boiler controls, to turn off power to the main gas valve from an easily accessible point inside the mechanical room door.
h. Fill the boiler with water, and pressure-test the boiler and system up to the rating of the relief valve. Clean the system with trisodium phosphate or equal cleaner; flush the system to remove all trash and dirt; and refill the system, including inhibitor as specified.

i. Bleed the gas line in a safe manner and energize the boiler controls.

3.2 TEST

a. Observe the ignition of the pilot and the main burner to make sure they are smooth and complete.
b. Check out the safety controls and verify that they are functioning properly.
c. Set the operating controls for the proper temperature.
d. Operate the boiler for no less than 2 hr., or return at the beginning of the heating season for completion of this test.
e. Provide a written report, through channels and on company letterhead, stating that ignition is proper, safety controls have been checked, and operating controls are set and functioning properly. The system is not complete until this report has been received.

3.3 OWNER'S INSTRUCTIONS

a. Provide the owner's operator with three copies of written instructions for operation and maintenance of the system. A letter shall be sent, through channels, listing the materials given to the operator, the operator's name, and the date of this transmittal.
b. The operator shall be instructed carefully in the methods, and necessity, for blowing down the boiler to remove the accumulated mud on a regular basis.

END OF SECTION 15811

SECTION 15812
****
HIGH-PRESSURE STEAM BOILER
VERTICAL TUBELESS TYPE
POWER BURNER

PART ONE - GENERAL

1.1 The GENERAL and SPECIAL CONDITIONS, Section 15100, are included as a part of this Section as though written in full in this document.

1.2 Scope of the Work shall include the furnishing and complete installation of the equipment covered by this Section, with all auxiliaries, ready for owner's use.

PART TWO - PRODUCTS

2.1 a. Boiler package shall be vertical tubeless type, with power burner for gas/oil operation, designed for 125-psi (1060-kPa) service and with output capacity shown on the drawings. Boiler packages shall occupy minimum floor space. See drawings.

b. Boiler package unit shall be AGA-certified, UL-labeled, and ASME-stamped.

c. Burner shall be pressure-atomizing forced-draft type, designed for #2 (light) oil and natural gas. Oil pump shall be integral with burner or as a separate unit mounted on the boiler. Burner shall be full modulating type on either fuel, with flame safeguard controls to meet or exceed UL requirements. Controls shall be 110-volt single-phase and shall include an operating pressure control and a manual reset high-limit control. Provide the following trim:

1. Main gas shutoff cock.
2. Pilot gas cock, regulator, and solenoid valve.
3. Gas pressure regulator.
4. Motorized main gas valve.
5. Auxiliary motorized main gas valve.
6. Solenoid bleed valve between the two gas valves.
7. Motorized modulating valve (gas and oil).
8. Auxiliary oil valve, piped in series.

d. Boiler shall be three-pass design, with a bottom mud ring protected from the direct flame and provided with no less than two handholes for cleaning. Provide cleanout plates for cleaning the fire side.

e. The boiler shall be insulated with high-temperature castable refractory or fire brick 4 in. (100 mm) thick with jacket. Refractory shall be field-replaceable.

f. Boiler jacket shall be no less than 22-gauge (0.759-mm) steel with baked-enamel finish. Jacket shall be removable for access to fire side.

g. Boiler trim shall include the following:

1. ASME full-capacity relief valve.
2. Pressure gauge not less than 4 1/2 in. (120 mm) in diameter with operating pressure at center of gauge.
3. Water column with try cocks and red-line water gauge glass. Provide blowdown valve for gauge glass.
4. Low-water cutoff with boiler feed pump control, float type with blowdown.
5. Second low-water cutoff with alarm contact, alarm horn, red alarm light, alarm panel including horn silencer, etc., all as required for a complete system.
6. Blowdown valve(s) as required by ASME code and recommended by manufacturer.
7. Main steam valve installed on the steam outlet, full size of outlet. (Where more than one outlet is available, the largest shall be used.)

h. Provide a substantial structural steel base for the boiler. This base shall also support the boiler feedwater pump(s), pipe, and any controls.

2.2 COMBUSTION EQUIPMENT

a. Burner shall be forced-draft design, fully automatic, and UL-labeled for operation on gas, oil, or gas/oil, as noted on drawings.

b. The gas burner shall be controlled-pressure mixed type, UL-labeled for operation on either natural or LP gas. The gas train piping shall include:

1. Gas pressure regulator for the actual gas pressure supplied. Furnish additional regulator if required.
2. Lubricated-plug-type manual gas cock.
3. Primary motorized or diaphragm gas valve.
4. Secondary gas valve downstream of the primary gas valve.
5. Solenoid bleed (vent) valve between the two main valves.
6. Gas pressure gauge following the regulator, with high-pressure cutoff.
7. Gas pressure gauge on inlet of gas trainer.

c. The gas pilot shall be the premix type with automatic electric ignition. The pilot gas line shall have a pilot gas regulator and a pilot gas solenoid valve. The burner controls shall modulate the flow of gas to the burner as required to maintain the steam pressure.

d. The oil burner shall be forced-draft, pressure-atomizing design, UL-labeled for operation with No. 2 fuel oil. An oil pump shall be directly connected to the burner fan motor or shall be a separate unit mounted on the boiler package. Burner shall be fully piped, including the primary oil valve, the secondary shutoff valve, and the pressure-regulating device ahead of the nozzle. The oil burner shall be electronically monitored so that the primary oil valve cannot open until the gas pilot has been established and proved. The burner controls shall modulate the supply of oil to the burner as required to maintain the steam pressure.

e. Flame safeguard control, equal to Honeywell, shall be furnished and wired at the factory and shall monitor both gas and oil burners. Flame safeguard control shall require a prepurge and a postpurge of the combustion chamber. The flame control shall indicate by lights if the pilot is on and if the main oil or gas valves are open. The flame control shall monitor the pilot flame and prove a stable and adequate pilot before allowing either the main gas valve or the main oil valve to open. Flame control shall monitor the main flame so that if there is any interruption of the main flame, the burners shall shut down and go into postpurge. An alarm shall sound and a light shall indicate main flame failure. Provide silencer for the alarm that will leave the light on. A low-water indication on any of the low-water cutoffs shall cause the burners to shut down, purge, and alarm. The flame monitor shall be ultraviolet or lead sulfide photocell. The pilot flame shall go off after the main flame is established.

f. An electric panel shall be furnished and installed on the boiler, which shall include a fused disconnect switch for the burner motor and control circuit. A second fused disconnect switch shall be provided for the boiler feedwater pump(s), along with relay and/or starter required for the pump(s). The second low-water cutoff shall be provided with any relay(s), etc., located in this panel. The electric panel shall be fully factory-wired and -tested, requiring only connection of electrical power to the panel terminals.

2.3 FEEDWATER RETURN SYSTEM

a. Provide a feedwater system that shall allow storage of return water, shall heat the water, and shall deliver the feedwater to the boiler as required.

b. The feedwater tank shall be mounted in the top of the boiler steam chest so that the water will be heated by the steam. Provide a float-type make-up valve to add city water as needed.

c.( )The feedwater return system shall consist of two turbine-type pumps mounted on the base with alternator. The alternator shall cause both pumps to operate if one will not keep up with the demand. The pump(s) shall be controlled by the feedwater control on the boiler.

( )The feedwater system shall consist of a turbine-type pump controlled by the feedwater control.

d. Provide all piping, valves, and strainers required for a complete system. Provide all wiring and controls required.

e. The feedwater system shall be factory-mounted on the boiler base and the entire system factory-tested before shipment.

2.4 AUXILIARY EQUIPMENT

a. Provide high-pressure gas requlator, or additional regulator, where the gas pressure is above the manufacturer's standard rating.

b. Provide a boiler water chemical treatment system, including treatment tank, pump, piping, controls, etc., for a complete system. Provide the initial 50-pound (22-kg) drum of treatment.

2.5 ACCEPTABLE MANUFACTURERS

a. Boiler shall be the make and model number shown on the drawings or equivalent by International (Plaza-Pack).

PART THREE - EXECUTION

3.1 MANUFACTURER'S INSTRUCTIONS

a. Piping and wiring diagrams, drawings, and installation instructions shall be obtained from the manufacturer and followed in the installation of the equipment.

3.2 MECHANICAL SPACE AND INSTALLATION

a. The boiler shall be installed in a space reserved for the boiler and provided with combustion air and ventilation required by the code and as shown on the drawings.
b. The boiler vent pipe shall be UL-listed for gas/oil service and shall be the full size of the boiler outlet. Provide a UL-listed flue cap and install both pipe and cap as required by the code.
c. The boiler room shall be provided with a drain adequate for the blowdown, draining, and washdown of the boiler.
d. Provide city water for make-up to the boiler and the boiler feedwater system. Include a water make-up assembly which includes a backflow preventer of the reduced-pressure type, acceptable to the local authorities. Provide necessary hose connections for washing down the boiler.
e. Connect gas to the gas train with a gas cock installed above ground and outside the mechanical space, in addition to the gas cock in the boiler gas train. Install gas regulator, as required, to reduce the gas pressure to that required by the equipment. Extend the vent from the gas bleed valve to the outside through the roof or other acceptable location. (Do not connect gas to the equipment if the mechanical room is not ventilated as required by code!)
f. Install steam header as shown on the drawings. Install the relief valve and carry full size through the roof. See detail on drawings.
g. Make the proper electrical connections for the boiler and equipment. Provide a disconnect means for the boiler controls, to turn off power to the main burner gas valve from an easily accessible point inside the mechanical room door.
h. Fill the boiler with water, and pressure-test the boiler and system up to the rating of the relief valve. Clean the boiler and system with trisodium phosphate or equal cleaner and flush the system to remove all trash and dirt; then chemically clean the system as specified and refill the system with clean water and the proper boiler water treatment.
i. Bleed the gas line in a safe manner and energize the boiler controls.
j. When the installation is complete, the manufacturer shall provide the services of a competent, factory-trained service person to review the installation and make the initial start-up of the system. Any problems found shall be corrected and the system made ready for the test called for below.

3.3 TEST

a. Observe the ignition of the pilot and the main burner to make sure they are smooth and complete.
b. Check out the safety controls and verify that they are functioning properly.
c. Set the operating controls for the proper pressure.
d. Operate the boiler for no less than 2 hr.
e. Provide a written report, through channels and on company letterhead, stating that burner operation is proper, safety controls have been checked, and operating controls are set and functioning properly. The system is not complete until this report has been received.

3.4 OWNER'S INSTRUCTIONS

a. Provide the owner's operator with three copies of written instructions for operation and maintenance of the system. A letter shall be sent, through channels, listing the materials given to the operator, the operator's name, and the date of this transmittal.
b. The operator shall be instructed carefully in the methods, and necessity, for blowing down the boiler to remove the accumulated mud on a regular basis.

END OF SECTION 15812

SECTION 15813

****

PACKAGED FIREBOX BOILER-BURNER UNIT

POWER BURNER - STEAM OR HOT WATER

PART ONE - GENERAL

1.1 The GENERAL and SPECIAL CONDITIONS, Section 15100, are included as a part of this Section as though written in full in this document.

1.2 Scope of the Work shall include the furnishing and complete installation of the equipment covered by this Section, with all auxiliaries, ready for owner's use.

PART TWO - PRODUCTS

2.1 BOILER PACKAGE

a. Provide a packaged firebox boiler-burner unit for use with natural (LP) gas, light oil, or combination gas/oil, with output capacity as shown on the drawings, to produce low-pressure steam (low-pressure hot water). The entire boiler-burner unit shall be UL-labeled.

b. Heating surface shall be no less than 5 square feet (0.465 m^2) per boiler horsepower (9809 watts). Heat release shall not exceed 55 MBH/ cubic foot (58,000 watts/m^3) of furnace volume.

c. Boiler shall be designed with an arched, upright crown sheet and two passes of fire tubes, in accordance with the ASME code for 15-psig (100-kPa) steam or 30-psig (200-kPa) water. Provide access to the front and rear tube sheets. Front flue doors shall be insulated and hinged. Provide refractory-lined rear access opening with an observation port.

d. Provide heavy-gauge, enameled steel jacket, with no less than 2 in. (50 mm) of high-temperature insulation. Floor of furnace shall be poured, high-temperature-insulating refractory. Rear wall of furnace shall be water-cooled. The entire unit shall be gastight.

e. Provide a flue gas thermometer, dial type, no less than 4 in. (100 mm) in diameter, mounted in the boiler flue outlet.

f. The following boiler trim shall be mounted, piped, and wired at the factory but may be removed for shipment if necessary.

( ) STEAM

1. ASME safety valve(s) for full capacity.
2. Steam gauge, 6 in. (150 mm) minimum, with siphon, shutoff cock, and test gauge connection.
3. Water column with red-line sight glass, try cocks, and blowdown valves.
4. Low-water cutoff and pump control for feedwater system.
5. Second low-water cutoff, manual water feeder, alarm contacts, alarm horn, alarm light, and control with silencer for the horn and manual reset.
6. Blowdown valves, as recommended by the manufacturer, for proper care of the boiler.

( ) HOT WATER

1. ASME relief valve(s) for full capacity.
2. Pressure gauge, 4 in. (100 mm) minimum.
3. Thermometer, 4 in. (100 mm) dial-type minimum.
4. Blowdown (drain) valves for washout of the mud legs, minimum of two per side and rear.

2.2 CONTROLS

a. All electrical equipment, motors, and control shall be wired to a control cabinet mounted on the boiler with terminals numbered. Provide a fused disconnect switch adjacent to the control cabinet.

b. Control cabinet shall include an electronic flame safeguard and programming control, equal to Honeywell or Fireye, to provide for prepurge, postpurge, pilot flame proving, main flame monitoring (gas or oil), low-water shutdown and alarm, and high-gas-pressure shutdown and alarm.

c. Provide terminal strips with numbers and identification nameplates. All wires shall be color-coded. Provide fuses for controls and for each circuit. Provide switch and transformer for controls. Provide control wiring for the boiler feedwater system pumps.

d. Provide burner on-off switch (selector switch for gas/oil), DC voltmeter indicating strength of flame signal, and indicating lights for the status of the burner, using low-voltage, long-life bulbs.

e. Provide motor starters and/or relays for all equipment controlled with the boiler.

2.3 COMBUSTION EQUIPMENT

( ) GAS BURNER

a. Gas burner shall be the annular port, flame retention, forced-draft type with gas-electric ignition, full modulation with low-fire start and manual high-fire potentiometer, lead sulfide (ultraviolet) flame detector cell, operating high-limit control, and manual reset high-limit safety control.

b. Gas train shall consist of:

1. Lubricated main gas cock.
2. Gas pressure regulator, system pressure to pressure required by burner.
3. Safety gas valve, motorized, normally closed.
( ) 4. Solenoid vent valve, normally open.
( ) 5. Gas pressure gauge.
6. Motorized primary gas valve, normally closed.
( ) 7. Modulating firing butterfly valve with control motor operator.
( ) 8. Push-to-test switch located in control cabinet.
9. Pilot gas cock, regulator, and solenoid valve.
10. Pressure gauge in gas main ahead of regulator.
11. Pressure gauge in gas train on leaving side of regulator.
12. Gas high-pressure safety cutout, on leaving side of regulator, to shut down boiler controls.

( ) OIL BURNER

a. Oil burner shall be pressure-atomizing forced-draft type, with gas-electric ignition, full modulation with low-fire start and manual high-fire potentiometer, lead sulfide (ultraviolet) flame detector cell, operating safety limit, and manual reset high-limit safety control.

b. Oil train shall include gate valve, oil filter with replaceable cartridge, check valve, oil pump driven by blower or as a separate unit mounted on the boiler, oil pressure gauge, main solenoid oil valve (normally closed), bypass pressure gauge, bypass metering valve, and check solenoid (normally closed).

( ) GAS/OIL COMBINATION

a. Include all items listed above for gas burner and the items listed for the oil burner.

b. Provide manual switch on the control cabinet to select gas or oil. Changeover shall not require adjustment of the air-fuel linkage.

( ) INSURANCE CONTROLS

a. Provide additional controls to comply with Factory Mutual (IRI) (FIA) insurance requirements.

2.4 ACCEPTABLE MANUFACTURERS

a. Boiler shall be the make and model number shown on the drawings or equivalent by Kewanee (Type C) or Burnham (Type 4F).

PART THREE - EXECUTION

3.1 MANUFACTURER'S INSTRUCTIONS

a. Piping and wiring diagrams, drawings, and installation instructions shall be obtained from the manufacturer and followed in the installation of the equipment.

3.2 MECHANICAL SPACE AND INSTALLATION

a. The boiler shall be installed in a space reserved for the boiler and provided with combustion air and ventilation required by the code and as shown on the drawings.

b. The boiler vent pipe shall be UL-listed for gas/oil service and shall be the full size of the boiler outlet. Provide a UL-listed flue cap and install both pipe and cap as required by the code.

c. The boiler room shall be provided with a drain adequate for the blowdown, draining, and washdown of the boiler.

d. Provide city water for make-up to the boiler and the boiler feedwater system. Include a water make-up assembly which includes a backflow preventer of the reduced-pressure type, acceptable to the local authorities. Provide necessary hose connections for washing down the boiler.

e. Install the low-water cutoffs and feed units in the following (descending) order:

1. First stage - bring on feedwater pump.
2. Second stage - low-water cutoff with automatic reset.
3. Third stage - low-water cutoff with manual reset and alarm circuit and system.
4. Fourth stage - automatic manual water feeder.

First stage shall be set at 1/2 in. (13 mm) below normal.

Second stage shall be set at 1 1/2 in. (30 mm) below normal water level.

Third stage shall be set at 2 1/2 in. (63 mm) below normal.

Fourth stage shall be set at 3 in. (75 mm) below normal water level.

f. Connect gas to the gas train with a gas cock installed above ground and outside the mechanical space, in addition to the gas cock in the boiler gas train. Install gas regulator as required to reduce the gas pressure to that required by the equipment. Extend the vent from the gas bleed valve to the outside through the roof or other acceptable location. (Do not connect gas to the equipment if the mechanical room is not ventilated as required by code!)

g. Install steam header as shown on the drawings. Install the relief valve and carry full size through the roof. See detail on drawings.

h. Make the proper electrical connections for the boiler and equipment. Provide a disconnect means for the boiler controls, to turn off power to the main burner gas valve from an easily accessible point inside the mechanical room door.

i. Fill the boiler with water, and pressure-test the boiler and system up to the rating of the relief valve. Clean the system with trisodium phosphate or equivalent cleaner, and flush the system to remove all trash and dirt; then chemically clean the system as specified and refill the system with clean water and the proper boiler water treatment.
j. Bleed the gas line in a safe manner and energize the boiler controls.
k. When the installation is complete, the manufacturer shall provide the services of a competent, factory-trained service person to review the installation and make the initial start-up of the system. Any problems found shall be corrected and the system made ready for the test called for below.

3.3 TEST
a. Observe the ignition of the pilot and the main burner to make sure they are smooth and complete.
b. Check out the safety controls and verify that they are functioning properly.
c. Set the operating controls for the proper pressure.
d. Operate the boiler for no less than 2 hr, or return at the beginning of the heating season for completion of this test.
e. Provide a written report, through channels and on company letterhead, stating that burner operation is proper, safety controls have been checked, and operating controls are set and functioning properly. The system is not complete until this report has been received.

( ) (THE SPEC WRITER MAY WISH TO REQUIRE A FACTORY FUEL-TO-STEAM OR -WATER TEST OF THE BOILER TO GUARANTEE THE PERFORMANCE AND/OR EFFICIENCY. IF SO, THE TEST SHOULD BE SPECIFIED HERE.
( ) SPEC WRITER MAY WISH TO REQUIRE A FIELD FUEL-TO-STEAM OR -WATER PERFORMANCE TEST OF THE BOILER, AS INSTALLED, TO GUARANTEE PERFORMANCE AND/OR EFFICIENCY. IF SO, THE TEST SHOULD BE SPELLED OUT HERE IN DETAIL.)

3.4 OWNER'S INSTRUCTIONS
a. After the system is in operation, the owner's operators shall be schooled in the proper operation of the boiler and associated equipment for no less than 16 hr by the factory-trained start-up personnel.
b. Provide the owner's operator with five bound copies of written instructions for operation and maintenance of the system. A letter shall be sent, through channels, listing the materials given to the operator, the operator's name, and the date of this transmittal.
c. The operator shall be instructed carefully in the methods, and necessity, for blowing down the boiler to remove the accumulated mud on a regular basis.

END OF SECTION 15813

SECTION 15820
****
PACKAGED SCOTCH MARINE BOILER-BURNER UNIT
HIGH-PRESSURE HOT WATER

PART ONE - GENERAL

1.1 The GENERAL and SPECIAL CONDITIONS, Section 15100, are included as a part of this Section as though written in full in this document.

1.2 Scope of the Work shall include the furnishing and complete installation of the equipment covered by this Section, with all auxiliaries, ready for owner's use.

PART TWO - PRODUCTS

2.1 BOILER PACKAGE

a. Provide a packaged Scotch Marine boiler-burner unit for use with natural (LP) gas, light oil, or combination gas/oil, with output capacity as shown on the drawings, to produce high-pressure hot water. The entire boiler-burner unit shall be UL-labeled.

b. Heating surface shall be no less than 5 square feet (0.465 m^2) per boiler horsepower (9809 watts). Heat release shall not exceed 55 MBH/ft^3 (2.05 MJ/m^3) of furnace volume.

c. Boiler shall have large-diameter fire tube and 3-in. (75-mm) return tubes arranged in a three- or four-pass design, with ASME stamp for 100-psig (690-kPa) working pressure and 150-psig (1035-kPa) test pressure. Provide access to the front and rear tube sheets. Front flue doors shall be insulated and hinged. Provide explosion relief opening and observation ports (front and rear).

d. Provide heavy-gauge, enameled steel jacket, with no less than 2 in. (50 mm) of high-temperature insulation.

e. Provide a flue gas thermometer, dial type, no less than 4 in. (100 mm) in diameter, mounted in the boiler flue outlet.

f. The following boiler trim shall be mounted, piped, and wired at the factory but may be removed for shipment if necessary.

1. ASME relief valve(s) for full capacity.
2. Pressure gauge, 4 in. (100 mm) minimum
3. Thermometer, 4 in. (100 mm) dial-type minimum.
4. Blowdown (drain) valves for washout of the mud.

2.2 CONTROLS

a. All electrical equipment, motors, and control shall be wired to a control cabinet mounted on the boiler with terminals numbered. Provide a fused disconnect switch adjacent to the control cabinet.

b. Control cabinet shall include an electronic flame safeguard and programming control, equal to Honeywell or Fireye, to provide for prepurge, postpurge, pilot flame proving, main flame monitoring (gas or oil), low-water shutdown and alarm, and high-gas-pressure shutdown and alarm.

c. Provide terminal strips with numbers and identification nameplates. All wires shall be color-coded. Provide fuses for controls and for each circuit. Provide switch and transformer for controls. Provide control wiring for the boiler feedwater system pumps.

d. Provide burner on-off switch (selector switch for gas/oil), DC voltmeter indicating strength of flame signal, and indicating lights for the status of the burner, using low-voltage, long-life bulbs.

e. Provide motor starters and/or relays for all equipment controlled with the boiler.

2.3 COMBUSTION EQUIPMENT

( ) GAS BURNER

a. Gas burner shall be the annular port, flame retention, forced-draft type with gas-electric ignition, full modulation with low-fire start and manual high-fire potentiometer, lead sulfide (ultraviolet) flame detector cell, operating high-limit control, and manual reset high-limit safety control.

b. Gas train shall consist of:

1. Lubricated main gas cock.
2. Gas pressure regulator, system pressure to pressure required by burner.
3. Safety gas valve, motorized, normally closed.

( ) 4. Solenoid vent valve, normally open.

5. Pressure-checking gauge.
6. Motorized primary gas valve, normally closed.

( ) 7. Modulating firing butterfly valve with control motor operator.

( ) 8. Push-to-test switch located in control cabinet.

9. Pilot gas cock, regulator, and solenoid valve.
10. Pressure gauge in gas main ahead of regulator.
11. Pressure gauge in gas train on leaving side of regulator.
12. Gas high-pressure safety cutout on leaving side of regulator, to shut down boiler controls.

( ) OIL BURNER

a. Oil burner shall be pressure-atomizing forced-draft type, with gas-electric ignition, full modulation with low-fire start and manual high-fire potentiometer, lead sulfide (ultraviolet) flame detector cell, operating safety limit, and manual reset high-limit safety control.

b. Oil train shall include gate valve, oil filter with replaceable cartridge, check valve, oil pump driven by blower or as a separate unit mounted on the boiler, oil pressure gauge, main solenoid oil valve (normally closed), bypass pressure gauge, bypass metering valve, and check solenoid (normally closed).

( ) GAS/OIL COMBINATION

a. Include all items listed above for gas burner and the items listed for the oil burner.

b. Provide manual switch on the control cabinet to select gas or oil. Changeover shall not require adjustment of the air-fuel linkage.

( ) INSURANCE CONTROLS

a. Provide additional controls to comply with Factory Mutual (FM) (IRI) (FIA) insurance requirements.

2.4 ACCEPTABLE MANUFACTURERS

a. Boiler shall be the make and model number shown on the drawings or equivalent by Kewanee (Classic III), Cleaver Brooks (CB), Superior (Mohawk).

PART THREE - EXECUTION

3.1 MANUFACTURER'S INSTRUCTIONS

a. Piping and wiring diagrams, drawings, and installation instructions shall be obtained from the manufacturer and followed in the installation of the equipment.

b. Should the manufacturer's drawings be in conflict with the contract drawings, the design engineer shall be contacted for a decision on which course to follow.

3.2 MECHANICAL SPACE AND INSTALLATION

a. The boiler shall be installed in a space reserved for the boiler and provided with combustion air and ventilation required by the code and as shown on the drawings. The boiler shall be mounted on a level concrete base.

b. The boiler vent pipe shall be UL-listed for gas/oil service and shall be the full size of the boiler outlet. Provide a UL-listed flue cap and install both pipe and cap as required by the code. The weight imposed on the boiler shall not be more than allowed by the manufacturer.

c. The boiler room shall be provided with a drain adequate for the blowdown, draining, and washdown of the boiler.

d. Provide city water for make-up to the boiler. Include a water make-up assembly which includes a backflow preventer of the reduced-pressure type, acceptable to the local authorities. Provide necessary hose connections for washing down the boiler.

e. Connect gas to the gas train with a gas cock installed above ground and outside the mechanical space, in addition to the gas cock in the boiler gas train. Install gas regulator as required to reduce the gas pressure to that required by the equipment. Extend the vent from the gas bleed valve to the outside through the roof or other acceptable location. (Do not connect gas to the equipment if the mechanical room is not ventilated as required by code.)

f. Install hot water supply and return connections as shown on the drawings. Install the relief valve and carry full size down to floor. See detail on drawings.

g. Make the proper electrical connections for the boiler and equipment. Provide a disconnect means for the boiler controls, to turn off power to the main burner gas valve from an easily accessible point inside the mechanical room door.

h. Fill the boiler with water, and pressure-test the boiler and system up to the rating of the relief valve. Clean the system with trisodium phosphate or equal cleaner and flush the system to remove all trash and dirt; then chemically clean the system as specified and refill the system with clean water and the proper boiler water treatment.

i. Bleed the gas line in a safe manner and energize the boiler controls.

j. When the installation is complete, the manufacturer shall provide the services of a competent, factory-trained service person to review the installation and make the initial start-up of the system. Any problems found shall be corrected and the system made ready for the test called for below.

3.3 TEST

a. Observe the ignition of the pilot and the main burner to make sure they are smooth and complete.

b. Check out the safety controls and verify that they are functioning properly.

c. Set the operating controls for the proper temperature.

d. Operate the boiler for no less than 2 hr, or return at the beginning of the heating season for completion of this test.

e. Provide a written report, through channels and on company letterhead, stating that burner operation is proper, safety controls have been checked, and operating controls are set and functioning properly. The system is not complete until this report has been received.

(THE SPEC WRITER MAY WISH TO REQUIRE A FACTORY FUEL-TO-WATER TEST OF THE BOILER TO GUARANTEE THE PERFORMANCE AND/OR EFFICIENCY. IF TEST IS REQUIRED, IT SHOULD BE SPECIFIED HERE.)
(THE SPEC WRITER MAY WISH TO REQUIRE A FIELD FUEL-TO-WATER PERFORMANCE TEST OF THE BOILER, AS INSTALLED, TO GUARANTEE PERFORMANCE AND/OR EFFICIENCY. IF TEST IS REQUIRED, IT SHOULD BE SPECIFIED IN DETAIL HERE.)

3.4 OWNER'S INSTRUCTIONS

a. After the system is in operation, the owner's operators shall be schooled in the proper operation of the boiler and associated equipment for no less than 16 hr by the factory-trained start-up personnel.
b. Provide the owner's operator with five bound copies of written instructions for operation and maintenance of the system. A letter shall be sent, through channels, listing the materials given to the operator, the operator's name, and the date of this transmittal.
c. The operator shall be instructed carefully in the methods, and necessity, for blowing down the boiler to remove the accumulated mud on a regular basis.

END OF SECTION 15820

SECTION 15821
****
PACKAGED SCOTCH MARINE BOILER-BURNER UNIT
LOW-PRESSURE STEAM

PART ONE - GENERAL

1.1 The GENERAL and SPECIAL CONDITIONS, Section 15100, are included as a part of this Section as though written in full in this document.

1.2 Scope of the Work shall include the furnishing and complete installation of the equipment covered by this Section, with all auxiliaries, ready for owner's use.

PART TWO - PRODUCTS

2.1 BOILER PACKAGE

a. Provide a packaged Scotch Marine boiler-burner unit for use with natural (LP) gas, light oil, or combination gas/oil, with output capacity as shown on the drawings, to produce low-pressure steam. The entire boiler-burner unit shall be UL-labeled.

b. Heating surface shall be no less than 5 square feet (0.465 m^2) per boiler horsepower (9809 watts). Heat release shall not exceed 175 MBH/ cubic foot (181,016 watts/m^3) of furnace volume.

c. Boiler shall be designed with a large-diameter fire tube and return fire tubes for three- or four-pass design. Boiler shall be ASME-stamped for 15-psig (100-kPa) steam. Provide access to the front and rear tube sheets. Flue doors shall be insulated and hinged or davited. Provide observation ports in front and rear.

d. Provide heavy-gauge, enameled steel jacket, with no less than 2 in. (50 mm) of high-temperature insulation.

e. Provide a flue gas thermometer, dial type, no less than 4 in. (100 mm) in diameter, mounted in the boiler flue outlet.

f. The following boiler trim shall be mounted, piped, and wired at the factory but may be removed for shipment if necessary.

1. ASME safety valve(s) for full capacity.
2. Steam gauge, 6 in. (150 mm) minimum, with siphon, shutoff cock and test gauge connection.
3. Water column with red-line sight glass, try cocks, and blowdown valves.
4. Low-water cutoff and pump control for feedwater system.

5. Second low-water cutoff with alarm contacts, alarm horn, alarm light, and control with silencer for the horn and manual reset.
6. Blowdown valves, as recommended by the manufacturer, for proper care of the boiler.

2.2 CONTROLS

a. All electrical equipment, motors, and control shall be wired to a control cabinet mounted on the boiler with terminals numbered. Provide a fused disconnect switch adjacent to the control cabinet.

b. Control cabinet shall include an electronic flame safeguard and programming control, equal to Honeywell or Fireye, to provide for prepurge, postpurge, pilot flame proving, main flame monitoring (gas or oil), low-water shutdown and alarm, and high-gas-pressure shutdown and alarm.

c. Provide terminal strips with numbers and identification nameplates. All wires shall be color-coded. Provide fuses for controls and for each circuit. Provide switch and transformer for controls. Provide control wiring for the boiler feedwater system pumps.

d. Provide burner on-off switch (selector switch for gas/oil), DC voltmeter indicating strength of flame signal, and indicating lights for the status of the burner, using low-voltage, long-life bulbs.

e. Provide motor starters and/or relays for all equipment controlled with the boiler.

2.3 COMBUSTION EQUIPMENT

( ) GAS BURNER

a. Gas burner shall be the annular port, flame retention, forced-draft type with gas-electric ignition, full modulation with low-fire start and manual high-fire potentiometer, lead sulfide (ultraviolet) flame detector cell, operating high-limit control, and manual reset high-limit safety control.

b. Gas train shall consist of:

1. Lubricated main gas cock.
2. Gas pressure regulator, system pressure to pressure required by burner.
3. Safety gas valve, motorized, normally closed.
4. ( ) Solenoid vent valve, normally open.
5. Gas pressure gauge.
6. Motorized primary gas valve, normally closed.
7. ( ) Modulating firing butterfly valve with control motor operator.
8. ( ) Push-to-test switch located in control cabinet.
9. Pilot gas cock, regulator, and solenoid valve.
10. Pressure gauge in gas main ahead of regulator.
11. Pressure gauge in gas train on leaving side of regulator.
12. Gas high-pressure safety cutout on leaving side of regulator, to shut down boiler controls.

( ) OIL BURNER

a. Oil burner shall be pressure-atomizing type, forced-draft, with gas-electric ignition, full modulation with low-fire start and manual high-fire potentiometer, lead sulfide (ultraviolet) flame detector cell, operating safety limit, and manual reset high-limit safety control.

b. Oil train shall include gate valve, oil filter with replaceable cartridge, check valve, oil pump driven by blower or as a separate unit mounted on the boiler, oil pressure gauge, main solenoid oil valve (normally closed), bypass pressure gauge, bypass metering valve, and check solenoid (normally closed).

( ) GAS/OIL COMBINATION

a. Include all items listed above for gas burner and the items listed for the oil burner.

b. Provide manual switch on the control cabinet to select gas or oil. Changeover shall not require adjustment of the air-fuel linkage.

( ) INSURANCE CONTROLS

a. Provide additional controls to comply with Factory Mutual (IRI) (FIA) insurance requirements.

2.4 ACCEPTABLE MANUFACTURERS

a. Boiler shall be the make and model number shown on the drawings or equivalent by Kewanee (Classic III), Cleaver Brooks (Model CB), Superior (Mohawk).

PART THREE - EXECUTION

3.1 MANUFACTURER'S INSTRUCTIONS

a. Piping and wiring diagrams, drawings, and installation instructions shall be obtained from the manufacturer and followed in the installation of the equipment.

b. Should there be any conflict between the manufacturer's drawings and the contract drawings, the design engineer shall be contacted for a decision as to which to follow.

3.2 MECHANICAL SPACE AND INSTALLATION

a. The boiler shall be installed in a space reserved for the boiler and provided with combustion air and ventilation required by the code and as shown on the drawings. A level concrete base shall be provided for the boiler.

b. The boiler vent pipe shall be UL-listed for gas/oil service and shall be the full size of the boiler outlet. Provide a UL-listed flue cap and install both pipe and cap as required by the code.

c. The boiler room shall be provided with a drain adequate for the blowdown, draining, and washdown of the boiler.

d. Provide city water for make-up to the boiler and the boiler feedwater system. Include a water make-up assembly which includes a backflow preventer of the reduced-pressure type, acceptable to the local authorities. Provide necessary hose connections for washing down the boiler.

e. Connect gas to the gas train with a gas cock installed above ground and outside the mechanical space, in addition to the gas cock in the boiler gas train. Install gas regulator as required to reduce the gas pressure to that required by the equipment. Extend the vent from the gas bleed valve to the outside through the roof or other acceptable location. (Do not connect gas to the equipment if the mechanical room is not ventilated as required by code!)

f. Install steam header as shown on the drawings. Install the relief valve and carry full size through the roof. See detail on drawings.

g. Make the proper electrical connections for the boiler and equipment. Provide a disconnect means for the boiler controls, to turn off power to the main burner gas valve from an easily accessible point inside the mechanical room door.

h. Fill the boiler with water, and pressure-test the boiler and system up to the rating of the relief valve. Clean the system with trisodium phosphate or equal cleaner and flush the system to remove all trash and dirt; then chemically clean the system as specified and refill the system with clean water and the proper boiler water treatment.

i. Bleed the gas line in a safe manner and energize the boiler controls.

j. When the installation is complete, the manufacturer shall provide the services of a competent, factory-trained service person to review the installation and make the initial start-up of the system. Any problems found shall be corrected and the system made ready for the test called for below.

3.3 TEST

a. Observe the ignition of the pilot and the main burner to make sure they are smooth and complete.

b. Check out the safety controls and verify that they are functioning properly.

c. Set the operating controls for the proper pressure.

d. Operate the boiler for no less than 2 hr, or return at the beginning of the heating season for completion of this test.

e. Provide a written report, through channels and on company letterhead, stating that burner operation is proper, safety controls have been checked, and operating controls are set and functioning properly. The system is not complete until this report has been received.

( ) (THE SPEC WRITER MAY WISH TO REQUIRE A FACTORY FUEL-TO-STEAM TEST OF THE BOILER TO GUARANTEE THE PERFORMANCE AND/OR EFFICIENCY. IF TEST IS REQUIRED IT SHOULD BE SPECIFIED HERE.)

( ) (THE SPEC WRITER MAY WISH TO REQUIRE A FIELD FUEL-TO-STEAM PERFORMANCE TEST OF THE BOILER, AS INSTALLED, TO GUARANTEE PERFORMANCE AND/OR EFFICIENCY. IF TEST IS REQUIRED, IT SHOULD BE SPECIFIED IN DETAIL HERE.)

3.4 OWNER'S INSTRUCTIONS

a. After the system is in operation, the owner's operators shall be schooled in the proper operation of the boiler and associated equipment for no less than 16 hr by the factory-trained start-up personnel.

b. Provide the owner's operator with five bound copies of written instructions for operation and maintenance of the system. A letter shall be sent, through channels, listing the materials given to the operator, the operator's name, and the date of this transmittal.

c. The operator shall be instructed carefully in the methods, and necessity, for blowing down the boiler to remove the accumulated mud on a regular basis.

END OF SECTION 15821

SECTION 15822
****

PACKAGED SCOTCH MARINE BOILER-BURNER UNIT
HIGH-PRESSURE STEAM

PART ONE - GENERAL

1.1 The GENERAL and SPECIAL CONDITIONS, Section 15100, are included as a part of this Section as though written in full in this document.

1.2 Scope of the Work shall include the furnishing and complete installation of the equipment covered by this Section, with all auxiliaries, ready for owner's use.

PART TWO - PRODUCTS

2.1 BOILER PACKAGE

a. Provide a packaged Scotch Marine boiler-burner unit for use with natural (LP) gas, light oil, or combination gas/oil, with output capacity as shown on the drawings, to produce high-pressure steam. The entire boiler-burner unit shall be UL-labeled.

b. Heating surface shall be no less than 5 square feet (0.465 m^2) per boiler horsepower (9809 watts). Heat release shall not exceed 175 MBH/cubic foot (6.5 MJ/m^3) of furnace volume.

c. Boiler shall be designed with a large-diameter fire tube, with 3 in (75-mm) return fire tubes for three or four passes. Boiler shall be ASME-stamped for 150-psig (1035-kPa) steam. Provide access to the front and rear tube sheets. Flue doors shall be insulated and hinged or davited. Provide observation ports in front and rear.

d. Provide heavy-gauge, enameled steel jacket, with no less than 2 in. (50 mm) of high-temperature insulation.

e. Provide a flue gas thermometer, dial type, no less than 4 in. (100 mm) in diameter, mounted in the boiler flue outlet.

f. The following boiler trim shall be mounted, piped, and wired at the factory but may be removed for shipment if necessary.

1. ASME safety valve(s) for full capacity.
2. Steam gauge, 6 in. (150 mm) minimum, with siphon, shutoff cock, and test gauge connection.
3. Water column with red-line sight glass, try cocks, and blowdown valves.
4. Low-water cutoff and pump control for feedwater system.
5. Second low-water cutoff with alarm contacts, alarm horn, alarm light, and control with silencer for the horn and manual reset.
6. Blowdown valves, as recommended by the manufacturer, for proper care of the boiler. (One or more fast-opening valves and one slow-opening valve.)

2.2 CONTROLS

a. All electrical equipment, motors, and control shall be wired to a control cabinet mounted on the boiler with terminals numbered. Provide a fused disconnect switch adjacent to the control cabinet.

b. Control cabinet shall include an electronic flame safeguard and programming control, equal to Honeywell or Fireye, to provide for prepurge, postpurge, pilot flame proving, main flame monitoring (gas or oil), low-water shutdown and alarm, and high-gas-pressure shutdown and alarm.

c. Provide terminal strips with numbers and identification nameplates. All wires shall be color-coded. Provide fuses for controls and for each circuit. Provide switch and transformer for controls. Provide control wiring for the boiler feedwater system pumps.

d. Provide burner on-off switch (selector switch for gas/oil), DC voltmeter indicating strength of flame signal, and indicating lights for the status of the burner, using low-voltage, long-life bulbs.

e. Provide motor starters and/or relays for all equipment controlled with the boiler.

2.3 COMBUSTION EQUIPMENT

( ) GAS BURNER

a. Gas burner shall be the annular port, flame retention, forced-draft type with gas-electric ignition, full modulation with low-fire start and manual high-fire potentiometer, lead sulfide (ultraviolet) flame detector cell, operating high-limit control, and manual reset high-limit safety control.

b. Gas train shall consist of:

1. Lubricated main gas cock.
2. Gas pressure regulator, system pressure to pressure required by burner.
3. Safety gas valve, motorized, normally closed.
( ) 4. Solenoid vent valve, normally open.
( ) 5. Gas pressure gauge between safety vlave and primary valve.
6. Motorized primary gas valve, normally closed.
( ) 7. Modulating firing butterfly valve with control motor operator.
( ) 8. Push-to-test switch located in control cabinet.
9. Pilot gas cock, regulator, and solenoid valve.
( ) 10. Pressure gauge in gas main ahead of regulator.
( ) 11. Pressure gauge in gas train on leaving side of regulator.
( ) 12. Gas high-pressure safety cutout on leaving side of regulator to shut down boiler controls.

( ) Indicates optional equipment to be specified if required.

( ) OIL BURNER

a. Oil burner shall be pressure-atomizing forced-draft type, with electric spark and/or gas ignition, full modulation with low-fire start and manual high-fire potentiometer, lead sulfide (ultraviolet) flame detector cell, operating safety limit, and manual reset high-limit safety control.

b. Oil train shall include gate valve, oil filter with replaceable cartridge, check valve, oil pump driven by blower or as a separate unit mounted on the boiler, oil pressure gauge, main solenoid oil valve (normally closed), bypass pressure gauge, bypass metering valve, and check solenoid (normally closed).

( ) GAS/OIL COMBINATION

a. Include all items listed above for gas burner and the items listed for the oil burner.

b. Provide manual switch on the control cabinet to select gas or oil. Changeover shall not require adjustment of the air-fuel linkage.

( ) INSURANCE CONTROLS

a. Provide additional controls to comply with Factory Mutual (FM) (IRI) (FIA) insurance requirements.

2.4 ACCEPTABLE MANUFACTURERS

a. Boiler shall be the make and model number shown on the drawings or equivalent by Kewanee (Classic III), Cleaver Brooks (Model CB), Superior (Mohawk).

PART THREE - EXECUTION

3.1 MANUFACTURER'S INSTRUCTIONS

a. Piping and wiring diagrams, drawings, and installation instructions shall be obtained from the manufacturer and followed in the installation of the equipment.

b. Should there be any conflict between the manufacturer's drawings and the contract drawings, the design engineer shall be contacted for a decision as to which to follow.

3.2 MECHANICAL SPACE AND INSTALLATION

a. The boiler shall be installed in a space reserved for the boiler and provided with combustion air and ventilation required by the code and as shown on the drawings. A level concrete base shall be provided for the boiler.

b. The boiler vent pipe shall be UL-listed for gas/oil service and shall be the full size of the boiler outlet. Provide a UL-listed flue cap and install both pipe and cap as required by the code.

c. The boiler room shall be provided with a drain adequate for the blowdown, draining, and washdown of the boiler.

d. Provide city water for make-up to the boiler and the boiler feedwater system. Include a water make-up assembly with a backflow preventer of the reduced-pressure type, acceptable to the local authorities. Provide necessary hose connections for washing down the boiler.

e. Connect gas to the gas train with a gas cock installed above ground and outside the mechanical space, in addition to the gas cock in the boiler gas train. Install gas regulator as required to reduce the gas pressure to that required by the equipment. Extend the vent from the gas bleed valve to the outside through the roof or other acceptable location. (Do not connect gas to the equipment if the mechanical room is not ventilated as required by code!)

f. Install steam header as shown on the drawings. Install the relief valve and carry full size through the roof. See detail on drawings.

g. Make the proper electrical connections for the boiler and equipment. Provide a disconnect means for the boiler controls, to turn off power to the main burner gas valve from an easily accessible point inside the mechanical room door.

h. Fill the boiler with water, and pressure-test the boiler and system up to the rating of the relief valve. Clean the boiler and system with trisodium phosphate or equal cleaner, and flush the system to remove all trash and dirt; then chemically clean the system as specified and refill the system with clean water and the proper boiler water treatment.

i. Bleed the gas line in a safe manner and energize the boiler controls.

j. When the installation is complete, the manufacturer shall provide the services of a competent, factory-trained service person to review the installation and make the initial start-up of the system. Any problems found shall be corrected and the system made ready for the test called for below.

3.3 TEST

a. Observe the ignition of the pilot and the main burner to make sure they are smooth and complete.

b. Check out the safety controls and verify that they are functioning properly.

c. Set the operating controls for the proper pressure.

d. Operate the boiler for no less than 10 hr, or return at the beginning of the heating season for completion of this test.

e. Provide a written report, through channels and on company letterhead, stating that burner operation is proper, safety controls have been checked, and operating controls are set and functioning properly. The system is not complete until this report has been received.

( ) (THE SPEC WRITER MAY WISH TO REQUIRE A FACTORY FUEL TO STEAM-TEST THE BOILER TO GUARANTEE THE PERFORMANCE AND/OR EFFICIENCY. IF SO, IT SHOULD BE ADDED HERE.)

( ) (THE SPEC WRITER MAY WISH TO REQUIRE A FIELD FUEL TO STEAM-PERFORMANCE-TEST THE BOILER AS INSTALLED TO GUARANTEE PERFORMANCE AND/OR EFFICIENCY. IF SO, IT SHOULD BE SPELLED OUT HERE IN DETAIL.)

3.4 OWNER'S INSTRUCTIONS

a. After the system is in operation, the owner's operators shall be schooled in the proper operation of the boiler and associated equipment for no less than 16 hr by the factory-trained start-up personnel.

b. Provide the owner's operator with five bound copies of written instructions for operation and maintenance of the system. A letter shall be sent, through channels, listing the materials given to the operator, the operator's name, and the date of this transmittal.

c. The operator shall be instructed carefully in the methods, and necessity, for blowing down the boiler to remove the accumulated mud on a regular basis.

END OF SECTION 15822

CHAPTER 13

Boiler Auxiliary Equipment

SECTION 15850
****
CONDENSATE RETURN UNIT
LOW PRESSURE - SIMPLEX

PART ONE - GENERAL

1.1 The GENERAL and SPECIAL CONDITIONS, Section 15100, are included as a part of this Section as though written in full in this document.

1.2 Scope of the Work shall include the furnishing and complete installation of the equipment covered by this Section, with all auxiliaries, ready for owner's use.

PART TWO - PRODUCTS

2.1 CONDENSATE RETURN UNIT

a. Condensate return unit shall be designed to store and return condensate to a low-pressure steam boiler. Unit shall be designed for a low suction elevation to allow maximum return pipe slope.

b. Unit shall consist of a cast iron receiver with a single vertical-shaft centrifugal pump mounted in the receiver or on the side with a low connection. Tank capacity shall be no less than that called for on drawings.

c. Pump shall be designed for the flow and head shown on the drawings and shall not require more power than shown.

2.2 AUXILIARY EQUIPMENT

a. Provide a float-operated water make-up unit to provide city water to the receiver, should the water level drop to just above the suction head required by the pump.

b. Provide a discharge line the full size of the pump outlet with a full-size check valve.

c. Provide a float-operated switch to operate the pump when the receiver is almost full. (OMIT THIS CONTROL IF THE PUMP IS CONTROLLED BY THE BOILER.)

d. Thermometer, adjustable dial type.

e. Gauge glass with cocks.

2.3 ACCEPTABLE MANUFACTURERS

a. The condensate return unit shall be the make and model shown on the drawings or equivalent by Weil (Series 4100), Underwood, Burks, Aurora (Series 221), Sterling (Series 4200).

PART THREE - EXECUTION

3.1 INSTALLATION

a. Install the condensate return unit, as shown on the drawings and as recommended by the manufacturer, on a level base. Clean out all trash and wash out the receiver and system.

b. Provide vent to the outside, if possible. If outside vent is not possible, provide a type M copper vent line and install no less than 6 ft. (2 m) of finned tube radiation in the vertical line to condense any exhaust steam. Carry the vent line up to the ceiling and then down to an acceptable point for water discharge.

c. Connect the system return lines to the receiver.

d. Connect water from the city water line through the vacuum breaker to the make-up float valve.

e. Connect electrical power and test the operation of the system.

END OF SECTION 15850

SECTION 15851

****

CONDENSATE RETURN UNIT

LOW PRESSURE - DUPLEX

PART ONE - GENERAL

1.1 The GENERAL and SPECIAL CONDITIONS, Section 15100, are included as a part of this Section as though written in full in this document.

1.2 Scope of the Work shall include the furnishing and complete installation of the equipment covered by this Section, with all auxiliaries, ready for owner's use.

PART TWO - PRODUCTS

2.1 CONDENSATE RETURN UNIT

a. Condensate return unit shall be designed to store and return condensate to a low-pressure steam boiler. Unit shall be designed for a low suction elevation to allow maximum return pipe slope.

b. Unit shall consist of a cast iron receiver with two vertical-shaft centrifugal pumps mounted in the receiver or on the side with a low connection. Tank capacity shall be no less than that called for on drawings.

c. Each pump shall be designed for the flow and head shown on the drawings and shall not require more power than shown.

2.2 AUXILIARY EQUIPMENT

a. Provide a float-operated water make-up unit to provide city water to the receiver, should the water level drop to just above the suction head required by the pump.

b. Provide an alternator of the mechanical (electrical) type to alternate operation of the pumps and to bring on both pumps if one pump is not able to keep up with the load.

c. Provide a discharge line the full size of the pump outlet with a full-size check valve for each pump, and connect the discharge to one header as shown.

d. Provide a float-operated switch to operate the pump when the receiver is almost full. (OMIT THIS CONTROL IF THE UNIT IS CONTROLLED BY THE BOILER.)

e. Dial-type thermometer.

f. Gauge glass with cocks.

2.3 ACCEPTABLE MANUFACTURERS

a. The condensate return unit shall be the make and model shown on the drawings or equivalent by Weil (Series 4100), Underwood, Burks, Aurora (Series 221), Sterling (Series 4200).

PART THREE - EXECUTION

3.1 INSTALLATION

a. Install the condensate return unit, as shown on the drawings and as recommended by the manufacturer, on a level base. Clean out all trash and wash out the receiver and system.

b. Provide vent to the outside, if possible. If outside vent is not possible, provide a type M copper vent line and install no less than 6 ft. (2 m) of finned tube radiation in the vertical line to condense any exhaust steam. Carry the vent line up to the ceiling and then down to an acceptable point for water discharge.

c. Connect the system return lines to the receiver.

d. Connect water from the city water line through the vacuum breaker to the make-up float valve.

e. Connect electrical power and test the operation of the system, including the alternator.

END OF SECTION 15851

SECTION 15855

****

CONDENSATE RETURN UNIT

HIGH PRESSURE - DUPLEX

PART ONE - GENERAL

1.1 The GENERAL and SPECIAL CONDITIONS, Section 15100, are included as a part of this Section as though written in full in this document.

1.2 Scope of the Work shall include the furnishing and complete installation of the equipment covered by this Section, with all auxiliaries, ready for owner's use.

PART TWO - PRODUCTS

2.1 CONDENSATE RETURN UNIT

a. Packaged condensate return unit shall be designed to store and heat returned condensate and to return the condensate to a medium- or high-pressure steam boiler. Unit shall be designed to provide a net positive suction head for the pump for the handling of very hot water.

b. Unit shall consist of a heavy galvanized steel receiver with two turbine-type boiler feed pumps mounted under the receiver, with separate suction lines to the tank. Tank capacity shall be no less than that called for on drawings.

c. Pumps shall each be designed for the flow and head shown on the drawings and shall not require more power than shown.

2.2 AUXILIARY EQUIPMENT

a. Provide a float-operated water make-up unit to provide city water to the receiver, should the water level drop to just above the suction head required by the pump.

b. Provide a prewired control panel with motor starters for each boiler feedwater pump, including an alternator of the mechanical (electrical) type to alternate operation of the pumps and to bring on both pumps if one pump is not able to keep up with the load. Provide a hand-off-auto selector switch for each pump. Provide a pilot light to indicate pump operation.

c. Provide a discharge line the full size of the pump outlet with a full-size check valve for each pump, and connect the discharge to one header as shown.

d. Provide a steam heater package to preheat the condensate by injection of steam. Provide ASME relief valve, pressure gauge, preheater tube, steam strainer, temperature-regulating valve with pressure reduction for high-pressure supply.

e. Dial-type thermometer.

f. Gauge glass with cocks.

g. Provide a magnesium anode.

2.3 ACCEPTABLE MANUFACTURERS

a. The condensate return unit shall be the make and model shown on the drawings or equivalent by Weil (Series 4400), Underwood, Burks, Aurora (Series 282), Sterling (Series 4200).

PART THREE - EXECUTION

3.1 INSTALLATION

a. Install the condensate return unit, as shown on the drawings and as recommended by the manufacturer, on a level base. Clean out all trash and wash out the receiver and system.

b. Provide vent to the outside, if possible. If outside vent is not possible, provide a type M copper vent line and install no less than 6 ft (2 m) of finned tube radiation in the vertical line to condense any exhaust steam. Carry the vent line up to the ceiling and then down to an acceptable point for water discharge.

c. Connect the system return lines to the receiver.

d. Connect water from the city water line through the vacuum breaker to the make-up float valve.

e. Connect electrical power and test the operation of the system, including the alternator.

f. Connect steam for the preheater with proper drip, valves, etc., as shown for steam main service. Operate the preheat system and adjust for proper temperature.

END OF SECTION 15855

SECTION 15860

****

DEAERATOR

SINGLE TANK

PART ONE - GENERAL

1.1 The GENERAL and SPECIAL CONDITIONS, Section 15100, are included as a part of this Section as though written in full in this document.

1.2 Scope of the Work shall include the furnishing and complete installation of the equipment covered by this Section, with all auxiliaries, ready for owner's use.

PART TWO - PRODUCTS

2.1 DEAERATOR TANK AND ACCESSORIES

a. Provide a packaged, spray-type, horizontal deaerator, with capacity as shown on the drawings (to match the boiler(s) specified). The system shall be a single-tank design, guaranteed to remove oxygen from the boiler feedwater to not more than 0.005 cc/liter in the effluent throughout all load conditions between 5 and 100%. Operating pressure shall be 5 psig (35 kPa), but unit shall be suitable for operation from 2 psig (14 kPa) to 15 psig (100 kPa).

b. Feedwater and condensate shall enter through a mixing device to heat the water and remove the noncondensable gases. All metals exposed to the gases shall be stainless steel. Provide automatic and manual vent valves to release the gases with a minimum amount of steam.

c. Deaerator shall have a storage capacity from minimum water level up to overflow shown on the drawings. Provide a manhole for access to the tank interior.

d. The deaerator tank shall be ASME-stamped for 50-psig (350-kPa) working pressure.

e. Trim for the deaerator shall include:

1. Inlet water-regulating valve.
2. External-float water-regulating valve with stainless steel float.
3. Steam-pressure-reducing valve to reduce steam pressure from system pressure to that required.
4. Overflow drainer with capacity of inlet system.
5. ASME relief valves of capacity required.
6. High-water-level alarm switch.
7. Low-water-level alarm switch.
8. Pressure gauge, 4-in. (100-mm) minimum diameter.
9. Thermometer, 4-in. (100-mm) dial type, with maximum and minimum set pointers.
10. Connections for make-up water to the water-regulating valve.
11. Steam supply to the reducing valve, return traps(s).
12. Pump suction with valves connected to the pumps.
13. Overflow and vent connections and control connections as required.

2.2 BOILER FEEDWATER PUMPS

a. Boiler feedwater pumps shall provide the capacity called for on the drawings with the head shown. Pumps shall be designed for operation with 250 deg. F (120 deg. C) water, with packing seals and recirculation orifices as recommended by the manufacturer.

b. Boiler feed pumps shall be mounted on a steel base, mounted under the deaerator tank and shipped as a unit with the tank.

c. Feed pumps shall be turbine (centrifugal) type with open, drip-proof, high-efficiency motors.

d. Pumps shall be direct-connected (flexible-connected) to the motors.

2.3 CONTROLS

a. All controls shall be factory-installed and wired to a control cabinet mounted on the unit.

b. Provide the necessary motor starters, indicating lights, numbered terminal blocks, switches, etc., to allow on-off-auto operation of each pump, with an alternator to alternate pump operation and to cause both pumps to operate if one pump is not able to maintain the load.

c. All wiring shall be color-coded and as required by the codes. Provide a control wiring diagram mounted permanently inside the control cabinet door. Submit five copies to the owner with the operating instructions.

2.4 FACTORY ENGINEER

a. Provide the services of a factory engineer to inspect the equipment after it has been installed, to verify proper installation and to place the equipment in operation and test performance.

2.5 ACCEPTABLE MANUFACTURERS

a. The deaerator package shall be the make and model number shown on the drawings or equivalent products by Cleaver Brooks (Spraymaster), Industrial (No-Ox), Kewanee.

PART THREE - EXECUTION

3.1 DRAWINGS

a. Obtain, from the manufacturer, drawings showing the recommended installation of the equipment and connections.
These drawings shall be used by the installer to produce a system acceptable to the manufacturer's engineer. Should these drawings be in conflict with the design drawings, the design engineer shall be consulted as to how to proceed.

3.2 INSTALLATION

a. Install the equipment level on a concrete base sloped so that water will not stand under or around the equipment.

b. Make all connections required, including service valves to isolate any equipment or component for service.

3.3 TEST

a. After equipment is installed and checked by the manufacturer's engineer, the equipment shall be placed in service and tested to show that it will produce water of the quality specified and that the pumps will produce the head and flow required within the motor rating.

3.4 INSTRUCTIONS

a. Instruct the owner's operators in the operation and maintenance of the system, and deliver to the owner five bound copies of instruction books, parts list, drawings, etc. Make a record of this transmittal, giving name of documents, name of person receiving same for the owner, name of factory engineer making inspection and giving instructions, and date.

END OF SECTION 15860

SECTION 15861
****
DEAERATOR
DUAL TANK WITH STORAGE

PART ONE - GENERAL

1.1 The GENERAL and SPECIAL CONDITIONS, Section 15100, are included as a part of this Section as though written in full in this document.

1.2 Scope of the Work shall include the furnishing and complete installation of the equipment covered by this Section, with all auxiliaries, ready for owner's use.

PART TWO - PRODUCTS

2.1 DEAERATOR TANK AND ACCESSORIES

a. Provide a packaged, spray-type, horizontal deaerator, with capacity as shown on the drawings (to match the boiler(s) specified). The system shall be a dual-tank design, guaranteed to remove oxygen from the boiler feedwater to not more than 0.005 cc/liter in the effluent throughout all load conditions between 5 and 100%. Operating pressure shall be 5 psig (35 kPa), but unit shall be suitable for operation from 2 psig (14 kPa) to 15 psig (100 kPa).

b. Feedwater and condensate shall first enter the vented storage portion (one-tank design) or a separate storage tank where it will mix and vent off some gases to the atmosphere. The feedwater will be pumped from this tank into the deaerator by the transfer pump(s) located under the tank.

c. The water transferred from the storage tank to the deaerator shall enter through a mixing device to heat the water and remove the noncondensable gases. All metals exposed to the gases shall be stainless steel. Provide automatic and manual vent valves to relieve the gases with a minimum amount of steam.

d. Deaerator shall have a storage capacity from minimum water level up to overflow shown on the drawings for the storage portion and capacity as required in the deaerator portion. Provide a manhole for access to both tank interiors.

e. The tank shall be ASME-stamped for 50-psig (350-kPa) working pressure.

f. Trim for the storage tank shall consist of external float-operated water make-up valve, overflow to drain, pump suction strainers, valves, etc., as required.

g. Trim for the deaerator shall include:

1. Inlet water-regulating valve.
2. External-float water-regulating valve with stainless steel float.
3. Steam-pressure-reducing valve to reduce steam pressure from system pressure to that required.
4. Overflow drainer with capacity of inlet system.

5. ASME relief valves of capacity required.
6. High-water-level alarm switch.
7. Low-water-level alarm switch.
8. Pressure gauge, 4-in. (100-mm) minimum diameter.
9. Thermometer, 4-in. (100-mm) dial type, with maximum and minimum set pointers.
10. Connections for make-up water to the water-regulating valve.
11. Steam supply to the reducing valve, return trap(s).
12. Pump suction with valves connected to the pumps.
13. Overflow and vent connections and control connections as required.

2.2 TRANSFER PUMPS

a. Provide transfer pump with standby pump to move the condensate from the storage tank to the deaerator tank, as required. Provide each pump with a suction strainer, suction valve, discharge check valve, and discharge valve.

b. Provide pressure gauges on suction and discharge of each pump.

c. Pumps shall be mounted on steel bases under the storage tank.

2.3 BOILER FEEDWATER PUMPS

a. Boiler feedwater pumps shall provide the capacity called for on the drawings with the head shown. Pumps shall be designed for operation with 250 deg. F (120 deg. C) water, with packing seals and recirculation orifices as recommended by the manufacturer.

b. Boiler feed pumps shall be mounted on a steel base, mounted under the deaerator tank, and shipped as a unit with the tank.

c. Feed pumps shall be turbine (centrifugal) type with open, drip-proof, high-efficiency motors.

d. Pumps shall be direct-connected (flexible-connected) to the motors.

2.4 CONTROLS

a. All controls shall be factory-installed and wired to a control cabinet mounted on the unit.

b. Provide the necessary motor starters, indicating lights, numbered terminal blocks, switches, etc., to allow on-off-auto operation of each pump, with an alternator to alternate pump operation (transfer and boiler feed) and to cause both pumps to operate if one pump is not able to maintain the load.

c. All wiring shall be color-coded and as required by the codes. Provide a control wiring diagram mounted permanently inside the control cabinet door. Submit five copies to the owner with the operating instructions.

2.5 FACTORY ENGINEER

a. Provide the services of a factory engineer to inspect the equipment after it has been installed, to verify proper installation, and to place the equipment in operation and test performance.

2.6 ACCEPTABLE MANUFACTURERS

a. The deaerator package shall be the make and model number shown on the drawings or equivalent products by Cleaver Brooks (Spraymaster), Industrial (No-Ox), Kewanee.

PART THREE - EXECUTION

3.1 DRAWINGS

a. Obtain, from the manufacturer, drawings showing the recommended installation of the equipment and connections. These drawings shall be used by the installer to produce a system acceptable to the manufacturer's engineer. Should these drawings be in conflict with the design drawings, the design engineer shall be consulted as to how to proceed.

3.2 INSTALLATION

a. Install the equipment level on a concrete base sloped so that water will not stand under or around the equipment.

b. Make all connections required, including service valves to isolate any equipment or component for service.

3.3 TEST

a. After equipment is installed and checked by the manufacturer's engineer, the equipment shall be placed in service and tested to show that it will produce water of the quality specified and that the pumps will produce the head and flow required within the motor rating.

3.4 INSTRUCTIONS

a. Instruct the owner's operators in the operation and maintenance of the system, and deliver to the owner five bound copies of instruction books, parts list, drawings, etc. Make a record of this transmittal, giving name of documents, name of person receiving same for the owner, name of factory engineer making inspection and giving instructions, and date.

END OF SECTION 15861

SECTION 15870
****
WATER SOFTENER

PART ONE - GENERAL

1.1 The GENERAL and SPECIAL CONDITIONS, Section 15100, are included as a part of this Section as though written in full in this document.

1.2 Scope of the Work shall include the furnishing and complete installation of the equipment covered by this Section, with all auxiliaries, ready for owner's use.

PART TWO - PRODUCTS

2.1 WATER SOFTENER
a. Provide a packaged water softener, consisting of a softener tank, valve, salt tank, etc., all as required for a complete system.
b. The softener system shall be provided by a manufacturer regularly engaged in the production of this equipment for use with boilers.
c. Provide the services of a field engineer to inspect the installation and start up the system.

2.2 SOFTENER TANK
a. Softener tank shall have no less than 50% freeboard, be designed for 100 psig (690 kPa), and be tested at 150 psig (1035 kPa). Tank shall be hot-dipped, galvanized, heavy steel construction and provided with lower distribution system to equally distribute the flow, requiring only one layer of supporting gravel.

2.3 BRINE TANK
a. Brine tank shall provide for a dry salt storage compartment and a saturated brine compartment, shall be equipped with a float-operated plastic brine valve, and shall allow for adjustment of the salt dosage without removing any parts.
b. Brine tank shall be constructed of fiberglass 3/16 in. (4.75 mm) thick and shall be provided with cover.

2.4 CAPACITY
a. Capacity shall be as shown on the drawings with the following information (to be filled in by design engineer).
1. Flow rate____ GPM (___ l/min)
2. Exchange capacity_____ kilograins.
3. Salt per regeneration ____ lbs. (kg)
4. Regenerations between refills ______ .
5. Water conditions in $CaCO_3$ equivalents:
_____ -ppm hardness
_____ -ppm ferrous iron

NOTE: SPEC WRITER SHOULD FILL IN THE CAPACITIES INDICATED BY THE BLANKS ABOVE.

2.5 MAIN OPERATING VALVE

a. Main operating valve shall be automatic, multiport diaphragm, slow-opening and -closing, pressure-actuated type. There shall be no dissimilar metals.

b. Controls shall have an adjustable duration of the various steps in regeneration and allow for push-button and complete manual operation. Regeneration shall be initiated by an electric time switch that shall allow selection of time for regeneration.

2.6 TEST KIT

a. A water test kit shall be provided for hardness testing. Provide a metal container for wall-mounted storage.

2.7 AUXILIARY EQUIPMENT

a. Provide pressure gauge on inlet and outlet of softener.

b. Provide test cock to sample water in and soft water out.

c. Provide a position indicator to show position of main valve.

d. Provide water meter to register the amount of soft water produced.

e. Provide an automatic reset head for the water meter to send a signal upon reaching a preset gallonage. The head shall automatically reset to this gallonage.

2.8 FACTORY ENGINEER

a. Provide the services of a factory engineer to inspect the equipment after it has been installed, to verify proper installation, and to place the equipment in operation and test performance.

2.9 ACCEPTABLE MANUFACTURERS

a. The softener package shall be the make and model number shown on the drawings or equivalent products by Cleaver Brooks or Kewanee.

PART THREE - EXECUTION

3.1 DRAWINGS

a. Obtain, from the manufacturer, drawings showing the recommended installation of the equipment and connections.
These drawings shall be used by the installer to produce a system acceptable to the manufacturer's engineer. Should these drawings be in conflict with the design drawings, the design engineer shall be consulted as to how to proceed.

3.2 INSTALLATION

a. Install the equipment level on a concrete base sloped so that water will not stand under or around the equipment.

b. Make all connections required, including service valves to isolate any equipment or component for service.

3.3 TEST

a. After equipment is installed and checked by the manufacturer's engineer, the equipment shall be placed in service and tested to show that it will produce water of the quality specified.

3.4 INSTRUCTIONS

a. Instruct the owner's operators in the operation and maintenance of the system, and deliver to the owner five bound copies of instruction books, parts list, drawings, etc. Make a record of this transmittal, giving name of documents, name of person receiving same for the owner, name of factory engineer making inspection and giving instructions, and date.

END OF SECTION 15870

SECTION 15875
****
BOILER WATER TREATMENT

PART ONE - GENERAL

1.1 The GENERAL and SPECIAL CONDITIONS, Section 15100, are included as a part of this Section as though written in full in this document.

1.2 Scope of the Work shall include the furnishing and complete installation of the equipment covered by this Section, with all auxiliaries, ready for owner's use.

PART TWO - PRODUCTS

2.1 WATER TREATMENT
a. Provide a boiler water treatment system for the steam boiler shown on the drawings, to inject controlled amounts of chemical into the boiler as required to prevent scaling.
b. The system shall be provided by a company regularly engaged in boiler installation and service or a company regularly engaged in the treatment of such systems.

2.2 EQUIPMENT
a. The equipment shall consist of a tank-mounted, positive displacement, variable-volume pump to draw chemicals from the tank and force them into the boiler, as recommended by the boiler manufacturer. The quantity of the chemical shall be determined by a qualified water treatment chemist for each boiler installation.
b. Provide a ball or gate valve and a check valve at each boiler connection. Connecting tube may be plastic or copper as required.

2.3 CHEMICALS
a. The chemical shall be formulated by a water treatment chemist to prevent scale formation in the boiler and any foaming of the boiler. An oxygen scavenger shall be included.

PART THREE - EXECUTION

3.1 INSTALLATION
a. Install the system completely. After the boiler and the system have been cleaned thoroughly and are ready to be filled, the treatment system shall provide the necessary treatment for the fill of the boiler.
b. After operation of several hours, the treatment level in the boiler shall be checked and adjusted to the proper level.
c. After operation for several days, the treatment level shall be checked and adjusted. Repeat in 30 days.
d. The installer shall be responsible for adjusting the proper amounts of chemicals for the boiler as operating and for the first 30 days. At that time, the owner should be responsible for the chemical treatment.
e. Instruct the owner in proper methods of testing and adjusting the water treatment equipment and chemicals.

END OF SECTION 15875

SECTION 15880
****
PRESSURE-POWERED PUMP
STEAM AND/OR AIR OPERATED

PART ONE - GENERAL

1.1 The GENERAL and SPECIAL CONDITIONS, Section 15100, are included as a part of this Section as though written in full in this document.

1.2 Scope of the Work shall include the furnishing and complete installation of the equipment covered by this Section, with all auxiliaries, ready for owner's use.

PART TWO - PRODUCTS

2.1 PRESSURE-OPERATED PUMP

a. Pressure-powered pump (pumping trap) shall consist of a receiver, a float-operated trip valve, and check valves, all designed to pump a fluid by means of a pressure applied to the surface of the liquid.
b. The receiver shall be of cast iron (low-pressure) or galvanized, heavy, fabricated steel with ASME stamp (high-pressure). The capacity of the receiver shall allow the pumping of approximately 7.5 gallons (1 cubic foot) (30 liters) for each cycle.
c. Float-operated trip valve shall be stainless steel or all bronze, designed for condensate operation. The trip valve shall allow the receiver to vent when the float is down. When the float rises to the trip point, the valve shall allow the pressure to be applied to the surface of the fluid, and the vent shall close. Steam consumption shall not exceed 3 lb (1.5 kg) per 1000 lb. (455 kg) of water.
d. Check valves shall be brass swing type with 3 in. (75 mm) (2 in. (50 mm) for smaller systems) on the receiver inlet and 2 in. (50 mm) on the outlet, sized for a flow rate of 90 gallons/min (340 liters/min).

2.2 AUXILIARY EQUIPMENT

a. Provide a cycle counter, operated by the float, to count the number of pump operations (and therefore the fluid volume moved).
b. Provide a steam-pressure-reducing valve on the steam connections to reduce the steam pressure to the pump rating. Provide ASME relief valve where steam supply pressure exceeds 125 psig (860 kPa) or the rating of the receiver, whichever is lower.
c. Provide a red-strip gauge glass on receiver to indicate the liquid level.
d. Outdoor installations should be provided with some means of draining the trap to prevent freezing in the event of steam failure. A Sarco Thermoton can be used for this.

2.3 ACCEPTABLE EQUIPMENT

a. The pressure-powered pump shall be the model shown on the drawings or equivalent by Sarco (PPC or PPF).

PART THREE - EXECUTION

3.1 INSTALLATION

a. Install the pressure-powered pump as directed by the manufacturer and generally as shown on the drawings. Provide a vented receiver, located no less than 12 in. (30 cm) above the top of the pump, to store the liquid while the pump is operating.

b. Provide ball or gate valve in inlet and discharge so that the pump and check valves can be isolated for servicing.

c. Carry the vent connection to a point where steam discharge will not create a problem (outside or to other appropriate areas). If vent is not carried to the outside, use copper line with sufficient sloped sections to condense the steam that will be discharged with the venting of the hot fluid.

d. Note that it is possible to operate pump on both steam and air pressure. The primary operating medium should have a higher pressure than the secondary medium. Each medium should be provided with spring-closed check valves.

3.2 TESTING

a. Operate the system and observe the filling of the receiver, the tripping of the valve, and the discharge of the fluid.

b. Time the cycle from the end of the pumping cycle to the trip of the valve, and determine that filling time will allow the required capacity, with allowance for the pumping time.

c. Time the cycle from the trip of the valve to the closing of the valve to determine the capacity of the pump cycle. The fill time and the pumping time together limit the capacity of the pump.

d. Divide the time for each cycle into 60 minutes and multiply times 7.5 gallons (30 liters) to determine capacity per hour.

END OF SECTION 15880

SECTION 15881

****

PRESSURE-POWERED CONDENSATE RETURN PUMP
STEAM AND/OR AIR OPERATED
ELECTRIC OR PNEUMATIC CONTROL

PART ONE - GENERAL

1.1 The GENERAL and SPECIAL CONDITIONS, Section 15100, are included as a part of this Section as though written in full in this document.

1.2 Scope of the Work shall include the furnishing and complete installation of the equipment covered by this Section, with all auxiliaries, ready for owner's use.

PART TWO - PRODUCTS

2.1 PRESSURE-POWERED PUMP

a. Pressure-powered pump shall consist of equalizing chamber, pumping chamber, float- or probe-operated control valve, and check valves; all designed to pump a fluid by means of a pressure applied to the surface of the liquid. A receiver and prefabricated stand shall be furnished as shown on the drawings. Capacity of the pump shall be no less than shown on the drawings.

b. The unit shall be fabricated from heavy steel and all chambers under pressure shall be ASME-stamped. Each chamber shall be provided with a red-line gauge glass with gauge valves as required for the pressure involved.

c. The activation of the control valve shall be by electric probes or by float-operated pneumatic system. All parts in the chambers shall be stainless steel or brass. The pump shall allow the pumping chamber to vent to the equalizing chamber when the liquid level is down, allowing the liquid to fill the pumping chamber by gravity from the receiver. When the liquid level rises, the controls shall apply air or steam pressure to the surface of the liquid, forcing it out. The vent shall close when the pressure is applied. Steam consumption shall not exceed 3 lb. (1.5 kg) per 1000 lb. (455 kg) of liquid pumped.

d. Check valves shall be swing type with regrindable seats, mounted horizontally and requiring low head for opening. Valves shall be rated for the operating pressure of the pumping medium.

2.2 AUXILIARY EQUIPMENT

a. Provide a counter, activated by the pumping pressure, to count the number of pumping cycles, allowing a determination of the total volume pumped between readings of the counter.

b. Provide a pressure regulator in the air line to the pump, and install a relief valve on the downstream side of the regulator. Provide a spring-loaded check valve on the discharge side.

c. Provide a pressure-reducing valve in the steam line supplying the pump, and install a relief valve on the downstream side. Provide a spring-loaded check valve on the discharge side.

d. Outdoor installations subject to freezing shall be provided with an automatic drain device to waste the liquid and prevent freeze-up in the event of failure of the pumping pressure.

e. Provide a duplicate control system for the pumping medium, to take over in case of failure of the main control system.

f. Provide replaceable magnesium anodes in each chamber to prevent internal corrosion of the steel.

PART THREE - EXECUTION

3.1 INSTALLATION

a. Install the system as recommended by the manufacturer and the drawings. Mount the unit on a substantial foundation and above any standing water.

b. Make the connections for the returning liquid to be pumped. Any hot condensate shall be trapped to prevent the return of live steam.

c. Provide the vent connections of the size recommended by the manufacturer, or extend vent the full size of the vent connection. The vent shall be carried to a point where the discharge of liquid or vapor will not be objectionable.

d. Provide the connections of the pumping medium (generally air and/or steam) to the control valve, with gate valves and unions for removal of the reducing stations for service. If air and steam are provided, install the check valves close together and allow sufficient pipe between the air check valve and the pressure regulator to prevent heat from the steam affecting the air regulator.

3.2 TESTING

a. Operate the system on each operating medium and measure the liquid pumped by each cycle. Determine the fill time for the pumping chamber and the overall cycle time. Calculate the overall pumping capacity for 1 hr. and compare against the design requirements.

b. Report the volume for each cycle and the cycle time to the engineer as required by Balancing and Testing, Section 15111.

c. Instruct the owner in the operation and the maintenance of the system.

END OF SECTION 15881

SECTION 15890

****

INDUCED-DRAFT FAN
IN-LINE MOUNTING

PART ONE - GENERAL

1.1 The GENERAL and SPECIAL CONDITIONS, Section 15100, are included as a part of this Section as though written in full in this document.

1.2 Scope of the Work shall include the furnishing and complete installation of the equipment covered by this Section, with all auxiliaries, ready for owner's use.

PART TWO - PRODUCTS

2.1 INDUCED DRAFT FAN

a. Fan shall be designed for induced-draft service for gas, oil, or gas/oil boilers to provide additional air for combustion for the boiler. Capacity shall be as shown on the drawings or as required to match the capacity of the boiler indicated.

b. Fan shall be a single-stage or two-stage propeller type, designed for operation in the hot and corrosive atmosphere normally present in boiler flue gases.

c. Fan shall be direct-connected to the motor shaft, but the motor shall be mounted in a separate chamber open to the atmosphere and isolated from the flue gases. Fan motor shall be cast iron construction, totally enclosed, and provided with long-life ball bearings for vertical or horizontal mounting. Motors for two-stage fans shall be provided with shaft extensions on both ends for mounting of the fans.

d. Housing shall be heavy-gauge welded steel, with flanged connections on both ends for mounting directly on boiler or in ductwork. Fan housing and motor enclosure shall be split with removable sections for easy removal of the fan(s) and motor.

2.2 ACCESSORY EQUIPMENT

a. Furnish fan and housing fabricated from special materials as shown on the drawings or as listed below.

b. Provide corrosion-resistant coating on inside or outside (or both inside and outside) of the fan and housing to protect the fan from severe conditions. Coating shall be outlined below.

c. Provide motor- (manual-) operated damper to reduce the fan capacity down to 40% of rated capacity. Provide controls or manual operators as shown on the drawings and outlined under the Section on controls.

d. Provide a pressure-sensing switch to act as a proving relay for the boiler, such that boiler controls cannot be energized until fan is operating properly.

2.3 ACCEPTABLE MANUFACTURERS

a. Induced-draft fan shall be the product of a company regularly manufacturing fans for this service, equal to the make and model shown on the drawings or approved equivalent by DeBothezat (Bifurcator).

PART THREE - EXECUTION

3.1 INSTALLATION

a. The induced-draft fan shall be installed above the boiler as recommended by the boiler manufacturer and the fan manufacturer for this application. The fan shall be supported by the boiler vent stack, and adequate additional support shall be provided as required to carry the weight of the unit. High-temperature gasket material shall be provided to reduce leakage into, and out of, the fan and connecting ductwork.

b. Provide all power and control wiring in proper conduit and of proper temperature rating as required by the electrical codes. Provide motor starter with heater trip elements to match the fan motor draw and ambient temperature during operation. Connection to the motor shall be made with flexible conduit to allow removal of the motor from the housing. Install the proving switch and make the proper changes in the boiler controls to prevent operation of the boiler unless the fan is producing the proper draft.

c. After installation is complete, operate the entire system, including the boiler, and prove that the installation is operating properly and that the draft is adequate but not in excess. The boiler shall be adjusted to produce minimum oxygen with no carbon monoxide in the flue gases by a qualified service person.

d. Read, record, and report the motor current draw to the engineer as required under Balancing and Testing, Section 15111.

END OF SECTION 15890

CHAPTER 14

HVAC Control Systems

EQUIPMENT Description of equipment.	SECTION NUMBER	NUMBER OF PAGES	PAGE NUMBER
HVAC CONTROLS Pneumatic control system for normal, uncomplicated system requiring pneumatically operated dampers, valves, etc. These Specifications allow a well-qualified contractor to install the system, but the design should be made by a qualified control engineer. Electrical components, such as interlocking, are included.	SECTION 15958	4 Pages	14-2
MICROPROCESSOR CONTROLS Microprocessor controls for a total energy management system, including time clock functions, duty cycling, optimization, and demand limiting. Includes remote communication through telephone modem. Solidine Specification.	SECTION 15962	7 Pages	14-6
ENERGY MANAGEMENT AND CONTROL SYSTEM Microprocessor controls for a total energy management system, including time clock functions, duty cycling, optimization, and demand limiting. Addition of slave units allows for expansion and use with larger systems. Control can be pneumatic or digital. Andover Specification.	SECTION 15964	6 Pages	14-13
HVAC ENERGY MANAGEMENT SYSTEM Energy management system for HVAC unit(s) including time scheduling, duty cycling, demand limiting, and optimization. Trane "Tracer" Specification.	SECTION 15966	4 Pages	14-19

SECTION 15958
****
HVAC CONTROLS

PART ONE - GENERAL

1.1 The GENERAL and SPECIFIC CONDITIONS, Section 15100, are included as a part of this Section as though written in full in this document.

1.2 Scope of the Work shall include the furnishing and complete installation of the equipment covered by this Section, with all auxiliaries, ready for owner's use.

PART TWO - PRODUCTS

2.1 CONTROL DEVICES

a. All control devices and products used in the control system shall be first-line products, manufactured for the application as used.

b. All electrical wiring for the control system shall be as specified in the Electrical Section of the Specifications and as required by local codes.

c. Pneumatic tubing shall be rigid copper where concealed or inaccessible in walls, etc., or where installed in mechanical rooms, open to public view, or where exposed to rodents. PVC tubing may be used above lay-in ceiling if permitted by code and if accessible for replacement. Tubing shall be virgin plastic of the best grade. All connectors shall be brass.

d. Control air compressor shall be duplex type with alternator. Provide pressure controls, relief valve, impulse drain valve, etc., as required for a complete system. Compressors shall be sized to operate no more than 50% of the time (25% each compressor). The ASME tank shall be sized to supply the system air for approximately 1 hr. Vibration pads shall be provided for each leg.

e. Refrigerated dryer shall be installed in the air supply. Provide impulse-type drain valve for the dryer.

f. Provide an oil-coalescing filter on the discharge of the dryer, followed by a final filter consisting of 24 in. (60 cm) x 2 in. (50 mm) steel pipe, with screwed cap on both ends for changing the filter elements. Provide three-valve bypass in air line to the filter. Filter elements shall consist of sanitary napkins, rolled and inserted to fill the pipe.

g. Pneumatic thermostats shall be two-pipe, nonbleed instruments, small in size, attractive in appearance, and with locking blank cover.

h. Electric thermostats shall be low-voltage, modulating type to control modulating devices, or low- or line-voltage type with heat anticipator for two-position controls. Provide locking covers.

i. Thermostat guards shall be plastic or metal covers to prevent tampering with the instrument. Provide substantial, locked, opaque cover, hinged to a base which is secured to the wall, not to the thermostat base.

j. Enthalpy selection system shall be entirely pneumatic and shall consist of one enthalpy transmitter in the outside air, one enthalpy transmitter in the return air, and a relay to select the lower of the two enthalpies. In operation, the signal from the two enthalpy transmitters shall be compared by the differential switching relay so that when the outside air enthalpy is lower than the return air enthalpy, the temperature control system shall modulate the outside, return, and relief dampers to supply up to 100% outside air for "free cooling." When the outside air enthalpy is higher than the return air, the system shall position to minimum outside air. The use of separate temperature and humidity transmitters to arrive at enthalpy will not be acceptable. Outside air transmitter shall not be damaged by operation during fog conditions. Provide Btu (watt) indication for both outside air and return air on the face of the control panel.

k. Pneumatic actuator housings shall be die-cast aluminum, motor stem shall be type 416 stainless steel, spring is to be tempered steel, diaphragm shall be ozone-resistant rubber, and the guide shall be brass. All damper actuators shall have pilot-positioning relays.

l. Control valves for fluids shall be two-position (On-Off), modulating two-position, three-way, or modulating three-way (mixing or divertor), as required for the application. Modulating valves shall be selected with the proper flow characteristics to allow control of the flow over as wide a range as is possible with a reasonable pressure drop (10 ft (3 kPa) of water unless noted otherwise).

2.2 ACCESSORY EQUIPMENT

a. Refrigerated air dryer shall be provided in the main supply air system to remove oil and moisture from the control air. Dryer shall cool the system air at the maximum compressor output from 120 to 40 deg. F (49 to 4.4 deg. C), with a pressure drop not to exceed 3 psi (20 kPa). Provide impulse-type automatic drain to expel all liquids from the dryer.

2.3 ACCEPTABLE MANUFACTURERS

a. Control equipment shall be manufactured by a company regularly engaged in production of this type equipment, as shown on the drawings, or equivalent equipment by Barber Colman, Honeywell, Johnson Service, Penn, Tour Anderson, or accepted equals.

b. Control air compressor shall be make and model number shown on the drawings, as manufactured by one of the companies above or Quincy.

c. Refrigerated air dryer shall be make and model number shown on the drawings, as manufactured by one of the companies above or Hankinson.

PART THREE - EXECUTION

3.1 INSTALLATION

a. All control equipment shall be installed as recommended by the manufacturer and as required for service in the field. No equipment shall be concealed or covered by other equipment unless adequate provisions are made for service and replacement.

b. All wiring and pneumatic tubing shall be run in neat, straight lines to present a finished appearance. Multiple runs shall be supported on brackets and spaced to give access to each line.

c. All wires shall be color-coded and numbered on both ends of each conductor for easy identification. Colors and numbers shall not change in the middle of a run, unless an accessible junction box is provided. Provide numbered terminal strips in all control panels.
d. All pneumatic tubing shall be identified by numbered tag, or color marking if in tube bundles, as is required for electrical wiring above.
e. Wiring diagrams shall be prepared for all electrical connections, showing the actual wire number and terminal identification as installed. Same shall be prepared for pneumatic tubing. No less than five copies of such diagrams shall be delivered to the owner.
f. Installation of all equipment shall be made by qualified mechanics familiar with control systems, forces involved, and their operation. Any work not neatly installed shall be removed and replaced.
g. All connections shall be made by technicians who are familiar with the operation of the equipment and the intent of the control designer.
h. After all equipment is mounted and connected, the control engineer shall inspect the system and verify the correct operation and connection of all equipment. Any equipment found to be installed improperly or connected incorrectly shall be changed as required. After the system is installed correctly, all instruments shall be calibrated and set points fixed at the correct setting.
i. After system operation has been checked and verified as correct, the control engineer shall notify the engineer, in writing and on company letterhead, that the system has been installed correctly, has been calibrated, and is functioning as designed.

3.2 TESTING

a. At the time of final review, the control contractor shall instruct the owner in the proper operation and maintenance of the system as installed and demonstrate how the system is designed to perform.
b. Any system found to be out of calibration or functioning improperly at this time shall be corrected immediately and the correct functions of the entire system demonstrated to the satisfaction of the engineer.

PART FOUR - CONTROL FUNCTIONS

4.1 SYSTEM OPERATION

a. All systems (each system) shall be started and stopped by a seven day time clock with reserve power of no less than 12 hr. Provide a manual mark-time switch (MTS) to override the clock for off-hour operation of the system. Provide a night low-limit (NLL) thermostat to bypass the clock in heating to maintain a minimum temperature in the building. Locate the NLL and the MTS as shown on the drawings or as directed by the engineer.
b. Each roof-mounted toilet exhaust fan shall be controlled with the air-handling unit serving the area, unless noted otherwise.

c. Ceiling-mounted exhaust fans shall be controlled with the lights in the area served, with delayed off-time operation as specified with the fan.

d. Water temperature delivered by hot water heating systems shall be reset by three-way valve as shown on drawings or by boiler reset controls if specified with the boiler. Water temperature shall be reset, and the reset schedule shall match the type of system.

e. On system start-up, the outside air dampers shall remain tightly closed until the return air temperature is within 5 deg. F (2 deg. C) of the design space temperature on either heating or cooling, unless the cooling system is operating on "free cooling."

(NOTE: THE DESIGN ENGINEER SHOULD CONTINUE TO SPELL OUT THE DESIRED OPERATION FOR EVERY COMPONENT OF THE SYSTEM.)

END OF SECTION 15958

SECTION 15962

****

MICROPROCESSOR CONTROLS

PART ONE - GENERAL

1.1 The GENERAL and SPECIAL CONDITIONS, Section 15100, are included as a part of this Section as though written in full in this document.

1.2 Scope of the Work shall include the furnishing and complete installation of the equipment covered by this Section, with all auxiliaries, ready for owner's use.

PART TWO - PRODUCTS

2.1 MICROPROCESSOR CONTROLLER

a. The microprocessor controller energy management system (EMS) shall provide the following functions:

1. Time of day scheduling
2. Duty cycling
3. Adaptive, variable analog-controlled duty cycling
4. Start and stop time optimization
5. Demand limiting
6. Minimum ON/OFF times
7. Maximum ON/OFF times
8. Temporary single-event programming
9. Holiday programming for 365 days
10. Monitoring and data logging for each input
11. Output channel run times
12. Override ON times
13. Demand kWh daily maximum/minimum and 15-minute averages.

2.2 CONNECTIONS

a. The EMS shall have the capability of being hard-wired on any channel or connected through power line carrier (PLC), or a combination of the two. The PLC shall send digitally coded radio frequency messages from the microprocessor control unit through the wiring system to programmable receiver relays located at the loads to be controlled.

2.3 EQUIPMENT

a. The controller shall be enclosed in a protective enclosure. The door shall be key-locking and have a viewing window to observe the display and the LED indicators when the door is closed. The door shall be removable.

b. The keyboard/microcomputer panel and standby battery shall be removable as one assembly, separate from the relay outputs. This shall allow separate installation of enclosure and relay wiring at job site, while allowing independent preprogramming of microcomputer panel before shipment to job site.

c. The power supply shall be 120 vac 60 Hz. The maximum power consumption shall be 80 watts. The relay contacts shall be rated at least 5 amperes at 24 vac, resistive and suppressed.

d. The operating range shall be 0 deg. F (-19 deg. C) to 125 deg. F (52 deg. C) and 5 to 90% rh, noncondensing.

e. The controller shall be provided with output relays to control up to 32 loads or groups of loads.
f. The controller shall provide the capability, if required, to remotely indicate normal operation or to remotely override critical loads in the event of a malfunction.
g. The controller shall be capable of monitoring up to 16 temperature sensors. All temperature sensors shall be solid state electronic devices with a range of -25 to 200 deg. F (-32 to 93 deg. C), and shall be capable of being installed up to 1000 ft from controller with 18 AWG shielded wiring.
h. The controller shall be capable of monitoring electrical power usage, utilizing either a watt transducer or a pulse-generating demand meter capable of being monitored in up to 16 demand limiting signals, each requiring the use of one of the 16 analog input channels. The first eight of the input channels shall be capable of storing daily peaks of demand or temperature for the previous 35 days.

2.4 PANEL FEATURES

a. The display shall provide visual indication of all system parameters. The display shall also display selected real-time data consisting of day/date/time, outside temperature, each inside temperature, each load's start and stop times, current demand, daily demand or temperature peaks or minimums for the past 35 days, cumulative kWh consumption, and 15-minute averages of analog values for previous 48 hr on two systems.
b. The controller shall be field-programmable. A front keyboard panel shall provide the operator interface.
c. Audible feedback of the keyboard shall provide a single tone for correct usage and a distinctive tone for incorrect usage.
d. The display shall provide LED indication of program parameters for each function, to aid in simplified programming.
e. The controller power shall be backed up with a rechargeable battery to maintain operator-programmable data, including date and time. The battery shall be trickle-charged continuously during daily operation. The fully charged battery backup shall be capable of providing a minimum of seven days' memory retention in event of a power failure.
f. The controller shall provide for application of a nonhold, timed manual bypass switch for each load.
g. The controller shall provide a speech synthesizer to verbally guide the operator in the programming procedure for verbal review of entered programs, verbal analog status, and verbal power failure annunciation after restoration of power.

2.5 SOFTWARE FEATURES

a. The controller shall provide a delay of 1 second between successive automatic starts and stops.
b. The controller shall provide a programmable minimum ON and OFF time for each load automatically started by the controller.
c. The controller shall operate from a 12-hr clock with AM or PM indication and the Gregorian calendar with month, day of month, and day of week indication.
d. Daylight savings time shall be automatically calculated and implemented. It shall be possible to disable the Daylight Savings Time function.
e. All temperature values and set points shall be displayed and entered in degrees Fahrenheit.

f. After completion of all program entries, entering date and time shall cause controller to examine its entire program and automatically place all connected loads in their correct ON or OFF states for the present time of day.

2.6 TIME OF DAY SCHEDULING

a. The Time of Day Scheduling function shall turn loads off and on according to a programmed time for each day, with 1 minute resolution.

b. Each load shall be capable of having 40 unique events (on or off) programmed for each day of the week and holidays.

c. The Time of Day Scheduling function shall allow up to 16 holiday periods. Each holiday period shall allow for up to 255 days. Holidays shall be programmable at least a year in advance.

d. The Time of Day Scheduling function shall allow for manual schedule override which will override scheduled on or off loads.

e. The Time of Day Scheduling function shall allow for single-event, temporary on/off programming, with automatic return to scheduled programming after the event.

2.7 DUTY CYCLING

a. The Duty Cycling function shall cycle loads off and on according to a programmed ON and OFF time for each assigned load.

b. The Duty Cycling function shall allow each load to be assigned to an ON-time interval of 1 to 255 minutes and an OFF-time interval of 1 to 255 minutes.

c. The Duty Cycling function shall be able to automatically rotate the loads to minimize electrical demand at all times by aligning the duty-cycled ON and OFF times in rotation.

d. The Duty Cycling function, together with the Analog Control function, shall allow any of the solid-state air temperature sensors (inside or outside) to be assigned to any load to allow individual high- and/or low-temperature set points to override a load that is duty-cycled off.

e. The Duty Cycling function shall be programmable to continuously vary the ON and OFF times in response to a selected analog temperature input, while evenly rotating up to eight loads assigned to that variable duty cycle program.

2.8 ANALOG CONTROL, FIXED AND VARIABLE

a. The controller shall have the capability to accept from 4 to 16 analog inputs such as kW, temperature, humidity, etc.

1. The Analog Control function shall allow for any one or several of the 16 analog inputs to be assigned to any load(s) to allow individual high- and/or low-analog set points to turn each load on or off.
2. The Variable Analog Control function shall allow the high and low set point of any analog input controlling a load to be continuously and linearly varied in response to a second analog input.

2.9 START TIME OPTIMIZATION

a. The Start Time Optimization function shall start assigned loads at the latest possible time which will permit the building's internal environmental conditions to reach the desired temperature by building occupancy time.

b. The Start Time Optimization function shall use outside air temperature, selected inside air temperatures, heating or cooling thermal efficiency factors, building loss factor, and building occupancy temperature set point to calculate optimum start time. The optimizing function shall be self-correcting over time.

2.10 STOP TIME OPTIMIZATION

a. The Stop Time Optimization function shall stop assigned loads at the earliest possible time that will permit the building's internal environmental conditions to be maintained until the scheduled occupancy stop time.

b. The Stop Time Optimization function shall use the outside air temperature, selected inside air temperatures, building loss factor, and the permissible internal environmental temperature range to calculate optimum stop time.

2.11 DEMAND CONTROL

a. The Demand Control function shall monitor and control electrical demand, shedding (turning off) and restoring (turning on) loads to maintain the peak demand below programmed peak-demand set points.

b. The Demand Control function shall provide for up to four or more time of day peak-demand set points.

c. The Demand Control function shall allow for any load to be assigned to be shed and restored on priority basis. Any number of loads may be assigned to a priority.

d. The Demand Control function shall allow for maximum OFF time assignable to each load to limit the amount of time a load may be shed.

e. The Demand Control function shall allow a minimum ON time assignable to each load to ensure a minimum ON time before the load may be reshed.

f. The Demand Control function shall allow a minimum OFF time assignable to each load to ensure a minimum OFF time before the load may be restarted.

g. The Demand Control function together with the Data Logging function shall allow for maintaining the following statistics: accumulated consumption (to be available on 16 inputs), past 35 days' demand peaks (to be available on at least the first 8 inputs), the time and date of occurrence, and the averaged kW for every 15 minutes of the previous 48 hr (to be available on at least the first 2 inputs).

h. The Demand Control function shall accept an input signal for electrical demand and consumption calculations from either a watt transducer or a pulse-generating demand meter and shall not require an end-of-interval signal. Up to 16 such demand-limiting signals shall be accepted by the controller and any load shall be able to be shed by any of the demand signals. Each demand-limiting input to the controller shall reduce by one the number of temperature-sensing or other inputs the controller can accept.

2.12 DATA LOGGING

a. The Data Logging function shall provide for the continuous accumulation of kWh, or any other analog accumulation, on any or all of the 16 analog input channels.

b. The Data Logging function shall provide for the recording of the daily maximum values or minimum values for the previous 35 days of the first eight of the analog inputs.

c. The Data Logging function shall provide, for the first 2 of the analog inputs, for the recording of the averaged analog data for every 15 minutes of the previous 48 hr.
d. The Data Logging function shall provide for recording of the accumulated ON time of each output channel and the accumulated override ON time of each output channel.

2.13 REMOTE COMMUNICATIONS (FOR FUTURE ADDITION)

a. The controller shall have the capability to have a field-addable communications module installed.
b. The communications module shall provide the capability to:
   1. Maintain all programming access of the controller.
   2. Remotely display and/or modify all data that may be entered at the controller's operator panel.
   3. Display all statistics available to the operator of the controller in easy-to-read, formatted reports.
   4. Provide a load control status report.
   5. Provide, for the current day, a report of temperatures, demands, and consumption data.
   6. Provide a report of the averaged analog data for every 15 minutes for the previous 48 hr.
   7. Provide a report of the accumulated ON time of all output channels.
   8. Provide a report of the accumulated override ON time for all output channels.
   9. Provide the ability to transfer and restore all programs.
   10. Provide the ability to transfer report data to a printed format via a computer terminal.
   11. Provide at least a 35-day history report of daily high or low temperatures, high or low demand, time that high or low temperature and demand occurred, and kWh consumption of all connected sensors.

c. Terminals for remote communication shall be connected directly to the remote communication module via standard RA-232-C electrical interface or remotely via telephone line with modem. Remote communication shall be capable of auto-answer operation when connected to telephone service.
d. The communication module shall provide the capability for the controller to dial out and report to a remote monitoring computer in response to any predetermined level of analog input (alarm).

2.14 LOCAL REPORT PRINTOUT

a. The controller shall have the capability to have a field-addable communications module installed that will provide for a printer connection at the installation site, for the purpose of generating data-logging and system status-report printouts automatically.

2.15 ACCESSORY EQUIPMENT

a. Provide heavy-duty lightning arresters installed at the point of entry into each building for any communication or control lines located outside the building.
b. Provide a central processing unit (CPU) to read, monitor, and communicate with the Master unit. The CPU shall be a personal computer with two disk drives and RAM of no less than 65K.

c. Provide a quality dot matrix printer having a 9.5-in. (240-mm) carriage and printing no less than 120 characters per second at 10 cpi. Printer shall be Apple or approved equivalent.

2.16 ACCEPTABLE MANUFACTURERS

a. The EMS shall be Solidyne, Series 3200A Master and necessary Slaves, or approved equivalent.

PART THREE - EXECUTION

3.1 INSTALLATION

a. Install the system as recommended by the manufacturer, using only equipment recommended or acceptable to the manufacturer.

b. Comply with all codes for electrical work. Run all power wiring in conduit. All sensor and control wiring located in mechanical rooms and other exposed areas shall be run in conduit. All wiring run inside walls shall be in conduit. All equipment located outside shall be in suitable weathertight enclosure.

c. Install all conduit, wiring, and cable, and install all equipment in first-class manner, using proper tools, equipment, hangers, and supports, and in locations as required for a neat, attractive installation. No material shall be exposed if it is possible to conceal it. Exposed materials shall be installed only with consent of the engineer.

d. Support all sensors as recommended by the manufacturer where inside equipment such as ductwork. Sensors in the space shall be in small, attractive housings designed for that purpose and mounted on an electrical junction box.

e. Extreme care shall be used in making connections to other equipment, such as boilers and chillers, to see that the safeties on this equipment are not inadvertently bypassed or overriden by the EMS.

f. All equipment having moving parts and controlled by the EMS shall be provided with warning labels no less than 2 in. (50 mm) in height, and in bright warning colors, stating that the equipment is remotely started by automatic controls. Such labels shall be posted clearly in the area of any moving parts, such as belts, fans, pumps, etc.

3.2 SOFTWARE

a. Load and debug the software to provide a complete operating EMS system, and operate the system to prove proper function of each system. Where necessary, the sensors shall be heated or cooled to demonstrate the correct function. Provide careful evaluation of the operation of chillers and boilers at part and full load under control of the EMS.

b. The EMS contractor shall review the programs with the engineer in the programming stage to make sure that the programmer understands the engineer's intent and that the program will carry out that intent.

c. Provide the engineer with a bound copy of the complete programming, as well as instruction books on reprogramming of the system, for future modifications of the system, if desirable.

d. Provide two bound copies of complete information on the equipment and all components, including programming, for the owner's use and records.

e. Instruct the owner and engineer in the proper operation of the system and the steps necessary to override or change any input data.

f. Provide the owner with five copies of complete wiring diagrams of the entire control system as installed, showing the location and the "as is" set point for every input and the location of every output. Where the EMS interlocks with other equipment, show enough of that equipment wiring diagram to allow the service department to follow the control sequence without having to pull the wiring diagram of the other equipment.

g. The EMS contractor shall monitor the system for one year, to verify performance of the system. This may be done by use of the modem and remote communications.

h. Remote communications require a telephone line to the EMS. The EMS contractor shall provide for the installation and upkeep of this line for the first year, as part of the contract. At the end of the first-year warranty period, the EMS contractor shall offer a service contract to continue to monitor the system and advise the owner of any alarms received.

PART FOUR - PROGRAMMED FUNCTIONS

4.1 a. Control functions shall be as follows:

(SPECIFICATION WRITER SHOULD INCLUDE HERE THE CONTROL FUNCTIONS REQUIRED OF THE EMS SYSTEM AS IT RELATES TO EVERY ITEM OF EQUIPMENT TO BE CONTROLLED OR MONITORED.)

END OF SECTION 15962

SECTION 15964

****

ENERGY MANAGEMENT AND CONTROL SYSTEM

PART ONE - GENERAL

1.1 The GENERAL and SPECIAL CONDITIONS, Section 15100, are included as a part of this Section as though written in full in this document.

1.2 Scope of the Work shall include the furnishing and complete installation of the equipment covered by this Section, with all auxiliaries, ready for owner's use.

PART TWO - PRODUCTS

2.1 a. Provide a microprocessor-based energy management and control system (EMS) to monitor the data environment (DE) and control the control functions (CF) specified below and/or shown on the drawings.

b. Analog output (AO), analog input (AI), digit
l output (DO), and
digital input (DI) represent a "point" where referred to in this Specification.

c. The EMS shall be composed of the system components listed below, all assembled in a rust-protected cabinet with a hinged and key-locked front cover, which shall contain a window that will allow observation of the parts of the panel requiring monitoring. All components not enclosed in the main panel shall be enclosed in matching panels of the same type. All panels shall be keyed alike and shall be connected as required to function as a single system.

2.2 MASTER

a. The Master shall function as the overall system coordinator, perform automated energy management functions, control peripheral devices, and perform all necessary calculations.

b. The Master shall have no less than 16K bytes of user-available memory, in addition to memory required by the system operation and diagnostics. Memory shall be expandable to 80K.

c. The Master shall have a battery-backed, uninterruptable real-time clock to provide time of day, month, year, and day of week. The clock shall automatically correct for daylight savings and leap years.

d. Power supply to the Master shall be 120 volts, 60 Hz power, and the system shall function down to 105 volts. Below this point, the system shall operate as in a power outage.

e. Power failure shall cause the internal clock, memory, and computation functions of the Master to continue to operate on the standard battery backup for no less than 3 hr.

f. Recovery from power failure shall automatically, and without operator attention, update all functions and resume operation based on the current time and status. The Master shall implement special building start-up strategies and log the time of outage and power restoration.

2.3 SLAVE PANELS

a. Slave panels shall be provided as necessary to connect the data-collecting system to the Master and to connect the field sensors to the control devices. The Slave shall be fully supervised to detect failures and shall report the status of all data points to the Master at least once each 10 seconds.

b. Spare capacity shall be provided equal to 10% of the installed I/O, but no less than two analog and two digital inputs shall be available.

c. Failure of a Slave panel shall cause the outputs to go to predetermined state, and the Slave shall transmit a failure message to the Master.

d. Power requirements for the Slave shall be same as for the Master.

e. The program shall bidirectionally transmit data to and from the Slaves.

2.4 MODEM

a. Provide a modem that is compatible with the local telephone utility being used by the owner to interface to the Master and to the operator's console via an RS-232-C port at 300 baud. Communication equipment shall be provided with surge protection as required to protect every component.

2.5 BULK LOADING

a. Provide a bulk-loading device for connection to the Master for uploading and downloading user application programs. This device shall remain with the owner at the completion of the installation.

2.6 I/O FUNCTIONS

a. Analog inputs shall be monitored with an analog/digital conversion. The digital value shall be held in a buffer for interrogation by the processor. Input range shall be 0 to 8.192 Vdc.

b. Pulse analog outputs shall accept the digital data and output a signal with a range of 0.1 to 25.5 seconds.

c. Digital outputs shall accept dry contact closure.

d. Digital outputs shall provide a contact closure for momentary and maintained programmable operation of field devices. Closure shall have a duration of 0.1 second to continuous. Contact rating shall be no less than 10 amps at 240 volts. Each digital output shall have a Hand-Off-Auto switch to manually obtain either ON or OFF state. Override switches shall be reported to the Master for an override signal to the operator.

e. Pulse accumulators shall have the same characteristics as a digital input, but with a buffer to totalize pulses between interrogations. Pulse rate shall be up to 10 pulses/second.

f. Temperature inputs shall originate with a thermistor and shall be monitored and buff an analog input. Automatic conversion of the signal to degrees Fahrenheit shall be performed without additional signal conditioning.

g. 4 to 20 mA output signal shall be provided for equipment requiring a current loop interface.

h. Signal conditioning shall be provided for all sensors requiring separate excitation, or provide external signal conditioners or transmitters.

2.7 INSTRUMENT AND CONTROLS

a. Thermistors shall be used for temperature measurement having a resistance of no less than 32 ohms and a temperature range of 32 to 300 deg. F (0 to 150 deg. C). Thermistor shall be encapsulated in epoxy or in stainless steel. Provide thermowells as recommended by the manufacturer for fluids. Provide louvered wooden shelters for outside air measurement as recommended by the manufacturer.

b. Pressure sensors shall be as recommended by the manufacturer, with a rating of at least 15% above the working pressure and accuracy of plus or minus 1% of full scale.

c. Pressure switches shall be snap-acting as recommended by the manufacturer, with repetitive accuracy of plus or minus 1% of range and with rating of 150% of the working pressure.

d. Watthour transducers shall have 1% accuracy for kW and kWh measurement of lead or lag power factor as recommended by the manufacturer.

e. Output devices shall consist of control relays, reed relays, contactors, solid-state relays, electric solenoid valves, electric/pneumatic transducers, all as recommended by the manufacturer.

f. Watthour meters shall have a pulse initiator utilizing light to indicate the pulse and consisting of form C contacts with current rating as recommended by the manufacturer. Provide demand register where required to allow demand limiting by the system.

2.8 SOFTWARE

a. Provide resident memory operating system. Bulk storage devices shall not be required to remain on line.

b. The operating system shall require no operator interaction to initialize and commence operations. The program shall provide operation and management of all devices and error detection and recovery from arithmetic and logical faults. Software shall allow user development and alterations of the programs. The system shall self-test and report any problems. All data shall be updated no less than every 10 seconds. Each point shall be given an identification of up to 8 characters.

c. Provide at least two levels of password protection. The lower level shall prevent the operator from changing any control action or modifying any programs.

2.9 PROGRAMMING

a. Provide programming as required for the functions listed below. All programs shall be as recommended by the manufacturer and as required for the necessary control.

b. Programs shall have capacity to include the following:

1. Scheduled start/stop program.
2. Optimized start/stop program.
3. Duty cycle program.
4. Demand-limiting program.
5. Day/night setback program.
6. Economizer program.
7. Ventilation and recirculation program.
8. Temperature control of modulating valve.
9. Enthalpy program using total heat content of air.

10. Reheat coil reset program.
11. Steam boiler optimization program.
12. Hot water boiler optimization program.
13. Hot water outside air reset program.
14. Chiller optimization program.
15. Chilled water temperature reset program.
16. Condenser water temperature reset program.
17. Chiller demand-limiting program.
18. Lighting control program.
19. Hot/cold deck temperature reset program.

2.10 SYSTEM REPORTS

a. The system shall have the capability of outputing reports through the communication channel on operator request or when triggered by the program as follows:

1. History report of at least 8 pages consisting of 8 variables over 32 preprogrammed time increments, ranging from 1 to 1999 minutes.
2. I/O reports of value and status of specified I/O.
3. Alarm reports consisting of up to 64 preprogrammed messages indicating an alarm condition.
4. Error reports of errors encountered in self-test and record of power interruptions. The program shall include the auto dialing of a series of phone numbers to report to a host computer the fact that there is an alarm situation.
5. Override report of those inputs and outputs that are currently overridden, and a list of overrides due to Hand-Off-Auto switch, as well as via operator's terminal.
6. Trace shall print a record of the operation of a given control sequence.
7. System integrity test shall be performed by the Master continuously. Any fault or errors shall be reported.

2.11 WIRE AND CABLE

a. Wire and cable connection for the system shall be as follows:

1. Communications cable shall be 22 AWG minimum, twisted and shielded in pairs, with shielding grounded approved for plenum location.
2. Coaxial cable shall be as recommended by the manufacturer.
3. Control wiring and sensor wiring shall be 22 AWG minimum with 600-volt insulation, shielded 2 or 3 wire, as recommended by the manufacturer. Digital functions do not require shielding. Wiring must be rated for plenum location when used in such areas.

2.12 ACCESSORY EQUIPMENT

a. Furnish a maintenance-free wet cell or gel cell battery to provide power for the EMS for a period of no less than 24 hr in case of extended power failure.

b. Provide heavy-duty lightning arresters installed at the point of entry into each building for any communication or control lines located outside the building.

c. Provide a central processing unit (CPU) to read, monitor, and communicate with the Master unit. The CPU shall be a personal computer with two disk drives and RAM of no less than 65K. CPU shall be Apple IIe or approved equivalent.

d. Provide a quality dot matrix printer having a 9.5-in. (240-mm) carriage and printing no less than 120 characters per second at 10 cpi. Printer shall be Apple, or approved equivalent.

2.13 ACCEPTABLE MANUFACTURERS

a. The EMS shall be Andover Model 256 Master and necessary Slaves or approved equivalent.

PART THREE - EXECUTION

3.1 INSTALLATION

a. Install the system as recommended by the manufacturer, using only equipment recommended or acceptable to the manufacturer.

b. Comply with all codes for electrical work. Run all power wiring in conduit. All sensor and control wiring located in mechanical rooms and other exposed areas shall be run in conduit. All wiring run inside walls shall be in conduit. All equipment located outside shall be in suitable weathertight enclosure.

c. Install all conduit, wiring, and cable, and install all equipment in first-class manner, using proper tools, equipment, hangers, and supports, and in locations as required for a neat, attractive installation. No material shall be exposed if it is possible to conceal it. Exposed materials shall be installed only with consent of the engineer.

d. Support all sensors as recommended by the manufacturer where inside equipment such as ductwork. Sensors in the space shall be in small, attractive housings designed for that purpose and mounted on an electrical junction box.

e. Extreme care shall be used in making connections to other equipment, such as boilers and chillers, to see that the safeties on this equipment are not inadvertently bypassed or overriden by the EMS.

f. All equipment having moving parts and controlled by the EMS shall be provided with warning labels no less than 2 in. (50 mm) in height, and in bright warning colors, stating that the equipment is remotely started by automatic controls. Such labels shall be posted clearly in the area of any moving parts, such as belts, fans, pumps, etc.

3.2 SOFTWARE

a. Load and debug the software to provide a complete operating EMS system, and operate the system to prove function of each system. Where necessary, the sensors shall be heated or cooled to demonstrate the correct function. Provide careful evaluation of the operation of chillers and boilers at part and full load under control of the EMS.

b. The EMS contractor shall review the programs with the engineer in the programming stage to make sure that the programmer understands the engineer's intent and that the program will carry out that intent.

c. Provide the engineer with a bound copy of the complete programming, as well as instruction books on reprogramming of the system for future modifications of the system, if desirable.

d. Provide two bound copies of complete information on the equipment and all components, including programming, for the owner's use and records.

e. Instruct the owner and engineer in the proper operation of the system and the steps necessary to override or change any input data.
f. Provide the owner with five copies of complete wiring diagrams of the entire control system as installed, showing the location and the "as is" set point for every input and the location of every output. Where the EMS interlocks with other equipment, show enough of that equipment wiring diagram to allow the service department to follow the control sequence without having to pull the wiring diagram of the other equipment.
g. The EMS contractor shall monitor the system monthly for one year to verify performance of the system. This may be done by use of the modem and remote communications.
h. Remote communications require a telephone line to the EMS. The EMS contractor shall provide for the installation and upkeep of this line for the first year as part of the contract. At the end of the first-year warranty period, the EMS contractor shall offer a service contract to continue to monitor the system and advise the owner of any alarms received.

PART FOUR - PROGRAMMED FUNCTIONS

4.1 CONTROL FUNCTIONS
a. Control functions performed by the EMS shall be as follows:

(SPECIFICATION WRITER SHOULD INCLUDE HERE THE CONTROL FUNCTIONS REQUIRED OF THE EMS AS IT RELATES TO EVERY ITEM OF EQUIPMENT TO BE CONTROLLED OR MONITORED.)

END OF SECTION 15964

SECTION 15966
****
HVAC ENERGY MANAGEMENT SYSTEM

PART ONE - GENERAL

1.1 The GENERAL and SPECIFIC CONDITIONS, Section 15100, are included as a part of this Section as though written in full in this document.

1.2 Scope of the Work shall include the furnishing and complete installation of the equipment covered by this Section, with all auxiliaries, ready for owner's use.

PART TWO - PRODUCTS

2.1 ENERGY MANAGEMENT SYSTEM

a. HVAC energy management system (EMS) shall have the capacity to control the loads and monitor the analog sensors as required to perform the functions listed later in this Specification.

b. The EMS shall include a control microprocessor with unit diagnostic and alarm capability, password-based data access system, battery for power failure backup with charger, manual override switches, relay status lights, and dry contact relays for load control, all mounted in an industrial steel cabinet.

c. The EMS shall be programmed from the face of the panel through a user keypad with LCD display. The system shall have the necessary binary outputs and analog and binary inputs as required to perform the functions listed in PART FOUR of this specification.

d. The system shall provide for an alarm for failure or loss of data and other information. The alarm shall be audible at the panel, with visual indication on the panel, and shall have provisions for remote indication.

e. Power supply for the EMS shall be 110 volts. Power supply for remote relays shall be provided by external power supply as required for the system.

f. The EMS shall be warranted for no less than one year from the date of written acceptance of the system, or the date on which the owner has beneficial use of the system, if not in writing.

2.2 SOFTWARE

a. The EMS shall provide for no less than the following routines:

1. Time of day scheduling
2. Duty cycling
3. Demand limiting
4. Remote scheduling override
5. Temperature control
6. Night setback
7. Optimum start and stop
8. Temperature-compensated duty cycling
9. Analog monitoring
10. Leap year compensation
11. Daylight savings compensation
12. Minimum ON and minimum OFF time for each load

b. Time of day scheduling shall be based on an eight-day time clock, seven workdays and one holiday. There shall be four day types for each load, with six timed events for each day type. There shall be a minimum of twelve holidays, each with a length of one to seven days.
c. Demand limiting shall read either a pulse meter, or the current and voltage through watts transducers, and shall shed the selected loads in priorities (priorities shall equal the number of loads), with equipment protected by minimum time off and temperature deadbands.
d. System shall monitor any analog input based on variable resistance, current, or voltage with day and night and high and low alarm limits.
e. The system shall provide for manual override of loads by use of a two-position switch with preassigned minimum time of 1 to 720 minutes. The system shall be capable of monitoring at least eight overridden inputs.
f. The system shall be able to monitor and independently generate trend prints (histories) for up to eight parameters, with a sampling period of 1 minute to 24 hr assignable to each load.
g. The system shall be able to compare up to four binary inputs and change control strategies based on logical statements.

2.3 WIRE AND CABLE

a. Wire and cable connections shall be made using the following:
   1. Communications cable shall be 22-gauge AWG minimum, twisted and shielded in pairs, with shielding grounding. Cables located in plenums shall be rated for that service.
   2. Coaxial cable shall be as recommended by the manufacturer of the EMS.
   3. Control wiring and sensor wiring shall be 22-gauge AWG minimum, with 600-volt insulation, shielded 2 or 3 wire, as recommended by the manufacturer of the EMS equipment. Digital functions do not require shielding. Wiring located in plenums shall be rated for that service.

2.4 AUXILIARY EQUIPMENT

a. Furnish a maintenance-free wet cell battery to provide for the EMS for a period of no less than 24 hr, in case of power failure.
b. Provide heavy-duty lightning arresters installed at the point of entry into each building for any communication or control lines located outside the building.
c. Provide a central processing unit (CPU) to read, monitor, and communicate with the Master unit. The CPU shall be a personal computer with two disk drives and RAM of no less than 65K. CPU shall be Apple IIe or approved equivalent.
d. Provide a telephone modem at the EMS and the computer for monitoring of the system. Modem shall be direct-connected auto-dial auto-answer type, compatible with the local telephone utility system.
e. Provide a quality dot matrix printer having a 9.5-in. (240-mm) carriage and printing no less than 120 characters per second at 10 cpi. Printer shall be Apple or approved equivalent.

2.5 ACCEPTABLE MANUFACTURERS

a. The EMS shall be Trane "Tracer" Master and necessary Slaves or approved equivalent.

PART THREE - EXECUTION

3.1 INSTALLATION

a. Install the system as recommended by the manufacturer, using only equipment recommended by, or acceptable to, the manufacturer.

b. Comply with all codes for electrical work. Run all power wiring in conduit. All sensors and control wiring located in mechanical rooms and other exposed areas shall be run in conduit. All wiring run inside walls shall be in conduit. All equipment located outside shall be in suitable weathertight enclosure.

c. Install all conduit, wiring, and cable, and install all equipment in a first-class manner, using proper tools, equipment, hangers, and supports, and in locations as required for a neat, attractive installation. No material shall be exposed if it is possible to conceal it. Exposed materials shall be installed only with consent of the engineer.

d. Support all sensors as recommended by the manufacturer where inside equipment such as ductwork. Sensors in the space shall be in small, attractive housings, designed for that purpose and mounted on an electrical junction box.

e. Extreme care shall be used in making connections to other equipment to see that the safeties on this equipment are not inadvertently bypassed or overridden by the EMS.

f. All equipment having moving parts and controlled by the EMS shall be provided with warning labels no less than 2 in. (50 mm) in height, and in bright warning color, stating that the equipment is remotely started by automatic controls. Such labels shall be posted clearly in the area of any moving parts, such as belts, fans, pumps, etc.

3.2 SOFTWARE

a. Load and debug the software to provide a complete operating EMS system, and operate the system to prove proper function of each system. Where necessary, the sensors shall be heated or cooled to demonstrate the correct function. Provide careful evaluation of the operation of chillers and boilers at part and full load under control of the EMS.

b. The EMS contractor shall review the programs with the engineer in the programming stage to make sure that the programmer understands the engineer's intent and that the program will carry out that intent.

c. Provide the engineer with a bound copy of the complete programming, as well as instruction books on reprogramming of the system for future modifications of the system, if desirable.

d. Provide two bound copies of complete information on the equipment and all components, including programming, for the owner's use and records.

e. Instruct the owner and engineer in the proper operation of the system and the steps necessary to override or change any input data.

f. Provide the owner with five copies of complete wiring diagrams of the entire control system as installed, showing the location and the "as is" set point for every input and the location of every output. Where the EMS interlocks with other equipment, show enough of that equipment wiring diagram to allow the service department to follow the control sequence without having to pull the wiring diagram of the other equipment.

g. The EMS contractor shall monitor the system monthly for one year to verify performance of the system. This may be done by use of the modem and remote communications.

h. Remote communications require a telephone line to the EMS. The EMS contractor shall provide for the installation and upkeep of this line for the first year as part of the contract. At the end of the first-year warranty period, the EMS contractor shall offer a service contract to continue to monitor the system and advise the owner of any alarms received.

PART FOUR - PROGRAMMED FUNCTIONS

4.1 CONTROL FUNCTIONS

a. Control functions performed by the EMS shall be as follows:

(SPECIFICATION WRITER SHOULD INCLUDE HERE THE CONTROL FUNCTIONS REQUIRED OF THE EMS SYSTEM AS IT RELATES TO EVERY ITEM OF EQUIPMENT TO BE CONTROLLED OR MONITORED.)

END OF SECTION 15966

Cross References to Sourcebook of HVAC Details

The specification writer may find the following list of details and schedules as published in a companion volume, GUIDEBOOK OF HVAC DETAILS, to be of interest and help in the preparation of the drawings associated with the specifications.

CHAPTER THREE — CENTRAL STATION REFRIGERATION EQUIPMENT

****

PAGE NO. OF DETAIL	DETAIL NUMBER	DESCRIPTION OF DETAIL
1-3	D-15150-1	RECIPROCATING COMPRESSOR REFRIGERANT PIPING
1-4	D-15150-2	RECIPROCATING CHILLER - PIPED WITH EVAPORATOR ABOVE COMPRESSOR
1-5	D-15150-3	RECIPROCATING CHILLER - PIPED WITH EVAPORATOR BELOW COMPRESSOR
1-6	D-15151-1	WATER-COOLED CONDENSER PIPE CONNECTIONS
1-7	D-15152-1	CHILLER-EVAPORATOR PIPE CONNECTIONS
1-8	D-15152-2	MULTIPLE CHILLER PIPING DIAGRAM
1-9	D-15153	MULTIPLE CHILLER/BOILER PIPING DIAGRAM
1-10	D-15155-1	STEAM-POWERED ABSORPTION CHILLER PIPING HOOKUP
1-11	D-15155-2	HOT WATER-POWERED ABSORPTION CHILLER HOT WATER PIPING
1-12	D-15155-3	TYPICAL CONDENSER WATER PIPING FOR ABSORPTION CHILLER
1-13	D-15155-4	STEAM PIPING FOR ABSORPTION CHILLER
1-14	D-15160	WATER-COOLED CENTRIFUGAL CHILLER PIPING
1-15	D-15175-9	CONCRETE PAD FOR GROUND-MOUNTED EQUIPMENT
1-16	D-15180	COOLING TOWER PIPING - PROPELLER FAN TYPE

CHAPTER FOUR PACKAGED EQUIPMENT

****

****

CHAPTER SIX VENTILATION EQUIPMENT AND FANS

****

4-16	D-15421-1	DOME-TYPE BELT-DRIVEN CENTRIFUGAL ROOF EXHAUSTER
4-17	D-15421-2	DOME TYPE BELT-DRIVEN CENTRIFUGAL ROOF EXHAUSTER WITH SOUND CURB
4-18	D-15423	KITCHEN EXHAUST HOOD WITH MAKEUP AIR - NO HEAT IN MAKEUP AIR - NO DETAIL ON HOOD
4-19	D-15427-1	PROPELLER WALL EXHAUST FAN - DIRECT DRIVE - LIGHT DUTY
4-20	D-15427-2	PROPELLER WALL FAN - DIRECT DRIVE - HEAVY DUTY - INTAKE WITH FIXED LOUVER - DUCT CONNECTION
4-21	D-15428	PROPELLER-TYPE WALL FAN - BELT DRIVE - EXHAUST SHUTTER
4-22	D-15428-1	PROPELLER-TYPE WALL FAN - NO LOUVER OR SHUTTER SHOWN
4-23	D-15428-2	PROPELLER-TYPE WALL FAN WITH INSULATED MOVABLE LOUVERS

PAGE NO. OF SCHEDULE	SCHEDULE NAME	DESCRIPTION OF SCHEDULE
11-14	CEILING-MOUNTED CENTRIFUGAL FANS	AIR QUANTITY, STATIC, AND POWER REQUIRED
11-15	FAN, LIGHT, HEATER COMBINATION	AIR QUANTITY, STATIC, AND POWER REQUIRED
11-16	CABINET FANS	AIR QUANTITY, STATIC, AND POWER REQUIRED
11-17	IN-LINE CENTRIFUGAL FANS	AIR QUANTITY, STATIC, AND POWER REQUIRED
11-18	WALL-MOUNTED PROPELLER EXHAUST FANS	AIR QUANTITY, STATIC, AND POWER REQUIRED
11-19	ROOF-MOUNTED PROPELLER FANS	AIR QUANTITY, STATIC, AND POWER REQUIRED
11-20	ROOF-MOUNTED CENTRIFUGAL EXHAUST FANS	AIR QUANTITY, STATIC, AND POWER REQUIRED

11-21	ROOF VENTILATORS PROPELLER TYPE VERTICAL SHAFT	AIR QUANTITY, STATIC, AND POWER REQUIRED
11-22	HOOD EXHAUST FANS	AIR QUANTITY, STATIC AND POWER
11-23	UTILITY VENT SETS	AIR QUANTITY, STATIC AND POWER REQUIRED
11-24	UTILITY FANS - CENTRAL AIR HANDLING SYSTEMS	AIR QUANTITY, STATIC AND POWER REQUIRED
11-25	WALL-MOUNTED PROPELLER EXHAUST (SUPPLY) FANS	AIR QUANTITY, STATIC AND POWER REQUIRED

CHAPTER SEVEN VENTILATION EQUIPMENT AND SYSTEMS

****

PAGE NO. OF DETAIL	DETAIL NUMBER	DESCRIPTION OF DETAIL
4-24	D-15450-1	RIDGE-MOUNTED VENTILATOR - GRAVITY TYPE
4-25	D-15450-2	VENTILATOR ON FLAT ROOF - GRAVITY TYPE - MANUAL DAMPER - CEILING RING GRILLE
4-26	D-15450-3	VENTILATOR ON PITCHED ROOF - GRAVITY TYPE - MANUAL DAMPER
4-27	D-15450-4	HORIZONTAL RELEF CAP - ROOF MOUNTED
4-28	D-15455-1	SHUTTER-TYPE RELIEF DAMPERS - WALL DISCHARGE
4-29	D-15455-2	WALL RELIEF DAMPERS WITH BIRD SCREEN
5-3	D-15460-1	WALL LOUVER INTAKE - FIXED LOUVER - BIRD SCREEN - MANUAL DAMPER
5-4	D-15461	INSULATED MOVABLE DAMPERS MOUNTED BEHIND WEATHERPROOF WALL LOUVERS WITH GYM GUARD - INTAKE SERVICE FOR VENTILATION SYSTEM
5-5	D-15465-1	LOUVERED PENTHOUSE - INTAKE OR EXHAUST - MOTORIZED DAMPERS
5-6	D-15465-2	LOUVERED PENTHOUSE - RAISED DUCT FOR STORM PROTECTION OF FRESH AIR INTAKE
5-7	D-15468-1	INTAKE HOOD WITH PREFABRICATED ROOF

CHAPTER EIGHT AIR DISTRIBUTION

****

CHAPTER NINE HEATING SOURCES

****

8-17	D-15700-25	PIPE HANGER (OVERHEAD SUPPORT) - SPLIT RING TYPE
8-18	D-15700-27	STRAP ANCHOR FOR NONMETALLIC PIPE
8-19	D-15700-30	PIPE SADDLE FOR INSULATED HOT PIPE (NO VAPOR BARRIER)
8-20	D-15700-32	CONCRETE INSERT ANCHOR SUPPORT FOR PIPE OR EQUIPMENT
8-21	D-15700-34	EYE ROD HANGER - WELDED OR OPEN EYE
8-22	D-15700-36	SIDE BEAM CLAMPS FOR EQUIPMENT OR PIPE SUPPORT
8-23	D-15700-38	SIDE BEAM CLAMPS FOR EQUIPMENT OR PIPE SUPPORT
8-24	D-15700-40	BEAM CLAMPS FOR PIPE OR EQUIPMENT SUPPORT
8-25	D-15700-42	WELDED STEEL HANGER BRACKET
8-26	D-15700-44	SPRING ISOLATED HANGER FOR PIPE OR EQUIPMENT
8-27	D-15700-46	SPRING/ELASTOMERIC HANGER FOR PIPE OR EQUIPMENT
8-28	D-15731	PIPE INSULATION UNDERGROUND - FOAMGLASS AND MASTIC WITH GLASS FABRIC AND MASTIC COVERING
8-29	D-15735	PRESSURE GAUGE WITH TEST GAUGE CONNECTION
8-30	D-15735-1	THERMOMETER INSTALLATION IN INSULATED PIPE
8-31	D-15736	EXPANSION TANK CONNECTIONS
8-32	D-15739	STEAM SEPARATOR - LOW- AND MEDIUM-PRESSURE STEAM (REMOVES MOISTURE AND TRASH) - SHOP FABRICATED
8-33	D-15748	CONDENSATE METER FOR HOT RETURN WATER

****

CHAPTER FOURTEEN HVAC CONTROL COMPONENTS

****

Index

About the Author

Frank E. Beaty, Jr. graduated from Clemson University with a B.S. in Mechanical Engineering and served as a sales engineer with Westinghouse Electric Corporation and a design engineer with James A. Evans Consulting Engineers before establishing his own firm, specializing in heating, ventilation, and air conditioning; plumbing; and fire protection. His consulting and design experience has included projects ranging from design of facilities for construction of nuclear reactors, water and sewage treatment plants, and large industrial plants to schools, churches, libraries, dormitories, and office buildings. Beaty is presently a staff engineer with the Energy Management Department of the University of Alabama at Birmingham.